Mathematics of Planet Earth

Volume 15

Series Editors

Dan Crisan, Imperial College London, London, UK

Ken Golden, University of Utah, Salt Lake City, USA

Darryl D. Holm, Imperial College London, London, UK

Mark Lewis, University of Victoria, Victoria, Canada

Yasumasa Nishiura, Tohoku University, Sendai, Japan

Joseph Tribbia, National Center for Atmospheric Research, Boulder, USA

Jorge Passamani Zubelli, Instituto de Matemática Pura e Aplicada, Rio de Janeiro, Brazil

This series provides a variety of well-written books of a variety of levels and styles, highlighting the fundamental role played by mathematics in a huge range of planetary contexts on a global scale. Climate, ecology, sustainability, public health, diseases and epidemics, management of resources and risk analysis are important elements. The mathematical sciences play a key role in these and many other processes relevant to Planet Earth, both as a fundamental discipline and as a key component of cross-disciplinary research. This creates the need, both in education and research, for books that are introductory to and abreast of these developments.

Springer's MoPE series will provide a variety of such books, including monographs, textbooks, contributed volumes and briefs suitable for users of mathematics, mathematicians doing research in related applications, and students interested in how mathematics interacts with the world around us. The series welcomes submissions on any topic of current relevance to the international Mathematics of Planet Earth effort, and particularly encourages surveys, tutorials and shorter communications in a lively tutorial style, offering a clear exposition of broad appeal.

Responsible Editors:

Martin Peters, Heidelberg (martin.peters@springer.com)
Robinson dos Santos, São Paulo (robinson.dossantos@springer.com)

Additional Editorial Contacts:

Donna Chernyk, New York (donna.chernyk@springer.com)
Masayuki Nakamura, Tokyo (masayuki.nakamura@springer.com)

the diffusion constant, and W_t denotes d-dimensional Brownian motion. The data $X_t^\dagger$ is being generated from a second-order SDE of the same type, i.e.,

$$dU_t^\dagger = g(X_t^\dagger, U_t^\dagger)dt + \sqrt{\sigma}dW_t^\dagger, \tag{1.2a}$$

$$dX_t^\dagger = U_t^\dagger dt \tag{1.2b}$$

with the drift $g(x, u)$ being unknown.

Note that for f, F, g independent of position, the system corresponds to a first-order diffusion process where only the integral of the process is being observed.

We assume that $g(x, u)$ is such that the reference process $(U_t^\dagger, X_t^\dagger)$ is ergodic with invariant distribution π. The optimal parameter θ_* can then be determined by minimizing the ergodic least squares functional

$$\mathcal{L}(\theta) = \frac{1}{2}\mathbb{E}_\pi \left[\|g(X, U) - f(X, U) - F(X, U)\theta\|^2 \right], \tag{1.3}$$

which has a unique minimizer provided the Fisher information matrix

$$\mathbb{E}_\pi[F(X, U)^T F(X, U)] \tag{1.4}$$

has full rank. Moreover, we assume that the unique minimizer satisfies

$$g(x, u) = f(x, u) + F(x, u)\theta_* . \tag{1.5}$$

However, the gradient of $\mathcal{L}(\theta)$ cannot be computed since $g(x, u)$ is unknown. Therefore we investigate online estimation techniques in this paper based on observed positions $X_t^\dagger$.

While standard online estimation techniques, such as stochastic gradient descent (SGD) (Sirignano and Spiliopoulos 2017) and Kalman filtering (Nüsken et al. 2019; Reich 2023), can be applied in case both $(X_t^\dagger, U_t^\dagger)$ from (1.2) are being observed, the estimation problem becomes more challenging in case only positions $X_t^\dagger$ are being observed and velocities need to be estimated using finite differencing (Pedersen et al. 2016; Brückner et al. 2020), such as,

$$\tilde{U}_{n+1/2} := \frac{X_{t_{n+1}}^\dagger - X_{t_n}^\dagger}{\tau} \tag{1.6}$$

with $\tau > 0$ denoting the sampling interval, $t_n = \tau n$, for all $n \geq 0$.

As demonstrated previously (Ditlevsen and Sorensen 2004; Gloter 2006; Ferretti et al. 2020), naive application of standard maximum likelihood-based (MLE) estimators (Kutoyants 2013; Pavliotis 2016) leads to a systematic bias. In order to correct for the arising estimation bias, appropriate corrections terms to the MLE estimator have been proposed in Gloter (2006), Brückner et al. (2020), Albrecht et al. (2024) and Ferretti et al. (2020). An alternative approach has been to treat the continuous time velocity process as a latent variable and calculate the MLE estimator using an

Chapter 1
Parameter Estimation for Partially Observed Second-Order Diffusion Processes

Jan Albrecht and Sebastian Reich

Abstract Estimating parameters of a diffusion process given continuous-time observations of the process via maximum likelihood approaches or, online, via stochastic gradient descent or Kalman filter formulations constitutes a well-established research area. It has also been established previously that these techniques are, in general, not robust to perturbations in the data in the form of temporal correlations of the driving noise. While the subject is relatively well understood and appropriate modifications have been suggested in the context of multi-scale diffusion processes and their reduced model equations, we consider here an alternative but related setting where a diffusion process in positions and velocities is only observed via its positions. In this note, we propose a simple modification to standard stochastic gradient descent and Kalman filter formulations, which eliminates the arising systematic estimation biases. The modification can be extended to standard maximum likelihood approaches and avoids computation of previously proposed correction terms.

1.1 Introduction

We consider the problem of online learning parameters $\theta \in \mathbb{R}^{d_\theta}$ of second-order stochastic differential equations (SDEs)

$$\mathrm{d}U_t = f(X_t, U_t)\mathrm{d}t + F(X_t, U_t)\theta\mathrm{d}t + \sqrt{\sigma}\mathrm{d}W_t, \qquad (1.1a)$$

$$\mathrm{d}X_t = U_t\mathrm{d}t \qquad (1.1b)$$

from incoming observations $X_t^\dagger \in \mathbb{R}^d$, $t \geq 0$, of an underlying reference process. Here $f : \mathbb{R}^d \times \mathbb{R}^d \to \mathbb{R}^d$ and $F : \mathbb{R}^d \times \mathbb{R}^d \to \mathbb{R}^{d \times d_\theta}$ are given functions, $\sigma > 0$ is

J. Albrecht
Institut für Physik und Astronomie, Universität Potsdam, Potsdam, Germany

S. Reich (✉)
Institut für Mathematik, Universität Potsdam, Potsdam, Germany
e-mail: sebastian.reich@uni-potsdam.de

© The Author(s) 2026

B. Chapron et al. (eds.), *Stochastic Transport in Upper Ocean Dynamics IV*, Mathematics of Planet Earth 15, https://doi.org/10.1007/978-3-032-12749-5_1

Contents

current and the fast wavy component are accordingly decomposed in terms of smooth in time resolved component and unresolved highly oscillating random field. The slow current is expressed as a two-dimensional evolution equation, potentially incorporating strong noise. The fast wavy components are associated with the random current but include their own noise contributions as well.

Mattéo Nex, Quentin Jamet, Etienne Mémin and Florian Sévellec in their paper entitled Chap. 8 highlight the constraints imposed by small ensemble sizes on capturing the Reynolds stress tensor and diagnosing its associated work. They employ the Location Uncertainty (LU) framework, which introduces stochastic variability into the governing equations to represent unresolved turbulent effects and compare deterministic and stochastic estimates of Reynolds stress work. The LU framework improves statistical convergence and stabilizes higher-order moments, leading to a more accurate representation of energy transfer statistics. Key contributions include a novel formulation of energy transfer equations and enhanced statistical diagnostics of Reynolds stress work within the LU framework with the study providing valuable dynamical diagnostics for identifying energetic patterns and regions of stability, offering new insights into eddy–mean flow interactions in the Gulf Stream.

The work of **Tom Protin, Valentin Resseguier, Momme Hell, Bertrand Chapron and Ronan Fablet** entitled Chap. 9 proposes a method to model the propagation and directional spread of ocean surface swell systems and follows the development of the Particle-in-Cell for Efficient Swell (PiCLES) wave modeling framework. They study how this model spreads an initial energy spectrum in space and time. Sequential importance resampling techniques are considered to control how the discrete distribution matches the true distribution after a time step without increasing the ensemble size. Tests are performed and compared to analytical solutions driven by the expected diffuse geometric optic behavior.

Finally, the STUOD Organising Committee would again like to acknowledge the financial support received from the European Research Council (ERC) under the European Union's Horizon 2020 Research and Innovation Programme (ERC, Grant Agreement No. 856408) for providing funds to cover the travel expenses of the invited speakers, catering costs and administrative support as well as the in-kind support from INRIA for hosting at the Inria Centre in Rennes, France.

STUOD Organising Committee

<table>
<tr><td>Plouzané, France</td><td>Prof. Bertrand Chapron</td></tr>
<tr><td>London, UK</td><td>Prof. Dan Crisan</td></tr>
<tr><td>London, UK</td><td>Prof. Darryl Holm</td></tr>
<tr><td>Rennes, France</td><td>Prof. Etienne Mémin</td></tr>
<tr><td>London, UK</td><td>Dr. Jane-Lisa Coughlan</td></tr>
<tr><td>July 2025</td><td></td></tr>
</table>

differential equation stirred by an additive random forcing term that is δ-correlated in time. The proposed linear dynamics generalize previous works for power-law spectra in the context of hydrodynamic turbulence to more general spectra, possibly non-radial, including sea wavenumber spectra such as the JONSWAP spectrum. They present simulations of the correlation structure of the solution to these dynamics in 2D, whose spectrum converges towards the JONSWAP spectrum.

The work of **Daniel Goodair** entitled Chap. 3 extends recent existence and uniqueness results for maximal solutions of SPDEs through an improved blow-up criterion. He shows that solutions exist until blow-up in the larger spaces of the variational framework. The result is applied through a bootstrap of regularity to show that solutions of 2D and 3D Stochastic Navier-Stokes Equations retain the higher order regularity of the initial condition on their time of existence.

Darryl D. Holm and Oliver D. Street in their paper entitled Chap. 4 extend the theoretical Euler-Poincaré framework for modelling ocean mixed layer dynamics. Through a symmetry-broken Lie group invariant variational principle, they derive a generalised Green-Naghdi equation with time dependent bathymetry, a complete Coriolis force, and inhomogeneity of the thermal buoyancy. The nature of the model derived lends it a potential future application to wave dynamics generated by changes to the bathymetry.

The work of **Maël Jaouen, Benjamin Dufée, Etienne Mémin and Gilles Tissot** entitled Chap. 5 proposes a novel approach based on reproducing kernel Hilbert spaces (RKHS), embedding dynamical system observables into time-evolving RKHSs. Their framework leads to reconstruction of trajectories with time-invariant linear combinations of ensemble members. The aim of the present paper is to construct an ensemble data assimilation method formulated in this time-evolving RKHS. They assess the approach using a multilayer quasi-geostrophic ocean model with synthetic sea surface height (SSH) observations derived from satellite tracks. Results show significantly improved reconstruction accuracy and robustness compared to a classical ensemble filter.

Long Li, Etienne Mémin and Bertrand Chapron in their paper entitled Chap. 6 introduce a stochastic coupled Ekman–Stokes model (SCESM) developed within the physically consistent Location Uncertainty framework, explicitly incorporating random turbulent fluctuations and surface wave effects. The SCESM integrates established parameterizations for air–sea fluxes, turbulent viscosity, and Stokes drift, and its performance is rigorously assessed through ensemble simulations compared against observations from the LOTUS field experiment. A performance ranking analysis quantifies the impact of different model components, highlighting the critical role of explicit uncertainty representation in both oceanic and atmospheric dynamics for accurately capturing system variability. The fully coupled stochastic framework provides a foundation for advancing boundary layer parameterizations in large-scale climate models.

The work of **Louis Marié, Etienne Mémin and Bertrand Chapron** entitled Chap. 7 explores a consistent derivation of the wave action conservation principle for stochastic flows. The study highlights a stochastic form of the wavefront Hamilton-Jacobi equation. Within this framework, the usual slow component of the underlying

The scientific programme of the four-day workshop featured an inspirational range of talks, theoretical and applied sessions, as well as networking opportunities. In particular, it showcased topics on: Data assimilation and stochastic modelling, Data models, SPDEs, Numerics for ocean models, Physics models, Stochastic Waves, New capabilities with SWOT data as well as covering new and future STUOD developments. Several members of the STUOD External Advisory Board gave invited talks: Prof. Sebastian Reich (University of Potsdam), Prof. Rosemary Morrow (Laboratoire d'Études en Géophysique et Océanographie Spatiales), Prof. Baylor Fox-Kemper (Brown University), Prof. Peter Korn (Max Planck Institute for Meteorology) and Prof. Arnaud Debussche (ENS Rennes). The programme also included individual presentations by the STUOD Principal Investigators and postdoctoral researchers that overall provided opportunities for investigators, at both early and established stages of their career, to foster future research collaborations and enable the next generation of researchers.

Workshop attendees on 23 September 2024

The following is a brief description of the **nine** contributions included in the proceedings:

The submitted manuscripts include the paper by **Jan Albrecht and Sebastian Reich** entitled Chap. 1 propose a simple modification to standard stochastic gradient descent and Kalman filter formulations that eliminates the arising systematic estimation biases. The modification can be extended to standard maximum likelihood approaches and avoids computation of previously proposed correction terms.

Geoffrey Beck, Charles-Edouard Bréhier, Laurent Chevillard, Ricardo Grande and Wandrille Ruffenach in their paper entitled Chap. 2 propose linear dynamics that can generate a given sea wavenumber spectrum via a linear partial

Preface

This volume contains the Proceedings of the **5th Stochastic Transport in Upper Ocean Dynamics Annual Workshop** held on **23–26 September 2024**. The workshop is a core part of the Stochastic Transport in Upper Ocean Dynamics (STUOD) project, which is supported by a European Research Council (ERC) Synergy Grant and is led by four Principal Investigators that bring together three world-class institutions: Prof. Bertrand Chapron, French Research Institute for Exploitation of the SEA (IFREMER), Prof. Dan Crisan, Imperial College London (ICL), Prof. Darryl Holm, Imperial College London (ICL) and Prof. Etienne Mémin, National Institute for Research in Digital Science and Technology (INRIA).

The project aims to deliver new capabilities for assessing variability and uncertainty in upper ocean dynamics and provides decision makers a means of quantifying the effects of local patterns of sea level rise, heat uptake, carbon storage and change of oxygen content and pH in the ocean. The project will make use of multimodal data and will enhance the scientific understanding of marine debris transport, tracking of oil spills and accumulation of plastic in the sea.

As in previous years, the 5th STUOD Annual Workshop 2024 focused on a range of fundamental topic areas, including:

1. Observations at high resolution of upper ocean properties such as temperature, salinity, topography, wind, waves, and velocity.
2. Large-scale numerical simulations.
3. Data-based stochastic equations for upper ocean dynamics that quantify simulation errors.
4. Stochastic data assimilation to reduce uncertainty.

Each chapter in this volume explores one or more of these thematic areas. They present mathematical frameworks along with novel methodological and numerical advancements aimed at supporting and advancing future research within the STUOD project. The workshop was held in the hybrid mode and brought together early-career academics, postgraduate students, senior members of the community and other invited guests from across the world that included the UK, France, Romania, Italy, Germany and the USA.

Editors
Bertrand Chapron
Institut Français de Recherche pour
l'Exploitation de la Mer
Plouzané, France

Darryl D. Holm 
Imperial College London
London, UK

Jane-Lisa Coughlan
Imperial College London
London, UK

Dan Crisan 
Imperial College London
London, UK

Etienne Mémin
Inria—Institut National de Recherche en
Sciences et Technolologies du Numérique
Rennes, France

ISSN 2524-4264   ISSN 2524-4272 (electronic)
Mathematics of Planet Earth
ISBN 978-3-032-12748-8   ISBN 978-3-032-12749-5   (eBook)
https://doi.org/10.1007/978-3-032-12749-5

Mathematics Subject Classification: 70L10, 35Q86, 00B25, 35R60, 60H15, 37M05

This work was supported by Horizon 2020 Framework Programme (856408).

Bertrand Chapron · Dan Crisan · Darryl D. Holm ·
Etienne Mémin · Jane-Lisa Coughlan
Editors

Stochastic Transport in Upper Ocean Dynamics IV

STUOD 2024 Workshop, Rennes, France,
September 23–26

Springer

expectation-maximization algorithm (Baltazar-Larios and Sørensen 2010). Viewing the SDEs as a partially observed system, the likelihood of θ could also be approximated using an extended Kalman filter (Särkkä 2013). Related phenomena have been investigated in the context of multi-scale diffusion (Ait-Sahalia et al. 2005; Papavasiliou et al. 2009; Bo and Celani 2013), which again can be corrected via an appropriately modified MLE estimator (Diehl et al. 2016).

While these correction terms can be applied to online learning, we follow an alternative approach in this note, which has been motivated by the related work (Abdulle et al. 2023; Pavliotis et al. 2025). Our approach can easily be implemented both for SGD (Sirignano and Spiliopoulos 2017) as well as Kalman filter (Nüsken et al. 2019; Reich 2023) formulations.

As demonstrated in the next section by means of a simple example, a discretized SGD of the form

$$\theta_{n+1} := \theta_n + \alpha_t F(X_{t_n}^{\dagger}, \tilde{U}_n)^{\mathsf{T}} \Delta I_n(\theta_n), \tag{1.7a}$$

$$\Delta I_n(\theta) := (\tilde{U}_{n+1/2} - \tilde{U}_{n-1/2}) - \left(f(X_{t_n}^{\dagger}, \tilde{U}_n) + F(X_{t_n}^{\dagger}, \tilde{U}_n)\theta \right) \tau, \tag{1.7b}$$

with $U_{t_{n+1/2}}$ being replaced by its estimated $\tilde{U}_{n+1/2}$ etc. and the symmetric approximation

$$\tilde{U}_n := \frac{1}{2} \left(\tilde{U}_{n+1/2} + \tilde{U}_{n-1/2} \right) = \frac{X_{t_{n+1}}^{\dagger} - X_{t_{n-1}}^{\dagger}}{2\tau} \tag{1.8}$$

will fail even in the limit $\tau \to 0$.[1] Here

$$\alpha_t = c_1/(c_2 + t) \tag{1.9}$$

denotes the learning rate for appropriate constants $c_i > 0$, $i = 1, 2$. While the centered approximation in (1.8) is the best approximation to U_n from the position measurements at the nearest neighbor time points $X_{t_{n-1}}, X_{t_n}, X_{t_{n+1}}$ alternative approximations to the velocity at t_n could also be employed.

The discussion from the following section will lead to an appropriate modification of (1.7), which removes the undesirable bias in the SGD estimator (1.7). See Sect. 1.3.

We extend the proposed methodology to Kalman filter based estimation in Sect. 1.4. The unbiasedness of the corrected SGD approach is demonstrated for a nonlinear second-order diffusion process in Sect. 1.5. The note closes with comments on further directions for research.

---

[1] For integers n, m, Eq. (1.6) defines the estimated velocity $\tilde{U}_{(n+m)+1/2}$ between time points $n + m$ and $n + m + 1$. To keep notation concise, we will write $\tilde{U}_{n+(2m+1)/2}$ instead of $\tilde{U}_{n+m+1/2}$. This means that $\tilde{U}_{n-1/2} = \tilde{U}_{(n-1)+1/2}$, $\tilde{U}_{n+3/2} = \tilde{U}_{(n+1)+1/2}$ etc.

1.2 Motivating Example

In order to demonstrate the difficulties the standard SGD implementation (1.7) encounters, we consider velocity data $U_t^\dagger$ being generated by the linear OU process

$$\mathrm{d}U_t^\dagger = -\theta_* U_t^\dagger \mathrm{d}t + \sqrt{\sigma}\mathrm{d}W_t^\dagger \tag{1.10}$$

for given diffusion constant $\sigma > 0$ and unknown drift parameter $\theta_* \geq 0$.

We will set $\theta_* = 0$ for some of the analysis to be conducted later in this section, which implies that solutions of (1.10) satisfy

$$U_t^\dagger = U_s^\dagger + \sqrt{\sigma}W_{t-s}^\dagger \tag{1.11}$$

for any $t \geq s$ and $s \geq 0$ and, consequently,

$$X_t^\dagger = X_s^\dagger + (t - s)U_s^\dagger + \sqrt{\sigma}\int_s^t W_{r-s}^\dagger \mathrm{d}r. \tag{1.12}$$

Here, we have introduced the shorthand $W_{t-s}^\dagger = W_t^\dagger - W_s^\dagger$.

We first recall some of the properties of SGD when a solution of (1.10) has in fact been fully observed. An application of SGD (Sirignano and Spiliopoulos 2017) to this problem yields

$$\mathrm{d}\theta_t = -\alpha_t U_t^\dagger \mathrm{d}I_t = -\alpha_t U_t^\dagger \left(\mathrm{d}U_t^\dagger + \theta_t U_t^\dagger \mathrm{d}t\right) \tag{1.13a}$$

$$= -\alpha_t U_t^\dagger \left((\theta_t - \theta_*)U_t^\dagger \mathrm{d}t + \sqrt{\sigma}\mathrm{d}W_t^\dagger\right). \tag{1.13b}$$

We call

$$I_t = \int_0^t \left(\mathrm{d}U_s^\dagger + \theta_s U_s^\dagger \mathrm{d}s\right) = \int_0^t \left((\theta_s - \theta_*)U_s^\dagger \mathrm{d}s + \sqrt{\sigma}\mathrm{d}W_s^\dagger\right) \tag{1.14}$$

the innovation and $\alpha_t U_t^\dagger$ the gain. It is obvious that I_t becomes a martingale at the optimal parameter choice $\theta_s = \theta_*$, $s \in [0, t]$.

It has been shown that $\theta_t \to \theta_*$ as $t \to \infty$ (Sirignano and Spiliopoulos 2017). In our context, a key observation is that the integral

$$\int_0^t \alpha_s U_s^\dagger \left(\mathrm{d}U_s^\dagger + \theta_t U_s^\dagger \mathrm{d}s\right) \tag{1.15}$$

also becomes a martingale for any $t > 0$ if and only if $\theta_s = \theta_*$ for all $s \in [0, T]$.

In terms of numerical implementations of the SGD formulation (1.13), we consider the discrete time SGD method

$$\theta_{n+1} := \theta_n - \alpha_{t_n} U_{t_n}^\dagger \left(U_{t_{n+1}}^\dagger - U_{t_n}^\dagger + \frac{\theta_n}{2} \left(U_{t_{n+1}}^\dagger + U_{t_n}^\dagger \right) \tau \right). \tag{1.16}$$

This discrete-time SGD method still implies that θ_n converges to θ_* as $n \to \infty$ and $\tau \to 0$. More precisely, any finite sum

$$\sum_{n=0}^{N-1} U_{t_n}^\dagger \left(U_{t_{n+1}}^\dagger - U_{t_n}^\dagger + \frac{\theta_n}{2} \left(U_{t_{n+1}}^\dagger + U_{t_n}^\dagger \right) \tau \right) \tag{1.17}$$

becomes a martingale as $\tau \to 0$ for $T = N\tau$ fixed if and only if $\theta_n = \theta_*$.

However, the focus of this note is on the case when only positions $X_t^\dagger$, $t \geq 0$, can be observed and we recall the velocity approximations (1.6), where the sampling interval $\tau > 0$ can be made arbitrarily small. We note that (1.6) can also be viewed as a local time average, i.e.,

$$\tilde{U}_{n+1/2} = \frac{1}{\tau} \int_{t_n}^{t_{n+1}} U_t^\dagger \mathrm{d}t, \tag{1.18}$$

which provides a link to the related work (Gloter 2006; Cialenco and Huang 2020).

Using (1.8) instead of the true $U_{t_n}^\dagger$ etc. and symmetrizing the gain leads to the previously stated (1.7), which reduces here to

$$\theta_{n+1} := \theta_n - \alpha_{t_n} \tilde{U}_n \left(\tilde{U}_{n+1/2} - \tilde{U}_{n-1/2} + \theta_n \tilde{U}_n \tau \right) \tag{1.19a}$$

$$= \theta_n - \frac{\alpha_{t_n}}{2} \left(X_{t_{n+1}}^\dagger - X_{t_{n-1}}^\dagger \right) I_N \tag{1.19b}$$

with innovation

$$I_N = \sum_{n=1}^{N-1} \left(\frac{X_{t_{n+1}}^\dagger - 2X_{t_n}^\dagger + X_{t_{n-1}}^\dagger}{\tau^2} + \frac{\theta_n}{2} \frac{X_{t_{n+1}}^\dagger - X_{t_{n-1}}^\dagger}{\tau} \right) \tau. \tag{1.20}$$

We note that, while the discrete innovations I_N still converge to a martingale as $\tau \to 0$ for $T = N\tau$ fixed provided $\theta_n = \theta_*$ for all $n \geq 0$, this is no longer true for

$$\frac{1}{2} \sum_{n=1}^{N-1} \left(X_{t_{n+1}}^\dagger - X_{t_{n-1}}^\dagger \right) \left(\frac{X_{t_{n+1}}^\dagger - 2X_{t_n}^\dagger + X_{t_{n-1}}^\dagger}{\tau^2} + \frac{\theta_n}{2} \frac{X_{t_{n+1}}^\dagger - X_{t_{n-1}}^\dagger}{\tau} \right). \tag{1.21}$$

This violation of the martingale property is perhaps not surprising since (1.21) does not resemble a discretized Itô-Integral like (1.17) but rather a discretized Stratonovitch integral (Pavliotis 2016). We note that such an interpretation as a discretized stochastic integral concerns only the second-order differencing term in (1.21) and we may therefore set $\theta = \theta_* = 0$ in the subsequent derivations. Recall that

$$\frac{X_{t_{n+1}} - X_{t_n}}{\tau} = \tilde{U}_{n+1/2} = U_{t_n}^\dagger + \frac{\sqrt{\sigma}}{\tau} \int_{t_n}^{t_{n+1}} W_{t-t_n}^\dagger \, dt \tag{1.22}$$

and we therefore obtain

$$\frac{1}{2}\mathbb{E}\left[\left(\tilde{U}_{n+1/2} + \tilde{U}_{n-1/2}\right)\left(\tilde{U}_{n+1/2} - \tilde{U}_{n-1/2}\right)\right] = \frac{1}{2}\mathbb{E}\left[\tilde{U}_{n+1/2}^2 - \tilde{U}_{n-1/2}^2\right] \tag{1.23a}$$

$$= \frac{1}{2}\mathbb{E}\left[(U_{t_n}^\dagger)^2 - (U_{t_{n-1}}^\dagger)^2\right] \tag{1.23b}$$

$$= \frac{\tau\sigma}{2}, \tag{1.23c}$$

which implies an estimation bias induced by the discrete sum (1.21) even in the limit $\tau \to 0$.

In order to remove the resulting bias in the SGD formulation (1.19), we propose to simply shift the innovation term forward in time such that it does no longer overlap with the gain; thus regaining an Itô-type stochastic integral approximation. A related idea has been proposed and investigated in Pavliotis et al. (2025) in the context of multi-scale diffusion. This modification leads to the following implementation:

$$\theta_{n+1} := \theta_n + \alpha_{t_n}\tilde{U}_n\left(\tilde{U}_{n+5/2} - \tilde{U}_{n+3/2} - \theta_n\tilde{U}_{n+2}\tau\right). \tag{1.24}$$

Note that the innovation in the modified implementation (1.24) only contains positional data from time points that are larger than or equal to those in the gain.

Let us introduce the random variables

$$\tilde{\Xi}_n := \frac{1}{\sqrt{\sigma\tau}}\left(\tilde{U}_{n+1/2} - \tilde{U}_{n-1/2}\right) \tag{1.25}$$

for $n \geq 1$, which is a rescaling of the innovation $\Delta I_n(\theta_*)$, as defined in (1.7b), with θ_* again set to $\theta_* = 0$. It is crucial to observe that $\tilde{\Xi}_{n+2}$ is correlated with both $\tilde{U}_{n+5/2}$ and $\tilde{U}_{n+3/2}$, while being independent of $\tilde{U}_n$. Hence, provided $\tilde{\Xi}_n$ is a martingale for all $n \geq 1$, the SGD scheme (1.24) will become unbiased in the limit $\tau \to 0$.

Hence let us analyze the statistical properties of $\tilde{\Xi}_n$. We find that

$$\tilde{\Xi}_n = \frac{1}{\sqrt{\sigma\tau}}\left(\tilde{U}_{n+1/2} - \tilde{U}_{n-1/2}\right) \tag{1.26a}$$

$$= \frac{1}{\sqrt{\sigma\tau}}\left(U_{t_n}^\dagger - U_{t_{n-1}}^\dagger\right) + \frac{1}{\sqrt{\tau}}\left(\frac{\int_{t_n}^{t_{n+1}} W_{t-t_n}^\dagger \, dt - \int_{t_{n-1}}^{t_n} W_{t-t_{n-1}}^\dagger \, dt}{\tau}\right) \tag{1.26b}$$

$$= \frac{W_{t_n-t_{n-1}}^\dagger}{\sqrt{\tau}} + \frac{1}{\sqrt{\tau}}\left(\frac{\int_{t_n}^{t_{n+1}} W_{t-t_n}^\dagger \, dt - \int_{t_{n-1}}^{t_n} W_{t-t_{n-1}}^\dagger \, dt}{\tau}\right). \tag{1.26c}$$

It follows that $\tilde{\mathcal{E}}_n$ is Gaussian with mean zero and that $\tilde{\mathcal{E}}_{n+2}$ is indeed independent of $\tilde{U}_n$.

We recall the following two properties of integrated Brownian motion:

$$\mathrm{var}\left(\int_{t_n}^{t_{n+1}} W_{t-t_n}^{\dagger}\,\mathrm{d}t\right) = \frac{1}{3}\tau^3 \tag{1.27}$$

and

$$\mathbb{E}\left[W_{t_n-t_{n-1}}^{\dagger}\left(\int_{t_{n-1}}^{t_n} W_{t-t_{n-1}}^{\dagger}\,\mathrm{d}t\right)\right] = \int_{t_{n-1}}^{t_n} \mathbb{E}\left[(W_{t-t_{n-1}}^{\dagger})^2\right]\mathrm{d}t = \frac{1}{2}\tau^2. \tag{1.28}$$

Upon applying these results to (1.26c) when computing the variance of $\tilde{\mathcal{E}}_n$, it is found that

$$\mathrm{var}\,(\tilde{\mathcal{E}}_n) = \frac{2}{3}. \tag{1.29}$$

In other words, we obtain that $\tilde{\mathcal{E}}_n \sim \mathrm{N}(0, 2/3)$ in agreement with results from Gloter (2006) and Cialenco et al. (2024). This result carries over to $\theta_* \neq 0$ since the arising additional terms vanish as $\tau \to 0$.

We finally formally investigate the time-continuous limit of the modified discrete-time SGD formulation (1.24). We first note that

$$\mathbb{E}[\tilde{\mathcal{E}}_n \tilde{\mathcal{E}}_{n+k}] = \begin{cases} \frac{1}{6} & |k| = 1 \\ 0 & |k| > 1 \end{cases}, \tag{1.30}$$

which implies that

$$Z_t := \begin{cases} 0 & \text{if } t < \tau \\ \sum_{k=1}^{\lfloor t/\tau \rfloor} \sqrt{\tau}\,\tilde{\mathcal{E}}_k & \text{else} \end{cases} \tag{1.31}$$

converges to standard Brownian motion in the limit $\tau \to 0$. When using this result in the SGD formulation (1.24) for general θ_*, we obtain

$$\theta_{n+1} \approx \theta_n - \alpha_{t_n}\tilde{U}_n\left((\theta_n - \theta_*)\tilde{U}_{n+2}\tau + \sqrt{\sigma\tau}\,\tilde{\mathcal{E}}_{n+2}\right) \tag{1.32a}$$

$$\approx \theta_n - \alpha_{t_n}\tilde{U}_n\left((\theta_n - \theta_*)\tilde{U}_n\tau + \sqrt{\sigma\tau}\,\tilde{\mathcal{E}}_{n+2}\right) \tag{1.32b}$$

and the limit $\tau \to 0$ leads formally to the time-continuous formulation

$$\dot{\theta}_t = -\alpha_t U_t^{\dagger}\left((\theta - \theta_*)U_t^{\dagger} + \sqrt{\sigma}\dot{W}_t^{\dagger}\right), \tag{1.33}$$

which coincides with (1.13b).

Remark Independent of the drift estimation, it is well known that the diffusion term can be estimated from the variability of a process. In line with previous results

(Cialenco et al. 2024; Ferretti et al. 2020; Gloter 2006), our above calculations and equation (1.29) imply the following estimator for the diffusion constant σ from the quadratic variation of the approximated velocities $\tilde{U}_{n+1/2}$:

$$\sigma_N := \frac{3}{2\tau N} \sum_{n=1}^{N} \|\tilde{U}_{n+1/2} - \tilde{U}_{n-1/2}\|^2. \tag{1.34}$$

1.3 Unbiased SGD

The analysis from Sect. 1.2 leads to the following modification of the discrete-time SGD implementation (1.7):

$$\theta_{n+1} = \theta_n + \alpha_{t_n} F(X_{t_n}^\dagger, \tilde{U}_n)^\mathrm{T} \Delta I_{n+2}(\theta_n), \tag{1.35}$$

for $n \geq 1$ with the innovations $\Delta I_n(\theta)$ defined by (1.7b) and the learning rate α_{t_n} by (1.9) with $t = t_n = n\tau$. We recall that the key point of shifting the innovation forward in time is to retain an Itô-type approximation in the gain times innovation stochastic integral approximation. This is the property that maintains the desired unbiasedness of the resulting sequential learning scheme in the limit $\tau \to 0$.

We note that unbiased formulations are not unique since only the shifting of the second-order differencing term in the innovation is relevant for consistency. For example, (1.35) could be replaced by

$$\theta_{n+1} = \theta_n + \alpha_{t_n} F(X_{t_n}^\dagger, \tilde{U}_n)^\mathrm{T} \times \tag{1.36a}$$

$$\left(\tilde{U}_{n+5/2} - \tilde{U}_{n+3/2} - \left(f(X_{t_n}^\dagger, \tilde{U}_n) + F(X_{t_n}^\dagger, \tilde{U}_n)\theta_n\right)\tau\right). \tag{1.36b}$$

This freedom is also utilized when formulating the modified MLE estimator (1.41) further below in Sect. 1.6.

1.4 Unbiased Kalman Filter

The SGD method (1.35) provides an online point estimator for the parameter θ. However, it is also possible to implement an online Bayesian estimator using the Kalman filter (Nüsken et al. 2019; Reich 2023). The key idea is to treat the unknown parameter as a Gaussian random variable Θ_n with mean m_n and covariance matrix Σ_n and with trivial dynamics $\Theta_{n+1} = \Theta_n, n \geq 0$ (Särkkä 2013). The evolution equation for the mean and the covariance matrix follow the standard Kalman update formulas:

$$m_{n+1} := m_n + K_n \frac{\Delta I_{n+2}(m_n)}{\tau}, \tag{1.37a}$$

$$\Sigma_{n+1} := \Sigma_n - K_n F(X_{t_n}^\dagger, \tilde{U}_n)\Sigma_n \tag{1.37b}$$

with Kalman gain

$$K_n := \Sigma_n F(X_{t_n}^\dagger, \tilde{U}_n)^{\mathrm{T}} \left(F(X_{t_n}^\dagger, \tilde{U}_n)^{\mathrm{T}} \Sigma_n F(X_{t_n}^\dagger, \tilde{U}_n) + \frac{\sigma}{\tau} I \right)^{-1} \tag{1.38}$$

for $n \geq 1$. The iteration is initialized from the Gaussian prior

$$\Theta_1 \sim \mathrm{N}(m_{\mathrm{prior}}, \Sigma_{\mathrm{prior}}). \tag{1.39}$$

In contrast to the SGD formulation (1.35), the learning rate α_{t_n} no longer appears and has instead been submerged implicitly into the Kalman gain K_n. Indeed, the singular values of the covariance matrix Σ_n decay with rate $1/t_n$ under the already stated ergodicity assumption on (1.2) and the non-singularity of the Fisher information matrix (1.4). If the required diffusion constant σ is unknown, it can either be estimated via (1.34) or, alternatively, be treated as part of the prior covariance matrix Σ_{prior} and then set to $\sigma = 1$ in the Kalman gain (1.38).

1.5 Numerical Example

We consider the nonlinear second-order SDE

$$\mathrm{d}U_t = -X_t \mathrm{d}t + U_t \mathrm{d}t - \theta U_t^3 \mathrm{d}t + \sqrt{2}\mathrm{d}W_t, \tag{1.40a}$$

$$\mathrm{d}X_t = U_t \mathrm{d}t \tag{1.40b}$$

with unknown parameter $\theta \in \mathbb{R}$. The data $X_t^\dagger$ is generated with $\theta_* = 1$, while the SGD iteration starts from $\theta_0 = 2$. The sample interval is set to $\tau = 0.025$. The data is generated with a much smaller step-size of $h = \tau/100$. A total of $N = 10^5$ positions $X_{t_n}, n = 1, \ldots, N$ have been generated. The learning rate is given by $\alpha_{t_n} = 6.0/n$. See Fig. 1.1 middle panel for the time evolved estimates θ_n generated by the modified SGD (1.35). These estimates clearly converge to θ_*. In Fig. 1.1 we also display results from the standard SGD as well as the modified Kalman filter presented in Sect. 1.4. For the Kalman filter, a Gaussian prior was used with $m_{\mathrm{prior}} = 2$ and $\Sigma_{\mathrm{prior}} = 6$. The diffusion constant σ was assumed to be unknown and treated as part of the prior. While the standard SGD converges slightly faster than the modified SGD, it converges to a wrong parameter value. The estimates of the modified Kalman filter are very similar to estimates of the modified SGD. An unmodified Kalman filter (not shown) on the other hand leads to a similar estimation bias as the standard SGD.

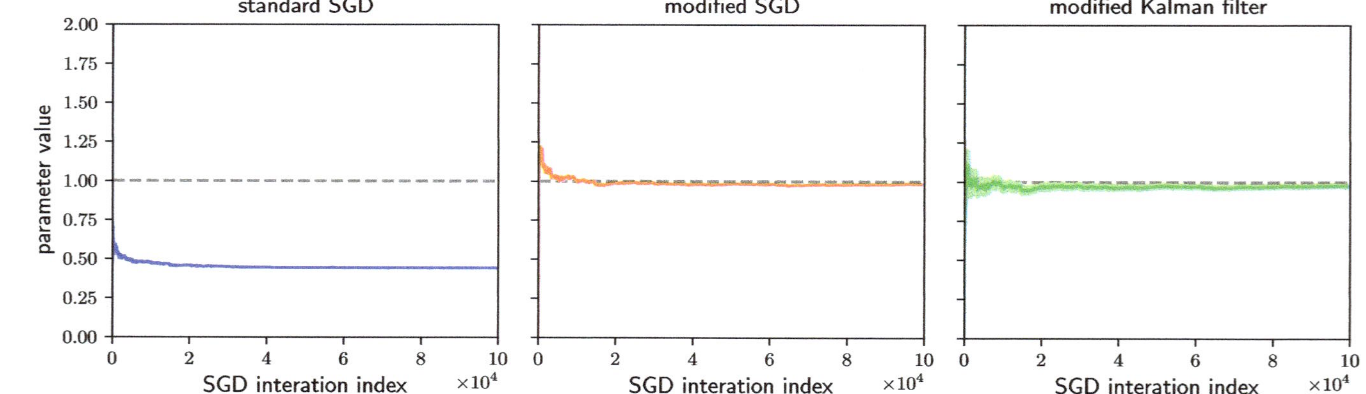

Fig. 1.1 Left panel: Estimated parameter value from standard SGD as a function of the iteration index. The iteration starts at $\theta_0 = 2$ and the reference value is $\theta_* = 1$. A systematic bias can clearly be seen. Middle panel: Same data is used in the modified SGD. The modified SGD algorithm converges to the correct reference value. Right panel: Same data is used in the unbiased Kalman filter with prior distribution $N(2, 1)$ and σ treated as part of the prior. The lightly colored lines indicate the 1σ-intervals of the Bayesian estimates

1.6 Conclusions

In this note, we have proposed a simple-to-implement modification of standard SGD and Kalman filter formulations for online parameter estimation from position data only. The essential idea is to establish an Itô-like modification that re-establishes the martingale property in the presence of temporal correlations. The approach has been inspired by related work (Abdulle et al. 2023; Pavliotis et al. 2025) on multi-scale diffusion and parameter estimation for the arising reduced equations. We finally note that a corrected MLE estimator can be based on the negative log-likelihood function

$$\mathcal{L}(\theta, \sigma) = \frac{1}{2\sigma} \sum_{n=1}^{N-3} \theta^{\mathrm{T}} F(X_{t_n}^{\dagger}, \tilde{U}_n)^{\mathrm{T}} F(X_{t_n}^{\dagger}, \tilde{U}_n)\theta \tau \tag{1.41a}$$

$$- \frac{1}{\sigma} \sum_{n=1}^{N-3} \theta^{\mathrm{T}} F(X_{t_n}^{\dagger}, \tilde{U}_n)^{\mathrm{T}} \left(\tilde{U}_{n+5/2} - \tilde{U}_{n+3/2} - f(X_{t_n}^{\dagger}, \tilde{U}_n)\tau \right) \tag{1.41b}$$

$$+ \frac{3}{4\tau\sigma} \sum_{n=1}^{N-3} \|\tilde{U}_{n+5/2} - \tilde{U}_{n+3/2}\|^2 + \frac{N-3}{2} \log \sigma. \tag{1.41c}$$

Compare this likelihood function to the contrast function presented in Gloter (2006), which is based on adding an appropriate correction term. Please also note that (1.41a)–(1.41b) replaces the ergodic least squares functional (1.3) using measured positions and estimated velocities and that (1.41c) delivers the estimator (1.34).

Future directions include the investigation of the proposed estimators in the context multi-scale diffusion when (1.1) arises from homogenization of an underlying multi-scale diffusion process and the inclusion of observation errors in the positions X_t. In the former case, a combination of the approaches taken here and in Pavliotis et al. (2025) emerges as a feasible starting point.

Acknowledgements This work has been partially funded by Deutsche Forschungsgemeinschaft (DFG)—Project-ID 318763901—SFB 1294.

References

Abdulle A, Garegnani G, Pavliotis G, Stuart A, Zanoni A (2023) Drift estimation of multiscale diffusions based on filtered data. Found Comput Math 23:33–84

Ait-Sahalia Y, Mykland PA, Zhang L (2005) How often to sample a continuous-time process in the presence of market microstructure noise. Rev Finan Stud 18:351–416

Albrecht J, Großmann R, Opper M (2024) Inferring parameter distributions in heterogeneous particle ensembles: a likelihood approach for second order Langevin models. arXiv:2411.08692

Baltazar-Larios F, Sørensen M (2010) Maximum likelihood estimation for integrated diffusion processes. In: Chiarella C, Novikov A (eds) Contemporary quantitative finance: essays in honour of Eckhard Platen, Berlin, Heidelberg, pp 407–423. Springer Berlin Heidelberg

Bo S, Celani A (2013) White-noise limit of nonwhite nonequilibrium processes. Phys Rev E 88:062150

Brückner D, Ronceray P, Broedersz C (2020) Inferring the dynamics of underdamped stochastic systems. Phy Rev Lett 125:058103

Cialenco I, Huang Y (2020) A note on parameter estimation for discretely sampled SPDEs. Stoch Dyn 20:2050016

Cialenco I, Kim H-J, Pasemann G (2024) Statistical analysis of discretely sampled semilinear SPDEs: a power variation approach. Stoch Part Diff Equ: Anal Comput 12:326–351

Diehl J, Friz P, Mai H (2016) Pathwise stability of likelihood estimators for diffusion via rough paths. Ann Appl Probab 26:2169–2192

Ditlevsen S, Sorensen M (2004) Inference for observations of integrated diffusion processes. Scand J Statist 31:417–429

Ferretti F, Chardès V, Mora T, Walczak AM, Giardina I (2020) Building general Langevin models from discrete datasets. Phys Rev X 10:031018

Gloter A (2006) Parameter estimation for a discretely observed integrated diffusion process. Scand J Stat 33:83–104

Kutoyants YA (2013) Statistical inference for ergodic diffusion processes. Springer Science & Business Media

Nüsken N, Reich S, Rozdeba PJ (2019) State and parameter estimation from observed signal increments. Entropy 21(5):505. https://doi.org/10.3390/e21050505

Papavasiliou A, Pavliotis G, Stuart A (2009) Maximum likelihood estimation for multiscale diffusions. Stoch Proc Appl 119:3173–3210

Pavliotis G (2016) Stochastic processes and applications. Springer, New York

Pavliotis GA, Reich S, Zanoni A (2025) Filtered data based estimators for stochastic processes driven by colored noise. Stoch Proc Appl 181:104558. ISSN 0304-4149

Pedersen JN, Li L, Gradinaru C, Austin RH, Cox EC, Flyvbjerg H (2016) How to connect time-lapse recorded trajectories of motile microorganisms with dynamical models in continuous time. Phys Rev E 94:062401

Reich S (2023) Frequentist perspective on robust parameter estimation using the ensemble Kalman filter. In: Chapron B, Crisan D, Holm D, Mémin E, Radomska A (eds) Stochastic transport in upper ocean dynamics. STUOD 2021. Mathematics of planet earth, vol 10, pp 237–258, Cham. Springer

Särkkä S (2013) Bayesian Filtering and Smoothing. Institute of Mathematical Statistics Textbooks. Cambridge University Press, Cambridge

Sirignano J, Spiliopoulos K (2017) Stochastic gradient descent in continuous time. SIAM J Finan Math 8:933–961

Chapter 2
A Linear Model of Dynamical Sea Wavenumber Spectra

Geoffrey Beck, Charles-Edouard Bréhier, Laurent Chevillard, Ricardo Grande, and Wandrille Ruffenach

Abstract We propose linear dynamics that can generate a given sea wavenumber spectrum via a linear partial differential equation stirred by an additive random forcing term that is δ-correlated in time. In particular, the correlation structure of the solution to this linear dynamics converges towards the target spectrum as time passes. The main linear mechanism is a transport in Fourier space, which models the transfer of energy from large scales towards small scales. The proposed linear dynamics generalize previous works for power-law spectra in the context of hydrodynamic turbulence to more general spectra, possibly non-radial, including sea wavenumber spectra such as the JONSWAP spectrum. Finally, we present simulations of the correlation structure of the solution to these dynamics in 2D, whose spectrum converges towards the JONSWAP spectrum. These simulations are based on a finite volume method in the Fourier domain and a splitting method in time, following a recently proposed numerical method by the same authors which improves on pseudospectral simulations.

G. Beck (✉)
IRMAR UMR 6625 and Centre Inria de l'Université de Rennes (MINGuS) and ENS Rennes, University of Rennes, Rennes, France
e-mail: geoffrey.a.beck@inria.fr

C.-E. Bréhier
LMAP, CNRS, E2S UPPA, Universite de Pau et des Pays de l'Adour, Pau, France
e-mail: charles-edouard.brehier@univ-pau.fr

L. Chevillard
Laboratoire de Physique, CNRS, Université Claude Bernard, ENS de Lyon, University of Lyon, Lyon, France
e-mail: laurent.chevillard@ens-lyon.fr

CNRS, ICJ UMR5208, Ecole Centrale de Lyon, INSA Lyon, Universite Claude Bernard Lyon 1, Université Jean Monnet, Villeurbanne, France

R. Grande
International School for Advanced Studies (SISSA), Trieste, Italy
e-mail: rgrandei@sissa.it

W. Ruffenach
UMR5672, LPENSL, CNRS, ENS de Lyon, Lyon cedex 07, France
e-mail: wandrille.ruffenach@ens-lyon.fr

© The Author(s) 2026

B. Chapron et al. (eds.), *Stochastic Transport in Upper Ocean Dynamics IV*, Mathematics of Planet Earth 15, https://doi.org/10.1007/978-3-032-12749-5_2

2.1 Introduction

2.1.1 Background

Turbulence is often understood as the intricate motion of vortices in fluid dynamics when the Reynolds number is large. From a phenomenological point of view, one of the main objects of study is the transfer of energy from such vortices to smaller and smaller vortices. This is known as energy cascade, a widely studied phenomenon since the pioneering work of Richardson, Kolmogorov and others (Tennekes and Lumley 1972; Frisch 1995).

In a more general setting, the word turbulence also refers to the chaotic behavior of systems having a large number of degrees of freedom. In the case of linear water wave systems, the surface elevation can be decomposed into independent oscillating waves (Fourier modes) whose evolution can be exactly computed. Gravity water waves equations are however nonlinear and thus Fourier modes do not evolve independently, but they interact with each other, generating new Fourier modes as time passes. Some of these interactions allow for the transfer of energy to higher and higher wavenumbers, which is reminiscent of Kolmogorov's work on energy cascades.

Due to the large number of interactions between these Fourier modes, a deterministic description of the evolution of the energy is not suitable and a stochastic approach is needed. Thus the natural object in wave turbulence is not the evolution of each individual mode but of its statistical average. Averaging allows one to carry out a kinetic theory of colliding waves in the same fashion as Boltzmann's kinetic theory for particles.

In the 1960s, Hasselmann heuristically derived a wave-kinetic equation for gravity water waves, which describes the effective evolution of the average energy of each Fourier mode of the system (Hasselmann 1962, 1963). At the same time, Zakharov made a connection between hydrodynamic turbulence and the interaction of waves in the general context of nonlinear dispersive Hamiltonian PDEs. Thus the kinetic theory for colliding waves became known as wave turbulence. The first rigorous derivation of a wave-kinetic equation was obtained by Deng and Hani (2023a, b) in the context of the cubic nonlinear Schrödinger equation (see also Buckmaster et al. 2021; Collot and Germain 2019; Grande and Hani 2024; Staffilani and Tran 2021) but analogous results for the gravity water waves system are not yet available due, in no small part, to its quasilinear nature. Even if rigorous derivation is not avaible, such wave kinetic equations are used in practical for weather forecasts by météo france (Wavewatch III, see Tolman 2014). Zakharov found stationary solutions to such wave kinetic equations which capture the mean energy transfer from small wavenumbers towards large wavenumbers once the wave system reaches a statistically stationary state. Such stationary solutions, also known as spectra, represent the correlation structure of the energy of the wave system.

Statistically homogeneous fields are a fundamental object in the phenomenology of homogeneous and isotropic turbulent flows (Batchelor 1953; Frisch 1995; Monin and Yaglom 2013). In this idealized situation, the underlying probability law of the turbulent velocity field is invariant by spatial translations and rotations, and has been repeatedly observed in laboratory flows and in simulations of the forced Navier-Stokes equations. Given a spatial field u which is statistically homogeneous, we call its wave number spectrum the function $S(k)$ such that

$$\mathbb{E}[\hat{u}(k)\overline{\hat{u}(k')}] = S(k)\,\delta_{k-k'}\ ,$$

where $\hat{u}$ is the spatial Fourier transform of u, k is the wave number and $\delta_{k-k'}$ stands for the Dirac delta distribution. In the context of the velocity field of a fluid with a large Reynolds number, it is experimentally observed that the fluid reaches a statistically stationary, homogeneous and isotropic state in the absence of boundary. In such cases, the wave number spectrum of the Eulerian velocity only depends on $|k|$. More precisely, according to Kolmogorov's 41 theory (Kolmogorov 1941), the wavenumber spectrum is a curve which may be roughly decomposed into three regions:

- Injection range: for small $|k|$, the shape of the curve $S(k)$ is determined by the forcing.
- Inertial range: for intermediate values of $|k|$, energy is transferred from large wavenumbers towards small wavenumbers. In this range, the spectrum is observed (and predicted by Kolmogorov's theory) to be a power law of the form $|k|^{-d-2H}$ with $H \approx 1/3$.
- Dissipative range: for large $|k|$, viscous dissipation damps the energy coming from small wavenumbers, effectively smoothing the spatial velocity profile.

Sea wavenumber spectra are often written as follows

$$S(k) = S_{\mathrm{r}}(\omega(k))S^{\cdot}(k), \tag{2.1}$$

where $\omega(k)$ is the dispersion relation of the system (Komen et al. 1994). For gravity water waves in the ocean with infinite depth, the dispersion relation is $\omega(k) = \sqrt{g|k|}$ where g is the gravity, whereas one has $\omega(k) = g|k|^2$ in the shallow water regime. Different dispersion relations are possible if one considers surface tension or other phenomena.

The angular part $S^{f}(k)$ of the spectrum (2.1) typically describes the energetic spreading of the waves as a function of the wind direction (Appel 1994; Elfouhaily et al. 1997). The most widely used radial spectrum $S_{\mathrm{r}}(\omega(\cdot))$ for fully developed sea is the JONSWAP (Joint North Sea Wave Observation Project) spectrum:

$$S_{\mathrm{r}}(\omega(k)) = \frac{\alpha g^2}{\omega(k)^5}\exp\left[-\frac{5}{4}\frac{\omega_p^4}{\omega(k)^4}\right]\gamma^{\exp\left[-\frac{(\omega(k)-\omega_p)^2}{2\sigma^2\omega_p^2}\right]} \tag{2.2}$$

where $\gamma = 3.3$, and the parameters α, ω_p, σ depend on the speed of the wind and its fetch (Hasselmann and Olbers 1973). The Pierson–Moskowitz spectrum corresponds to $\gamma = 1$. The choice $\gamma = 1$ and $\omega_p = 0$ leads to the Phillips spectrum,[1] which describes white capping. In intermediate range of $|k|$, the JONSWAP spectrum $S_r(\omega(k)) \sim \omega(k)^{-4}$ which corresponds to the Zakharov stationary solutions to the wave kinetic equation associated to the water waves equations, also known as the Hasselmann spectrum.

A great number of wave spectra are based on observations and do not (yet) admit a rigorous mathematical justification. In particular, in the context of wave turbulence, there is no rigorous derivation of such spectra for the gravity water waves system. Similarly, in the context of hydrodynamic turbulence, no mathematical proof of the regimes described in Kolmogorov's theory exists starting from the Navier-Stokes equation in 3D. For this reason, the construction of toy models which feature such spectra is of fundamental importance. A large number of such models are static, in the sense that the model is already in a statistically stationary state featuring the desired spectrum. Other models are dynamic but only in the form of discrete dynamical systems.

While turbulent transfers of energy have traditionally been associated to nonlinear interactions, the work of Colin de Verdière and Saint-Raymond show that some features of turbulence such as energy cascades and loss of regularity can also occur in the context of a linear dynamics (deVerdière et al 2019). The main result of this paper is the construction of a linear PDE driven by a random additive force whose solution converges to a statistically stationary state featuring the desired spectrum.

2.1.2 Main Results

We are ready to introduce our PDE in wavenumber space:

$$\begin{cases} \partial_t \widehat{u}(t, k) + c\, \partial_{|k|} \widehat{u}(k) + cV(k)\widehat{u}(t, k) = \widehat{\xi^\varphi}(t, k) & t > 0, k \in \mathbb{R}^d, |k| > \kappa > 0, \\ \widehat{u}(t, k) = 0 & t > 0, k \in \mathbb{R}^d, |k| \leq \kappa, \\ \widehat{u}(0, k) = 0. \end{cases}$$

$$(2.3)$$

where $\partial_{|k|} = \frac{k}{|k|} \cdot \nabla_k$ stands for the transport with respect to the radial variable and

$$V(k) = \partial_{|k|} \log(\sigma(k)) \quad \text{where} \quad \sigma(k) = |k|^{\frac{d-1}{2}} S(k)^{-\frac{1}{2}} \qquad (2.4)$$

where $k \mapsto S(k)$ is a fixed wavenumber spectrum satisfying Assumption 2.1.2 below.

The forcing term $(t, x) \mapsto \xi^\varphi(t, x)$ is a statistically homogeneous and stationary Gaussian field which is a white noise in time and colored in space. More precisely, its correlations satisfy

---

[1] Kuznetsov argues that the Phillips spectrum in deep water should be taken with $\omega \sim |k|$ instead of $\omega \sim \sqrt{|k|}$, as the most common slope breaks occur on 1D lines/ridges, not on 0D point/peaks, and that wave crests are propagating with preserved shape.

$$\mathbb{E}[\xi^{\varphi}(t, x)\xi^{\varphi}(s, x')] = C_{\varphi}(x - x')\,\delta_{t-s} \qquad \forall t, s \geq 0, \ x, x' \in \mathbb{R}^d, \tag{2.5}$$

where δ is the delta Dirac distribution and

$$C_{\varphi}(x) = \int_{\mathbb{R}^d_x} \varphi(x + y)\varphi(y)\mathrm{d}y, \quad \forall\, x \in \mathbb{R}^d.$$

Note that the correlations of the forcing term are distributions. In Sect. 2.2.1, we construct such a forcing term as the spatial convolution of a space-time white noise with the function φ. We make the following assumptions on the latter.

Assumption 2.1.1 The function φ is a real-valued, spectrally supported function. More precisely, we assume that its Fourier transform $\widehat{\varphi}$ is continuous and compactly supported away from the origin, i.e. there exist $\kappa_{\varphi} > \kappa > 0$ such that

$$\widehat{\varphi}(k) = 0 \text{ if } |k| < \kappa \text{ or } |k| > \kappa_{\varphi}.$$

We make the following assumptions on the wavenumber spectrum $k \mapsto S(k)$.

Assumption 2.1.2 $S : \mathbb{R}^d \to \mathbb{R}$ is a continuous function which is strictly positive in the region $\mathbb{R}^d \setminus \{0\}$. We further assume that $k \mapsto S(k)$ is integrable in the region $|k| > \kappa > 0$, with κ defined in Assumption 2.1.1. Moreover, for every $a \geq 0$, the following property holds:

$$\sup_{\substack{k \in \mathbb{R}^d, \\ |k| \geq \kappa}} \frac{S\left(\frac{|k|+a}{|k|}k\right)}{S(k)} < \infty. \tag{2.6}$$

Note that (2.3) must be interpreted in a weak sense since the source $\widehat{\xi^{\varphi}}(t, k)$ is delta-correlated in t and k. An important part of this work is thus to give a rigorous meaning to (2.3) and to define a notion of solution. In particular, we will show that the solution is a Gaussian field with an explicit correlation structure, which we obtain in Propositions 2.2.2 (in Fourier space) and 2.2.7 (in physical space).

Our main goal is to show that the solution to (2.3) exhibits the desired spectrum $S(k)$ as $t \to \infty$. The main result of this paper is the following:

Main Theorem (Description of the spectrum) *Under Assumptions 2.1.1 and 2.1.2, there exists a solution $\widehat{u}(t, k)$ to (2.3) whose inverse Fourier transform is a well-defined Gaussian field. Moreover, the solution has the following asymptotic correlations;*

$$\lim_{t \to \infty} \mathbb{E}[\widehat{u}(t, k)\overline{\widehat{u}(t, k')}] = S(k)\,\vartheta(k)\,\delta_{k-k'} \tag{2.7}$$

where

$$\vartheta(k) = \mathbb{1}_{|k|>\kappa} \int_{\kappa}^{|k|} \left|\widehat{\varphi}\left(s\frac{k}{|k|}\right)\right|^2 \frac{1}{S\left(s\frac{k}{|k|}\right)} \frac{\mathrm{d}s}{c}. \tag{2.8}$$

Equality (2.7) must be interpreted in the sense of distributions, i.e. for any $\phi, \phi' \in \widehat{\mathcal{H}}$ (defined in (2.10)),

$$\lim_{t \to \infty} \mathbb{E}[\langle \widehat{u}(t), \phi \rangle \, \overline{\langle \widehat{u}(t), \phi' \rangle}] = \int_{\mathbb{R}^d} S(k) \, \vartheta(k) \, \phi(k) \, \phi'(k) \, dk.$$

A more precise statement of this theorem may be found in Propositions 2.2.2 (in Fourier space), 2.2.4 (in physical space) and Corollaries 2.2.9 (asymptotic correlations). An example of energy cascades in the case $S(k) = |k|^{-(d+2H)}$ may be found in Proposition 2.2.10.

In Sect. 2.4, we present numerical simulations to illustrate the results in this theorem in the case where $S(k)$ is the JONSWAP spectrum (2.2).

Remark 2.1.3 If the spectrum S is radial away from the injection range, namely $S(k) = S_r(|k|)$ for all $|k| > \kappa_f$, then ϑ in (2.8) only depends on the angular variable away from the injection range, i.e.

$$\vartheta(k) = S_\theta \left(\frac{k}{|k|} \right) := \int_\kappa^{\kappa_\varphi} \left| \widehat{\varphi} \left(s \frac{k}{|k|} \right) \right|^2 \frac{1}{S\left(s \frac{k}{|k|} \right)} \frac{ds}{c}, \qquad \forall \, |k| > \kappa_\varphi.$$

Remark 2.1.4 If the spectrum S and $\widehat{\varphi}$ are radial, then $\vartheta(k) = \mathfrak{C}$ is constant for all $|k| > \kappa_\varphi$, where

$$\mathfrak{C} = \int_\kappa^{\kappa_\varphi} \frac{|\widehat{\varphi}(s\gamma)|^2}{c \, S(s\gamma)} \, ds,$$

for an arbitrary unit vector $\gamma \in \mathbb{S}^{d-1}$. In particular, this means that away from the injection range the solution $\widehat{u}$ to (2.3) has the following spectrum:

$$\lim_{t \to \infty} \mathbb{E}[\widehat{u}(t, k) \overline{\widehat{u}(t, k')}] = \mathfrak{C} \, S(k) \, \delta_{k-k'} \qquad \forall \, |k| > \kappa_\varphi.$$

Remark 2.1.5 For a continuous function $k \in \mathbb{R}^d \mapsto S(k) \in \mathbb{R}$ which is strictly positive in $\mathbb{R}^d \setminus \{0\}$, it suffices to check (2.6) for a sufficiently large $|k| > K_a$. If S is a radial function, a sufficient condition for (2.6) to hold is for S to be non-increasing. Combining these observations, (2.6) is easily shown to hold in the case of the JONSWAP spectrum (2.2).

Remark 2.1.6 The Gaussianity of the forcing (2.5) is a natural modeling choice in order to guarantee that the solution to the linear PDE (2.3) is Gaussian. The latter is a desirable property in a toy model, as velocity fields in turbulent fluids have been empirically shown to be well-approximated by Gaussian fields, see Tennekes and Lumley (1972) and also Wilczek et al. (2011) for estimates on the deviation from Gaussianity.

2.1.3 Outline

In Sect. 2.1.4 we introduce the notation used throughout this article. In Sect. 2.2 we present a rigorous analysis of the model (2.3) in Fourier and physical space, including detailed definitions of the forcing and notion of solution. In particular, Sect. 2.2.3 contains the asymptotic analysis of the correlations, including the proof of (2.7). In Sect. 2.3 we present the numerical method used to conduct numerical simulations of our model (2.3). Finally, in Sect. 2.4 we present and discuss said simulations in the cas of the JONSWAP spectrum in dimension $d = 2$.

2.1.4 Notation

Given two functions $\phi, \phi' \colon \mathbb{R}^d \to \mathbb{C}$, define the convolution product $\phi * \phi'$ and the correlation product $\phi \star \phi'$ by

$$\phi * \phi'(x) = \int_{\mathbb{R}^d} \phi(x - y)\phi'(y)\mathrm{d}y, \qquad \phi \star \phi'(x) = \int_{\mathbb{R}^d} \phi(x + y)\phi'(y)\mathrm{d}y.$$

One has

$$\int_{\mathbb{R}^d} (\varphi * \phi)\phi' = \int_{\mathbb{R}^d} \phi(\varphi \star \phi') \tag{2.9}$$

for any function $\varphi, \phi, \phi' \colon \mathbb{R}^d \to \mathbb{C}$ such that the above integrals are well-defined.

The indicator function of a set B is denoted by $\mathbb{1}_B$. We let $\mathbb{R}^+ = [0, \infty)$.

Given $\mathbb{K} = \mathbb{R}$ or $\mathbb{K} = \mathbb{C}$, let $L^2(\mathbb{R}^d_x; \mathbb{K})$, resp. $L^2(\mathbb{R}^+_t \times \mathbb{R}^d_x; \mathbb{K})$, denote the spaces of square-integrable functions $\phi \colon x \in \mathbb{R}^d \mapsto \phi(x) \in \mathbb{K}$, resp. $\phi \colon (t, x) \in \mathbb{R}^+ \times \mathbb{R}^d \mapsto \phi(t, x) \in \mathbb{K}$. The inner products in $L^2(\mathbb{R}^d_x; \mathbb{C})$ and $L^2(\mathbb{R}^+_t \times \mathbb{R}^d_x; \mathbb{C})$ are denoted by $(\cdot, \cdot)_{L^2_x}$ and $(\cdot, \cdot)_{L^2_{t,x}}$ and defined by

$$(\phi, \phi')_{L^2_x} = \int_{\mathbb{R}^d} \phi(x)\overline{\phi'(x)}\mathrm{d}x, \quad \forall\, \phi, \phi' \in L^2(\mathbb{R}^d_x; \mathbb{C}),$$

$$(\Phi, \Phi')_{L^2_{t,x}} = \int_0^{+\infty} \int_{\mathbb{R}^d} \phi(t, x)\overline{\phi'(t, x)}\mathrm{d}t\mathrm{d}x, \quad \forall\, \Phi, \Phi' \in L^2(\mathbb{R}^+_t \times \mathbb{R}^d_x; \mathbb{C}),$$

and the associated norms are denoted by $\| \cdot \|_{L^2_x}$ and $\| \cdot \|_{L^2_{t,x}}$ respectively.

The Fourier transform of an integrable function $\phi \colon \mathbb{R}^d_x \to \mathbb{C}$ is defined by

$$\widehat{\phi}(k) = \mathcal{F}\phi(k) := \int_{\mathbb{R}^d_x} e^{-2\pi i k \cdot x}\phi(x)\mathrm{d}x, \quad \forall\, k \in \mathbb{R}^d.$$

Whenever defined, the inverse Fourier transform of an integrable function $\widehat{\phi} \colon \mathbb{R}^d_k \to \mathbb{C}$ is

$$\phi(x) = \widehat{\phi}^{\vee}(x) = \mathcal{F}^{-1}\widehat{\phi}(x) := \int_{\mathbb{R}_k^d} e^{2\pi i x \cdot k}\, \widehat{\phi}(k)\, dk, \quad \forall\, x \in \mathbb{R}^d.$$

It is well know that the Fourier transform is an isometry from $L^2(\mathbb{R}_x^d)$ to $L^2(\mathbb{R}_k^d)$:

$$(\widehat{\phi}, \widehat{\phi'})_{L_k^2} = (\phi, \phi')_{L_x^2}, \quad \forall\, \phi, \phi' \in L^2(\mathbb{R}_x^d; \mathbb{C}).$$

Let $(\Omega, \mathcal{F}, \mathbb{P})$ denote the probability space. Given $\mathbb{K} = \mathbb{R}$ or $\mathbb{K} = \mathbb{C}$, let $L^2(\Omega; \mathbb{K})$ be the space of square-integrable random variables with values in $\mathbb{K}$.

The equation studied in this article is driven by Gaussian real-valued space-time white noise $\xi(t, x)$ for $(t, x) \in \mathbb{R}^+ \times \mathbb{R}^d$. Formally, this Gaussian noise is characterized by its mean $\mathbb{E}[\xi(t, x)] = 0$ and its covariance

$$\mathbb{E}[\xi(t, x)\xi(t', x')] = \delta(t - t')\delta(x - x'),$$

where δ stands for the Dirac pulse distribution. The formulation above is formal and shows that white noise needs to be understood in a distributional sense. A more rigorous interpretation consists in considering a Gaussian process

$$\begin{aligned} \varXi : L^2(\mathbb{R}_t^+ \times \mathbb{R}_x^d; \mathbb{R}) &\to L^2(\Omega; \mathbb{R}), \\ \Phi &\mapsto \varXi(\Phi) = \langle \xi, \Phi \rangle, \end{aligned}$$

that satisfies

- for any $N \in \mathbb{N}$ and any $\Phi_1, \ldots, \Phi_N \in L^2(\mathbb{R}_t^+ \times \mathbb{R}_x^d; \mathbb{R})$, $\big(\langle \xi, \Phi_1 \rangle, \ldots, \langle \xi, \Phi_N \rangle\big)$ is a $\mathbb{R}^d$-valued Gaussian random variable,
- for any $\Phi \in L^2(\mathbb{R}_t^+ \times \mathbb{R}_x^d; \mathbb{R})$,

$$\mathbb{E}[\langle \xi, \Phi \rangle] = 0,$$

- for any $\Phi, \Phi' \in L^2(\mathbb{R}_t^+ \times \mathbb{R}_x^d; \mathbb{R})$,

$$\mathbb{E}[\langle \xi, \Phi \rangle \langle \xi, \Phi' \rangle] = (\Phi, \Phi')_{L_{t,x}^2}.$$

Such a Gaussian process is called an isonormal Gaussian process.

Setting

$$\langle \xi, \Phi \rangle = \langle \xi, \operatorname{Re}(\Phi) \rangle + i\langle \xi, \operatorname{Im}(\Phi) \rangle$$

for any complex-valued square-integrable function $\Phi \in L^2(\mathbb{R}_t^+ \times \mathbb{R}_x^d; \mathbb{C})$ extends the definition of ξ as an isonormal Gaussian process defined from $L^2(\mathbb{R}^+ \times \mathbb{R}^d; \mathbb{C})$ to $L^2(\Omega; \mathbb{C})$.

The Gaussian random variable $\varXi(\Phi) = \langle \xi, \Phi \rangle$ is sometimes written as a stochastic Wiener integral

$$\langle \xi, \Phi \rangle = \int_0^{+\infty} \int_{\mathbb{R}^d} \Phi(t, x)\, d\xi(t, x).$$

Note that it is sufficient to consider functions $\Phi = \psi \otimes \phi$, such that $\Phi(t, x) = \psi(t)\phi(x)$, given square-integrable functions $\psi \in L^2(\mathbb{R}_t^+; \mathbb{C})$ and $\phi \in L^2(\mathbb{R}_x^d; \mathbb{C})$. It is then possible to define the Fourier transform $\widehat{\xi}(t, k)$ of the space-time white noise $\xi(t, x)$ as follows: for any $\psi \in L^2(\mathbb{R}_t^+; \mathbb{C})$ and $\phi \in L^2(\mathbb{R}_k^d; \mathbb{C})$, set

$$\langle \widehat{\xi}, \psi \otimes \phi \rangle = \langle \xi, \psi \otimes \widehat{\phi} \rangle.$$

The notation $\widehat{\xi}(t, k)$ is formal, since this is a white noise, with mean $\mathbb{E}[\widehat{\xi}(t, k)] = 0$ and the covariance

$$\mathbb{E}[\widehat{\xi}(t, k)\overline{\widehat{\xi}(t', k')}] = \delta(t - t')\delta(k - k').$$

The above expression is a formal version of the property

$$\mathbb{E}[\langle \widehat{\xi}, \psi \otimes \phi \rangle \overline{\langle \widehat{\xi}, \psi' \otimes \phi' \rangle}] = \mathbb{E}[\langle \xi, \psi \otimes \widehat{\phi} \rangle \overline{\langle \xi, \psi' \otimes \widehat{\phi'} \rangle}]$$
$$= (\psi \otimes \widehat{\phi}, \psi' \otimes \widehat{\phi'})_{L^2_{t,x}}$$
$$= (\psi \otimes \phi, \psi' \otimes \phi')_{L^2_{t,k}}$$

which follows from the definition of the Fourier transform of the white noise and of the isometry property of the Fourier transform.

Being the Fourier transform of a real-valued white noise $\xi(t, x)$, a property $\overline{\widehat{\xi}(t, k)} = \widehat{\xi}(t, -k)$ is satisfied, in a weak sense, therefore it is not necessary to deal with $\mathbb{E}[\widehat{\xi}(t, k)\widehat{\xi}(t', k')]$ in the description of $\widehat{\xi}(t, k)$.

Define the subspace $\widehat{\mathcal{H}}$ of $L^2(\mathbb{R}_k^d)$ by

$$\widehat{\mathcal{H}} = \{\widehat{f} \in L^2(\mathbb{R}_k^d) \mid \widehat{f}(k) = 0 \text{ if } |k| < \kappa\}. \tag{2.10}$$

Let also $\mathcal{H}$ be the subspace of $L^2(\mathbb{R}_x^d)$ defined by

$$\mathcal{H} = \{f \in L^2(\mathbb{R}_x^d) \mid \widehat{f} \in \widehat{\mathcal{H}}\}. \tag{2.11}$$

Let us mention a useful change of variables formula which is used several times below: given a function $F : (k_1, k_2) \in \mathbb{R}_k^d \times \mathbb{R}_k^d \mapsto F(k_1, k_2) \in [0, \infty]$, one has

$$\int_{\mathbb{R}^d} F\left(k, \frac{|k| - ct}{|k|}k\right) \mathbb{1}_{|k| - ct > \kappa} dk = \int_{\mathbb{R}^d} \left(\frac{|k| + ct}{|k|}\right)^{d-1} F\left(\frac{|k| + ct}{|k|}k, k\right) \mathbb{1}_{|k| > \kappa} dk. \tag{2.12}$$

The identity (2.12) is due to a change of variables in spherical coordinates, letting $r = |k| \in \mathbb{R}^+$ and $\theta = \frac{k}{|k|} \in \mathbb{S}^{d-1}$ for all $k \in \mathbb{R}^d$. If μ denotes the surface measure on the sphere $\mathbb{S}^{d-1}$, using successively the changes of variables $k \mapsto (r, \theta) = (|k|, \frac{k}{|k|})$, $(r, \theta) \mapsto (r', \theta) = (r - ct, \theta)$ and $(r', \theta) \mapsto k' = r'\theta$, one obtains

$$
\begin{aligned}
\int_{\mathbb{R}^d} F\left(k, \frac{|k| - ct}{|k|}k\right) \mathbb{1}_{|k|-ct>\kappa} \mathrm{d}k &= \int_{\mathbb{R}^+ \times \mathbb{S}^{d-1}} F(r\theta, (r-ct)\theta)\, \mathbb{1}_{r-ct>\kappa}\, r^{d-1}\, \mathrm{d}r\, \mathrm{d}\mu(\theta) \\
&= \int_{\mathbb{R}^+ \times \mathbb{S}^{d-1}} F((r'+ct)\theta, r'\theta)\, \mathbb{1}_{r'>\kappa}\, (r'+ct)^{d-1}\, \mathrm{d}r\, \mathrm{d}\mu(\theta) \\
&= \int_{\mathbb{R}^d} \left(\frac{|k'|+ct}{|k'|}\right)^{d-1} F\left(\frac{|k'|+ct}{|k'|}k', k'\right) \mathbb{1}_{|k'|>\kappa}\, \mathrm{d}k'.
\end{aligned}
$$

2.2   Analysis of the Model

2.2.1   In the Fourier Domain

The objective of this section is to give a meaning to the model when considered in the Fourier domain. This formulation will be considered for the numerical approximation. Looking for the solution as a Gaussian random field $(t, k) \in \mathbb{R}^+ \times \mathbb{R}^d \mapsto \widehat{U}(t, k)$ is not convenient, indeed there is no regularization mechanism in the dynamics, and at any time $t \geq 0$ the mapping $k \mapsto \widehat{U}(t, k)$ would have the same irregularity of white noise: more precisely one would have

$$
\mathbb{E}[\widehat{U}(t, k)\, \overline{\widehat{U}(t, k')}] = \Gamma(t, k)\, \delta(k - k'),
$$

for a function $(t, k) \in \mathbb{R}^+ \times \mathbb{R}^d \mapsto \Gamma(t, k)$. As in the description of the white noise $\xi(t, x)$ and of its Fourier transform $\widehat{\xi}(t, k)$ in Sect. 2.1.4, it is more convenient to consider, at each time $t \geq 0$, the Gaussian random variable $\widehat{U}(t) : L^2(\mathbb{R}^d_k; \mathbb{C}) \to L^2(\Omega; \mathbb{C})$, i.e. the complex-valued random variables $\langle \widehat{U}(t), \phi \rangle$ for all $\phi \in L^2(\mathbb{R}^d_k; \mathbb{C})$. It is even sufficient to deal with functions ϕ that belong to the subspace $\widehat{\mathcal{H}}$. In the sequel, we show how $\widehat{U}(t)$ is defined and we justify the following identity: for all $t \in \mathbb{R}^+$ and all $\phi, \phi' \in \widehat{\mathcal{H}}$ one has

$$
\mathbb{E}[\langle \widehat{U}(t), \phi \rangle \overline{\langle \widehat{U}(t), \phi' \rangle}] = \int_{\mathbb{R}^d} \Gamma(t, k)\phi(k)\overline{\phi'(k)}\mathrm{d}k.
$$

In order to define the process $\left(\widehat{U}(t)\right)_{t\geq 0}$, several ingredients need to be introduced: a Wiener process $\left(\widehat{W}(t)\right)_{t\geq 0}$, the linear operator $\widehat{A}$, the semigroup $\left(\widehat{\Pi}_t\right)_{t\geq 0}$ and its adjoint $\left(\widehat{\Pi}_t^\star\right)_{t\geq 0}$.

For any $t \in \mathbb{R}^+$ and any $\phi \in L^2(\mathbb{R}^d_k)$, set

$$
\langle \widehat{W}(t), \phi \rangle = \langle \widehat{\xi}, \mathbb{1}_{[0,t]} \otimes \phi \rangle. \tag{2.2.1}
$$

where $\left(\mathbb{1}_{[0,t]} \otimes \phi\right)(s, k) = \mathbb{1}_{[0,t]}(s)\, \phi(k)$. For all $t, t' \in \mathbb{R}^+$ and any $\phi, \phi' \in L^2(\mathbb{R}^d_k)$, it follows that

$$\mathbb{E}[\langle \widehat{W}(t), \phi\rangle \overline{\langle \widehat{W}(t'), \phi\rangle}] = \mathbb{E}[\langle \widehat{\xi}, \mathbb{1}_{[0,t]} \otimes \phi\rangle \overline{\langle \widehat{\xi}, \mathbb{1}_{[0,t']} \otimes \phi'\rangle}]$$
$$= (\mathbb{1}_{[0,t]} \otimes \phi, \mathbb{1}_{[0,t']} \otimes \phi')_{L^2_{t,k}}$$
$$= (t \wedge t')\,(\phi, \phi')_{L^2_k},$$

where $t \wedge t' = \min(t, t')$. The above means that $\widehat{W}$ behaves like Brownian motion in the time variable t and like white noise in the wave number variable k. The identity above is a rigorous version of the formal expression for the correlations

$$\mathbb{E}[\widehat{W}(t, k)\overline{\widehat{W}(t', k')}] = (t \wedge t')\,\delta(k - k').$$

Next let us introduce the linear operator $\widehat{A}$ given by

$$\widehat{A} : \widehat{f} \longmapsto (\widehat{A}\widehat{f})(k) = c\,\mathrm{div}\left(\frac{k}{|k|}\,\widehat{f}(k)\right) + c\,\partial_{|k|}\log\big(\sigma(k)\big)\widehat{f}(k) - c\,\frac{d-1}{|k|}\,\widehat{f}(k)$$
$$= c(\partial_{|k|}\widehat{f})(k) + cV(k)\widehat{f}(k)$$

$$(2.2.2)$$

with domain

$$D(\widehat{A}) := \{\widehat{f} \in \widehat{\mathcal{H}} \mid \partial_{|k|}\widehat{f} \in \widehat{\mathcal{H}}, \ V\widehat{f} \in \widehat{\mathcal{H}} \ \text{ and } \ \widehat{f}_{|_{|k|=\kappa}} = 0\},$$

where $\widehat{f}_{|_{|k|=\kappa}}$ stands for the trace

$$\widehat{f}_{|_{|k|=\kappa}}(\kappa\theta) := \lim_{\varepsilon \to 0^+}\widehat{f}((\kappa + \varepsilon)\theta), \quad \forall \theta \in \mathbb{S}^{d-1}$$

which is well defined since $\widehat{f} \in \widehat{\mathcal{H}}$ and $\partial_{|k|}\widehat{f} \in \widehat{\mathcal{H}}$.

2.2.1.1   Well-Posedness in Fourier Domain

Lemma 2.2.1 *Suppose that the function $k \mapsto S(k)$ satisfies Assumption 2.1.2. Then the operator $\widehat{A}$ generates a C^0-semigroup of $\widehat{\mathcal{H}}$ given by*

$$\big(\widehat{\Pi}_t\phi\big)(k) = \frac{\sigma\left(\frac{|k|-ct}{|k|}k\right)}{\sigma(k)}\,\phi\left(\frac{|k|-ct}{|k|}k\right)\mathbb{1}_{|k|-ct>\kappa}, \quad \forall k \in \mathbb{R}^d, \qquad (2.2.3)$$

whose dual operator is given by

$$\big(\widehat{\Pi}_t^\star\phi\big)(k) = \left(\frac{|k|+ct}{|k|}\right)^{d-1}\frac{\sigma(k)}{\sigma\left(\frac{|k|+ct}{|k|}k\right)}\,\phi\left(\frac{|k|+ct}{|k|}k\right)\mathbb{1}_{|k|>\kappa}, \quad \forall k \in \mathbb{R}^d.$$

$$(2.2.4)$$

More precisely, one has for all $\phi, \phi' \in \widehat{\mathcal{H}}$

$$(\widehat{\Pi}_t \phi, \phi')_{L_k^2} = (\phi, \widehat{\Pi}_t^\star \phi')_{L_k^2}. \tag{2.2.5}$$

Proof Our first goal is to show that for any $\widehat{\phi} \in \widehat{\mathcal{H}}$, the function $\widehat{\Pi}_t \widehat{\phi}$ defined by (2.2.3) is indeed an element of $\widehat{\mathcal{H}}$. It is straightforward to check that $\widehat{\Pi}_t \widehat{\phi}(k) = 0$ if $|k| < \kappa$. Applying the change of variable formula (2.12) with

$$F(k_1, k_2) = \left| \frac{\sigma(k_2)}{\sigma(k_1)} \, \widehat{\phi}(k_2) \right|^2,$$

one obtains

$$\int_{\mathbb{R}^d} |\widehat{\Pi}_t \widehat{\phi}(k)|^2 \, \mathrm{d}k = \int_{\mathbb{R}^d} \left| \frac{\sigma\left(\frac{|k|-ct}{|k|}k\right)}{\sigma(k)} \, \widehat{\phi}\left(\frac{|k|-ct}{|k|}k\right) \right|^2 \mathbb{1}_{|k|-ct>\kappa} \, \mathrm{d}k$$

$$= \int_{\mathbb{R}^d} \left(\frac{|k|+ct}{|k|} \right)^{d-1} \left(\frac{\sigma(k)}{\sigma\left(\frac{|k|+ct}{|k|}k\right)} \right)^2 |\widehat{\phi}(k)|^2 \mathbb{1}_{|k|>\kappa} \, \mathrm{d}k$$

$$= \int_{\mathbb{R}^d} \frac{S\left(\frac{|k|+ct}{|k|}k\right)}{S(k)} |\widehat{\phi}(k)|^2 \, \mathrm{d}k,$$

and the desired result follows from (2.6) in Assumption 2.1.2.

It is straightforward to check $\left(\widehat{\Pi}_t\right)_{t\geq 0}$ is a C^0-semigroup, whose generator is given by

$$\partial_t [\widehat{\Pi}_t \widehat{\phi}] = -c\partial_{|k|} \widehat{\phi} - cV\widehat{\phi} = -\widehat{A}\widehat{\phi}, \text{ for any } \widehat{\phi} \in D(\widehat{A}).$$

Finally, (2.2.5) follows from (2.12) with

$$F(k_1, k_2) = \frac{\sigma(k_2)}{\sigma(k_1)} \phi(k_2)\overline{\phi'(k_1)}, \qquad \phi, \phi' \in \widehat{\mathcal{H}}.$$

$\square$

Lemma 2.2.1 guarantees that the initial value problem

$$\begin{cases} \partial_t \widehat{u}(t) + \widehat{A}\widehat{u}(t) = 0, & \forall\, t \geq 0, \\ \widehat{u}(0) = \widehat{u}_0, \end{cases} \tag{2.2.6}$$

is wellposed in $C^0([0, \infty), \widehat{\mathcal{H}})$, the space of space of time continuous function with value in the Hilbert space $\widehat{\mathcal{H}}$, for any initial condition $\widehat{u}_0 \in \widehat{\mathcal{H}}$, and that its solution is given by $\widehat{u}(t) = \widehat{\Pi}_t \widehat{u}_0$. This is by definition the mild solution of (2.2.6).

We are ready to introduce our stochastic model, given by the following stochastic evolution equation

$$\begin{cases} d\widehat{U}(t) + \widehat{A}\,\widehat{U}(t)dt = \widehat{\varphi}d\widehat{W}(t), & \forall\, t \geq 0, \\ \widehat{U}(0) = 0. \end{cases} \tag{2.2.7}$$

A formal mild solution of the stochastic evolution equation is given by the expression

$$\widehat{U}(t) = \int_0^t \widehat{\Pi}_{t-s}\big(\widehat{\varphi}(\cdot)d\widehat{W}(s)\big), \quad \forall\, t \geq 0, \tag{2.2.8}$$

which is interpreted in a weak sense: for any $\phi \in \widehat{\mathcal{H}}$, $\langle \widehat{U}(t), \phi \rangle$ is given by

$$\langle \widehat{U}(t), \phi \rangle = \int_0^t \langle d\widehat{W}(s), \widehat{\varphi}(\cdot)\widehat{\Pi}_{t-s}^\star \phi \rangle, \quad \forall\, t \geq 0. \tag{2.2.9}$$

Note that the expression above is well-defined since $\widehat{\varphi}(\cdot)\widehat{\Pi}_{t-s}^\star \phi \in \widehat{\mathcal{H}}$ as consequence of $\widehat{\varphi} \in \widehat{\mathcal{H}} \cap L_k^\infty$ and the Hölder inequality.

2.2.1.2 Correlation Structure in Fourier Domain

Proposition 2.2.2 *Suppose that φ and S satisfy Assumptions 2.1.1 and 2.1.2, respectively. Then the process defined by (2.2.8)–(2.2.9) satisfies*

$$\mathbb{E}[\langle \widehat{U}(t), \phi \rangle \overline{\langle \widehat{U}(t), \phi' \rangle}] = \int_{\mathbb{R}^d} \Gamma(t, k)\phi(k)\overline{\phi'(k)}dk, \quad \forall \phi, \phi' \in \widehat{\mathcal{H}}, \tag{2.2.10}$$

where

$$\begin{aligned} \Gamma(t, k) &= \int_0^t \left| \widehat{\varphi}\left(\frac{|k| - cs}{|k|}k \right) \right|^2 \frac{S(k)}{S\left(\frac{|k|-cs}{|k|}k \right)} \mathbb{1}_{|k|-cs>\kappa}ds \\ &= S(k) \int_{|k|-ct}^{|k|} \left| \widehat{\varphi}\left(s\frac{k}{|k|} \right) \right|^2 \frac{1}{S\left(s\frac{k}{|k|} \right)} \mathbb{1}_{s>\kappa}\frac{ds}{c}. \end{aligned} \tag{2.2.11}$$

Moreover, $\sup_{t\geq 0} \Gamma(t, \cdot) \in L^1(\mathbb{R}^d)$.

Proof From (2.2.9), one obtains

$$\mathbb{E}[\langle \widehat{U}(t), \phi \rangle \overline{\langle \widehat{U}(t), \phi' \rangle}] = \int_0^t \langle \widehat{\varphi}(\cdot)\widehat{\Pi}_{t-s}^\star \phi, \widehat{\varphi}(\cdot)\widehat{\Pi}_{t-s}^\star \phi' \rangle ds. \tag{2.2.12}$$

Combining (2.2.12) and (2.2.4), and applying the change of variables formula (2.12) with

$$F(k_1, k_2) = |\widehat{\varphi}(k_2)|^2 \left(\frac{|k_1|}{|k_2|} \right)^{d-1} \left(\frac{\sigma(k_2)}{\sigma(k_1)} \right)^2 \phi(k_1)\overline{\phi'(k_1)},$$

one obtains

$$\mathbb{E}[\langle \widehat{U}(t), \phi \rangle \overline{\langle \widehat{U}(t), \phi' \rangle}] = \int_0^t \langle \widehat{\varphi}(\cdot) \widehat{\Pi}^{\star}_{t-s} \phi, \widehat{\varphi}(\cdot) \widehat{\Pi}^{\star}_{t-s} \phi' \rangle \, ds$$

$$= \int_0^t \int_{\mathbb{R}^d} |\widehat{\varphi}(k)|^2 \left(\frac{|k| + cs}{|k|} \right)^{2(d-1)} \left(\frac{\sigma(k)}{\sigma\left(\frac{|k|+cs}{|k|}k\right)} \right)^2 \phi\left(\frac{|k| + cs}{|k|}k \right) \overline{\phi'\left(\frac{|k| + cs}{|k|}k \right)} \mathbb{1}_{|k|>\kappa} \, dk \, ds$$

$$= \int_0^t \int_{\mathbb{R}^d} \left| \widehat{\varphi}\left(\frac{|k| - cs}{|k|}k \right) \right|^2 \left(\frac{|k|}{|k| - cs} \right)^{d-1} \left(\frac{\sigma\left(\frac{|k|-cs}{|k|}k\right)}{\sigma(k)} \right)^2 \phi(k) \, \overline{\phi'(k)} \mathbb{1}_{|k|-cs>\kappa} \, dk \, ds$$

$$= \int_0^t \int_{\mathbb{R}^d} \left| \widehat{\varphi}\left(\frac{|k| - cs}{|k|}k \right) \right|^2 \frac{S(k)}{S\left(\frac{|k|-cs}{|k|}k\right)} \phi(k) \, \overline{\phi'(k)} \mathbb{1}_{|k|-cs>\kappa} \, dk \, ds.$$

Setting

$$\Gamma(t, k) = \int_0^t \left| \widehat{\varphi}\left(\frac{|k| - cs}{|k|}k \right) \right|^2 \frac{S(k)}{S\left(\frac{|k|-cs}{|k|}k\right)} \mathbb{1}_{|k|-cs>\kappa} \, ds.$$

By Assumption 2.1.1,

$$\mathrm{supp}(\widehat{\varphi}) \subset \{ k \in \mathbb{R}^d \mid \kappa \leq |k| \leq \kappa_\varphi \}.$$

Since S is continuous, cf. Assumption 2.1.2, there exists a positive constant $c(\kappa, \kappa_\varphi)$ such that

$$\inf_{0 \leq s \leq \min\{t, \frac{|k|-\kappa}{c}\}} S\left(\frac{|k| - cs}{|k|}k \right) \mathbb{1}_{|k|>\kappa} \geq \inf_{\kappa \leq |k'| \leq \kappa_\varphi} S(k') := c(\kappa, \kappa_\varphi) > 0.$$

Define

$$\Phi\left(t, \frac{k}{|k|} \right) = \int_0^t \left| \widehat{\varphi}\left(s \, \frac{k}{|k|} \right) \right|^2 ds.$$

Since $\widehat{\varphi}$ is continous, the function $\Phi(t, \cdot) : \mathbb{S}^{d-1} \to \mathbb{R}^+$ is continuous. As a result,

$$\Gamma(t, k) \leq \frac{S(k)}{c(\kappa, \kappa_\varphi)} \int_0^t \left| \widehat{\varphi}\left(s \, \frac{k}{|k|} \right) \right|^2 ds \leq \frac{S(k)}{c(\kappa, \kappa_\varphi)} \Phi\left(t, \frac{k}{|k|} \right) \lesssim \frac{S(k)}{c(\kappa, \kappa_\varphi)} \Phi\left(\kappa_\varphi, \frac{k}{|k|} \right).$$

Thanks to the continuity of $\Phi(\kappa_\varphi, \cdot)$, there exists some constant $C > 0$ such that $\Phi\left(\kappa_\varphi, \frac{k}{|k|} \right) \leq C$ for all $k \in \mathbb{R}^d$. In particular,

$$\sup_{t \geq 0} \Gamma(t, k) \leq \frac{C}{c(\kappa, \kappa_\varphi)} S(k),$$

which is in $L^1(\mathbb{R}^d)$ by Assumption 2.1.2.

The second line in (2.2.11) follows from the change of variables $s \mapsto |k| - cs$. $\quad\square$

2.2.2 In the Spatial Domain

In this section, we consider a physical space formulation of the model (2.2.7). One advantage of this point of view is that the mild solution may be viewed as a Gaussian random field, i.e. the random variable $U(t, x)$ will be well-defined for all $t \in \mathbb{R}^+$ and $x \in \mathbb{R}^d$. More precisely, we consider the initial value problem:

$$\begin{cases} dU(t) + AU(t)dt = dW^\varphi(t), & \forall\, t \geq 0, \\ U(0) = 0. \end{cases} \tag{2.2.13}$$

which corresponds to (2.2.7) in physical space.

In order to make sense of (2.2.13), we introduce a Wiener process $\big(W(t)\big)_{t \geq 0}$ defined as follows:

$$\langle W(t), \phi \rangle = \langle \xi, \mathbb{1}_{[0,t]} \otimes \phi \rangle, \qquad \text{for } t \in \mathbb{R}^+ \text{ and } \phi \in L_x^2. \tag{2.2.14}$$

Since the space-time white noise ξ is assumed to be real-valued, $W(t)$ is also real-valued and it suffices to deal with functions $\phi \in L_x^2$ which are-valued.

For any $t, t' \in \mathbb{R}^+$ and $\phi, \phi' \in L_x^2$, one has

$$\mathbb{E}[\langle W(t), \phi \rangle \langle W(t'), \phi' \rangle] = (t \wedge t')(\phi, \phi')_{L_x^2}.$$

Note that the Wiener process $\big(\widehat{W}(t)\big)_{t \geq 0}$ defined by (2.2.1) can be seen as the Fourier transform of the Wiener process $\big(W(t)\big)_{t \geq 0}$ defined by (2.2.14) above, indeed one has

$$\langle \widehat{W}(t), \phi \rangle = \langle W(t), \widehat{\phi} \rangle.$$

In the evolution equation (2.2.7) considered in the Fourier domain, the forcing is given by $\widehat{\varphi}(\cdot)\dot{W}(t)$, therefore in the physical domain it should be given by the convolution $W^\varphi(t) = \varphi * W(t)$, which is defined as follows: for all $t \in \mathbb{R}^+$ and $\phi \in L^2(\mathbb{R}_x^d)$

$$\langle W^\varphi(t), \phi \rangle = \langle W(t), \varphi \star \phi \rangle, \tag{2.2.15}$$

where $\varphi \star \phi$ denotes the correlation product defined in Sect. 2.1.4. For any $t, t' \in \mathbb{R}^+$ and $\phi, \phi' \in L_x^2$, one has

$$\begin{aligned} \mathbb{E}[\langle W^\varphi(t), \phi \rangle \langle W^\varphi(t'), \phi' \rangle] &= \mathbb{E}[\langle W(t), \varphi \star \phi \rangle \langle W(t'), \varphi \star \phi' \rangle] \\ &= (t \wedge t')(\varphi \star \phi, \varphi \star \phi')_{L_x^2} \\ &= (t \wedge t') \int_{\mathbb{R}_x^d} \int_{\mathbb{R}_x^d} C_\varphi(x - x')\phi(x)\phi(x')dxdx', \end{aligned}$$

where the mapping C_φ is given by

$$C_\varphi(x) = \int_{\mathbb{R}^d_x} \varphi(x+y)\varphi(y)\mathrm{d}y = (\varphi \star \varphi)(x), \quad \forall\, x \in \mathbb{R}^d.$$

Note that the Fourier transform of C_φ is given by $\widehat{C}_\varphi(k) = |\widehat{\varphi}(k)|^2$ for all $k \in \mathbb{R}^d$.

One may interpret $\langle W^\varphi(t), \phi \rangle$ as the integral $\int_{\mathbb{R}^d_x} W^\varphi(t,x)\phi(x)\mathrm{d}x$, where $\left(W^\varphi(t,x)\right)_{t \geq 0, x \in \mathbb{R}^d}$ is a spatially homogeneous Gaussian random field which satisfies

$$\mathbb{E}[W^\varphi(t,x)W^\varphi(t',x')] = (t \wedge t')C_\varphi(x - x'), \quad \forall\, t, t' \in \mathbb{R}^+, \forall\, x, x' \in \mathbb{R}^d.$$

Next, let us introduce the operator A in (2.2.13), which corresponds to (2.2.2) in Fourier space:

$$A = \mathcal{F}^{-1}\widehat{A}\mathcal{F} \tag{2.2.16}$$

that is to say

$$Af = \mathcal{F}^{-1}[\widehat{A}\widehat{f}], \quad \forall f \in D(A) := \{f \in \mathcal{H} \,|\, \widehat{f} \in D(\widehat{A})\}.$$

It turns out that the operator A may be written as a pseudo-differential operator of the form:

$$Af(x) = \int_{\mathbb{R}^d_k} e^{2\pi i k \cdot x}\, a(x,k)\, \mathbb{1}_{|k|>\kappa}\, \widehat{f}(k)\, dk,$$

with symbol

$$a(x,k) = -c\,\frac{2\pi i\, k \cdot x + (d-1)}{|k|} + c\, V(k),$$

for regular function f.

Remark 2.2.3 When $S(|k|) = |k|^\alpha$ for some $\alpha \in \mathbb{R}$, $V(k) = \alpha\,|k|^{-1}$ and the principal symbol of $a(x,k)$ is $-c\,\frac{2\pi i\, k\cdot x}{|k|}\,k$ which is a degree 0 symbol. See Apolinário et al. (2023, Proposition 4.6) and the comments therein.

Let us introduce the C^0-semigroups

$$\left(\Pi_t\right)_{t \geq 0} = \left(\mathcal{F}^{-1}\widehat{\Pi}_t\mathcal{F}\right)_{t \geq 0} \quad \text{and} \quad \left(\Pi_t^\star\right)_{t \geq 0} = \left(\mathcal{F}^{-1}\widehat{\Pi}_t^\star\mathcal{F}\right)_{t \geq 0} \tag{2.2.17}$$

of linear operators on $\mathcal{H}$. Lemma 2.2.1 shows that A generates the C^0-semigroup $\left(\Pi_t\right)_{t \geq 0}$ and $\left(\Pi_t^\star\right)_{t \geq 0}$ is the dual of $\left(\Pi_t\right)_{t \geq 0}$ in the sense that

$$(\Pi_t\phi, \phi')_{L^2_x} = (\phi, \Pi_t^\star\phi')_{L^2_x}, \tag{2.2.18}$$

for all $t \in \mathbb{R}^+$ and $\phi, \phi' \in \mathcal{H}$ as a consequence of the fact that the Fourier transform is an isometry from L^2_x to L^2_k.

2.2.2.1   Wellposedness in Space Domain

In the spirit of the expression of the solution (2.2.8) in the Fourier domain, we claim that there exists a mild solution to (2.2.13) given by

$$U(t) = \int_0^t \Pi_{t-s} dW^\varphi(s), \quad \forall\, t \in \mathbb{R}^+, \tag{2.2.19}$$

which as to be interpreted in a weak sense: for all $\phi \in \mathcal{H}$ one has

$$\langle U(t), \phi \rangle = \int_0^t \langle dW^\varphi(s), \Pi_{t-s}^\star \phi \rangle, \quad \forall\, t \in \mathbb{R}^+. \tag{2.2.20}$$

The definition above can be used to derive a more convenient expression for $\langle U(t), \phi \rangle$: one has with (2.9)

$$\langle U(t), \phi \rangle = \int_0^t \langle dW(s), \varphi \star \left(\Pi_{t-s}^\star \phi \right) \rangle,$$

where the last stochastic integral above can be written as

$$\int_0^t \langle dW(s), \varphi \star \left(\Pi_{t-s}^\star \phi \right) \rangle = \int_0^{+\infty} \langle dW(s), \Phi_t(s, \cdot) \rangle \rangle = \langle \xi, \Phi_t \rangle = \int_0^{+\infty} \int_{\mathbb{R}^d} \Phi_t(s, y) d\xi(s, y),$$

with $\Phi_t(s, y) = \mathbb{1}_{[0,t]}(s)\left(\varphi \star \left(\Pi_{t-s}^\star \phi \right) \right)(y)$ for all $s \in \mathbb{R}^+$ and $y \in \mathbb{R}^d$. In particular, for all $s \in [0, t]$ and all $y \in \mathbb{R}^d$, one has

$$\Phi_t(s, y) = \int_{\mathbb{R}^d} \varphi(x + y)\left(\Pi_{t-s}^\star \phi \right)(x) dx = \langle \varphi(\cdot + y), \Pi_{t-s}^\star \phi \rangle = \langle \Pi_{t-s}\varphi(\cdot + y), \phi \rangle$$

where the last equality follows by duality, cf. (2.2.18). We may therefore introduce the kernel G defined by:

$$G(t, x, y) = (\Pi_t \varphi(\cdot + y))(x), \quad \forall\, t \in \mathbb{R}^+, \forall\, x, y \in \mathbb{R}^d. \tag{2.2.21}$$

Combining the results above, one obtains

$$\langle U(t), \phi \rangle = \int_0^t \int_{\mathbb{R}^d} \langle G(t - s, \cdot, y), \phi \rangle d\xi(s, y) = \langle \int_0^t \int_{\mathbb{R}^d} \langle G(t - s, \cdot, y) d\xi(s, y), \phi \rangle.$$

The last equality is justified by the following result.

Proposition 2.2.4 *The function*

$$U(t, x) = \int_0^t \int_{\mathbb{R}^d} G(t - s, x, y) d\xi(s, y), \tag{2.2.22}$$

is a well-defined Gaussian random field which is a mild solution to (2.2.13) in the sense of (2.2.19)–(2.2.20). Moreover, for any multi-index β, any $t_1, t_2 > 0$ and any $x_1, x_2 \in \mathbb{R}^d$ there exists a (locally uniform) constant $C > 0$ such that

$$\mathbb{E}|\partial_x^\beta U(t_1, x_1) - \partial_x^\beta U(t_2, x_2)|^2 \leq C\left(|t_1 - t_2| + |x_1 - x_2|^2\right). \tag{2.2.23}$$

It remains to justify that the computations above are valid. Properties of the mapping G are given in Lemma 2.2.5 below. Moreover, one needs to show that the Gaussian random field $(U(t, x))_{t \geq 0, x \in \mathbb{R}^d}$ is well-defined and to study its regularity properties.

Lemma 2.2.5 *Suppose that φ and S satisfy Assumptions 2.1.1 and 2.1.2, respectively, and let $G(t, x, y)$ be defined by (2.2.21). Then the following statements hold:*

(i) The Fourier transform in the x-variable of G is given by

$$\widetilde{G}(t, k, y) = e^{-2\pi i \frac{|k|-ct}{|k|} k \cdot y} \frac{\sigma\left(\frac{|k|-ct}{|k|}k\right)}{\sigma(k)} \varphi\left(\frac{|k| - ct}{|k|}k\right) \mathbb{1}_{|k|-ct > \kappa}. \tag{2.2.24}$$

(ii) The Fourier transform in the y-variable of G is given by

$$\widehat{G}(t, x, k) = e^{-2\pi i \frac{|k|+ct}{|k|} k \cdot x} \left(\frac{|k| + ct}{|k|}\right)^{d-1} \frac{\sigma(k)}{\sigma\left(\frac{|k|+ct}{|k|}k\right)} \overline{\widehat{\varphi}(k)} \mathbb{1}_{|k| > \kappa}. \tag{2.2.25}$$

(iii) For each multi-index β and for fixed $(t, x) \in \mathbb{R}^+ \times \mathbb{R}^d$, $\partial_x^\beta G(t, x, \cdot) \in L^2(\mathbb{R}^d)$.
(iv) For all $y \in \mathbb{R}^d$, the mapping $G_y : (t, x) \in \mathbb{R}^+ \times \mathbb{R}^d \mapsto G(t, x, y)$ is a solution to

$$\begin{cases} (\partial_t + A)G_y = 0, & \forall t > 0, \\ G(0, x, y) = \varphi(x + y), & \forall x, y \in \mathbb{R}^d \times \mathbb{R}^d. \end{cases}$$

Proof (i) Using (2.2.3), (2.2.17) and (2.2.21), we obtain

$$G(t, x, y) = \int_{\mathbb{R}^d} e^{2\pi i\, k \cdot x}\, e^{-2\pi i \frac{|k|-ct}{|k|} k \cdot y} \frac{\sigma\left(\frac{|k|-ct}{|k|}k\right)}{\sigma(k)} \varphi\left(\frac{|k| - ct}{|k|}k\right) \mathbb{1}_{|k|-ct > \kappa}\, dk, \tag{2.2.26}$$

from which the Fourier transform in the x-variable follows.

(ii) In order to compute the Fourier transform of G in the y variable, it suffices to perform the change of variables (2.12) on the right-hand side of (2.2.26). More precisely,

$$G(t, x, y) = \int_{\mathbb{R}^d_k} e^{2\pi i k \cdot y}\, e^{2\pi i \frac{|k|+ct}{|k|} k \cdot x} \left(\frac{|k| + ct}{|k|}\right)^{d-1} \frac{\sigma(k)}{\sigma\left(\frac{|k|+ct}{|k|}k\right)} \widehat{\varphi}(k)\, \mathbb{1}_{|k| > \kappa}\, dk,$$

from which $\widehat{G}(t, x, k)$ follows.

(iii) By (2.6), there exists a constant $C(t) > 0$ such that

$$\sup_{k \in \mathbb{R}^d} \frac{S\left(\frac{|k|+ct}{|k|}k\right)}{S(k)} \le C(t).$$

By the continuity of S, cf. Assumption 2.1.2, the function $t \mapsto C(t)$ is also continuous.

Differentiating (2.2.25) and using the Plancherel theorem, we obtain

$$\left\|\partial_x^\beta G(t, x, \cdot)\right\|_{L^2(\mathbb{R}_y^d)}^2 = \left\|\partial_x^\beta \widehat{G}(t, x, \cdot)\right\|_{L^2(\mathbb{R}_k^d)}^2$$

$$\le \int_{\mathbb{R}_k^d} \left(\frac{|k|+ct}{|k|}\right)^{2(d-1)+2|\beta|} \frac{\sigma(k)^2}{\sigma\left(\frac{|k|+ct}{|k|}k\right)^2} |\widehat{\varphi}(k)|^2\, \mathbb{1}_{|k|>\kappa}\, dk$$

$$\le \int_{\mathbb{R}_k^d} \left(\frac{|k|+ct}{|k|}\right)^{(d-1)+2|\beta|} \frac{S\left(\frac{|k|+ct}{|k|}k\right)}{S(k)} |\widehat{\varphi}(k)|^2\, \mathbb{1}_{|k|>\kappa}\, dk$$

$$\le C(t) \left(1 + \frac{ct}{\kappa}\right)^{d-1+2|\beta|} \|\varphi\|_{L^2}^2 \tag{2.2.27}$$

(iv) Note that

$$\left(\partial_t + c\, \partial_{|k|}\right)\left[(|k| - ct)\frac{k}{|k|}\right] = 0.$$

Applying the operator $\partial_t + c\, \partial_{|k|}$ to both sides of (2.2.24), it easily follows that:

$$\left(\partial_t + c\, \partial_{|k|}\right)\widecheck{G}(t, k, y)$$

$$= e^{-2\pi i \frac{|k|-ct}{|k|} k \cdot y} \sigma\left(\frac{|k|-ct}{|k|}k\right) \varphi\left(\frac{|k|-ct}{|k|}k\right) \mathbb{1}_{|k|-ct>\kappa} \left(\partial_t + c\, \partial_{|k|}\right)\left[\sigma(k)^{-1}\right]$$

$$= e^{-2\pi i \frac{|k|-ct}{|k|} k \cdot y} \sigma\left(\frac{|k|-ct}{|k|}k\right) \varphi\left(\frac{|k|-ct}{|k|}k\right) \mathbb{1}_{|k|-ct>\kappa}\, c\, \partial_{|k|}\left[\sigma(k)^{-1}\right]$$

$$= \widetilde{G}(t, k, y)\, c\, \sigma(k)\, \partial_{|k|}\left[\sigma(k)^{-1}\right].$$

Finally, it suffices to use the observation that

$$\sigma(k)\, \partial_{|k|}\left[\sigma(k)^{-1}\right] = -\sigma(k)^{-1}\, \partial_{|k|}\sigma(k) = -\partial_{|k|}\left(\log\sigma(k)\right) = -V(k).$$

$\square$

Proof of Proposition 2.2.4 The fact that $U(t, x)$ in (2.2.22) is a well-defined Gaussian random variable follows from point (iii) in Lemma 2.2.5 and the Itô formula. Similarly, the proof that $U(t, x)$ solves (2.2.13) is a straight-forward application of (iv) in

Lemma 2.2.5. Finally, property (2.2.23) follows from the representation (2.2.22) and Lemma 2.2.5; the proof is similar to that of Apolinário et al. (2023, Corollary 4.9) and we omit it. $\qquad\square$

Remark 2.2.6 Introducing an appropriate notion of weak solutions to (2.2.13), one may prove that weak solutions are unique, which implies that (2.2.22) is the unique solution to (2.2.13) in this class. See Sect. 4.1.1. in Apolinário et al. (2023) for additional details.

2.2.2.2 Correlation Structure in Space Domain

By Proposition 2.2.4, $U(t, x)$ in (2.2.22) is a well-defined Gaussian random field. Next we compute its two-point correlation.

Proposition 2.2.7 *Suppose that φ and S satisfy Assumptions 2.1.1 and 2.1.2, respectively. Then the solution $U(t, x)$ to (2.2.13) has the following correlation structure*

$$\mathbb{E}[U(t, x)U(t, x')] = \int_{\mathbb{R}^d} e^{2\pi i k \cdot (x - x')} \Gamma(t, k) dk, \qquad (2.2.28)$$

where $\Gamma(t, k)$, given by (2.2.11), is absolutely integrable.

Proof Using the expression (2.2.22), the Plancherel theorem and (2.2.25), one obtains

$$
\begin{aligned}
\mathbb{E}[U(t, x)U(t, x')] &= \int_0^t \int_{\mathbb{R}^d} G(t - s, x, y)G(t - s, x', y) ds dy \\
&= \int_0^t \int_{\mathbb{R}^d} \widehat{G}(t - s, x, k)\overline{\widehat{G}(t - s, x', k)} ds dk \\
&= \int_0^t \int_{\mathbb{R}^d} e^{2\pi i \frac{|k| + ct}{|k|} k \cdot (x - x')} \left(\frac{|k| + ct}{|k|} \right)^{2(d-1)} \frac{|\sigma(k)|^2}{\left| \sigma\left(\frac{|k| + ct}{|k|} k \right) \right|^2} |\widehat{\varphi}(k)|^2 \, \mathbb{1}_{|k| > \kappa} dk ds.
\end{aligned}
$$

Using the change of variable formula (2.12), one then obtains

$$\mathbb{E}[U(t, x)U(t, x')]$$

$$
\begin{aligned}
&= \int_0^t \int_{\mathbb{R}^d} e^{2\pi i k \cdot (x - x')} \left| \widehat{\varphi}\left(\frac{|k| - cs}{|k|} k \right) \right|^2 \left(\frac{|k|}{|k| - cs} \right)^{d-1} \left(\frac{\sigma\left(\frac{|k| - cs}{|k|} k \right)}{\sigma(k)} \right)^2 \mathbb{1}_{|k| - cs > \kappa} dk ds \\
&= \int_0^t \int_{\mathbb{R}^d} e^{2\pi i k \cdot (x - x')} \left| \widehat{\varphi}\left(\frac{|k| - cs}{|k|} k \right) \right|^2 \frac{S(k)}{S\left(\frac{|k| - cs}{|k|} k \right)} \mathbb{1}_{|k| - cs > \kappa} dk ds \\
&= \int_{\mathbb{R}^d} e^{2\pi i k \cdot (x - x')} \Gamma(t, k) dk,
\end{aligned}
$$

where we use formula (2.2.11) in the last step. The use of the Fubini theorem is allowed since $\Gamma(t, \cdot) \in L^1(\mathbb{R}^d)$ by Proposition 2.2.2. $\qquad\square$

Note that the above result coincides with (2.2.10). Indeed, for all $\phi, \phi' \in \widehat{\mathcal{H}}$, (2.2.28) yields

$$
\mathbb{E}[\langle \widehat{U}(t), \phi \rangle \overline{\langle \widehat{U}(t), \phi' \rangle}] = \mathbb{E}[\langle U(t), \widehat{\phi} \rangle \overline{\langle U(t), \widehat{\phi'} \rangle}]
$$

$$
= \mathbb{E}\left[\int_{\mathbb{R}^d_x \times \mathbb{R}^d_{x'}} \widehat{\phi}(x)\overline{\widehat{\phi'}(x')}\, U(t,x)\, \overline{U(t,x')}\, \mathrm{d}x\, \mathrm{d}x' \right]
$$

$$
= \int_{\mathbb{R}^d_x \times \mathbb{R}^d_{x'}} \widehat{\phi}(x)\overline{\widehat{\phi'}(x')} \int_{\mathbb{R}^d_k} e^{2\pi i k\cdot(x-x')}\Gamma(t,k)\mathrm{d}k\, \mathrm{d}x\, \mathrm{d}x'
$$

$$
= \int_{\mathbb{R}^d_k} \phi(k)\overline{\phi'(k)}\, \Gamma(t,k)\mathrm{d}k.
$$

The two formulations of the problem studied above are thus consistent.

2.2.3 Asymptotic Behavior

2.2.3.1 Limiting Correlation Structure

Under Assumptions 2.1.1 and 2.1.2, one may use the dominated convergence theorem on the expression (2.2.11) to take the limit as $t \to \infty$, which yields:

$$
\Gamma(\infty, k) := \lim_{t \to \infty} \Gamma(t, k) = \mathbb{1}_{|k|>\kappa}\, S(k) \int_\kappa^{|k|} \left| \widehat{\varphi}\left(s\frac{k}{|k|} \right) \right|^2 \frac{1}{S\left(s\frac{k}{|k|} \right)} \frac{\mathrm{d}s}{c}. \tag{2.2.29}
$$

If $\widehat{\varphi}$ and S are radial, one has

$$
\frac{\Gamma(\infty, k)}{S(k)} = \int_\kappa^{|k|} \left| \widehat{\varphi}\left(s\frac{k}{|k|} \right) \right|^2 \frac{1}{S\left(s\frac{k}{|k|} \right)} \frac{\mathrm{d}s}{c} = \mathfrak{C} \qquad \text{for all } |k| \geq \kappa_\varphi. \tag{2.2.30}
$$

As a result, the ratio $\frac{\Gamma(\infty,k)}{S(k)}$ is constant for all $|k| \geq \kappa_\varphi$.

Remark 2.2.8 For the purpose of numerical simulations, we will normalize $\widehat{\varphi}$ in such a way that the ratio $\frac{\Gamma(\infty,k)}{S(k)} = 1$ for large $|k|$, cf. (2.4.9).

Corollary 2.2.9 *Suppose that φ and S satisfy Assumptions 2.1.1 and 2.1.2, respectively. Then the solution $U(t,x)$ to (2.2.13) has the following asymptotic correlation structure*

$$
\lim_{t \to \infty} \mathbb{E}[U(t,x)U(t,x')] = \int_{\mathbb{R}^d} e^{2\pi i k\cdot(x-x')} S(k)\, \vartheta(k)\, \mathrm{d}k, \tag{2.2.31}
$$

where ϑ was defined in (2.8).

Proof By Proposition 2.2.7,

$$\mathbb{E}[U(t, x)U(t, x')] = \int_{\mathbb{R}^d} e^{2\pi i k \cdot (x-x')} \Gamma(t, k) dk,$$

where $\Gamma(t, k)$ is given in (2.2.11).

By Proposition 2.2.2, $\sup_{t \geq 0} \Gamma(t, \cdot) \in L^1(\mathbb{R}^d)$ and we may thus use the dominated convergence to take the limit $t \to \infty$ inside the integral, namely

$$\lim_{t \to \infty} \mathbb{E}[U(t, x)U(t, x')] = \int_{\mathbb{R}^d} e^{2\pi i k \cdot (x-x')} \Gamma(\infty, k) dk.$$

Finally, by (2.2.29),

$$\Gamma(\infty, k) = S(k)\,\vartheta(k)$$

for ϑ as in (2.8). $\qquad\square$

2.2.3.2 Loss of Regularity

In the case $S(k) = |k|^{-(d+2H)}$, our next result shows that the energy cascade results in a *loss of regularity* phenomenon at infinite time.

Proposition 2.2.10 *Suppose that $S(k) = |k|^{-(d+2H)}$ for $H \in (0, 1)$ and suppose that φ satisfies Assumption 2.1.1. Then the solution $U(t, x)$ to (2.2.13) converges in law to zero-mean Gaussian field $U_\infty(x)$ with correlation structure given by*

$$\mathbb{E}[U_\infty(x)\, U_\infty(x')] = \int_{\mathbb{R}^d_k} e^{2\pi i\, k \cdot (x-x')} |k|^{-d-2H} \, \vartheta(k)\, dk \qquad (2.2.32)$$

where

$$\vartheta(k) = \mathbb{1}_{|k|>\kappa} \int_\kappa^{|k|} s^{d+2H} \left| \widehat{\varphi}\left(s \frac{k}{|k|}\right) \right|^2 ds. \qquad (2.2.33)$$

Moreover, a loss of regularity *phenomenon takes place: while $x \mapsto U(t, x)$ is smooth for each finite time $t \geq 0$, $x \mapsto U_\infty(x)$ is only α-Hölder continuous for any $0 < \alpha < H$, namely*

$$\mathbb{E}|U_\infty(x) - U_\infty(x')|^2 \leq \frac{C}{H(1-H)} |x - x'|^{2H} + C |x - x'|^2, \quad \forall x, x' \qquad (2.2.34)$$

where the constant $C > 0$ depends of H and d without blowing up when H tends to 0 or 1.

Proof Formulas (2.2.32) and (2.2.33) follow from Corollary 2.2.9 since $S(k) = |k|^{-d-2H}$ satisfies Assumption 2.1.2. Next we set

$$V(t, x) = \int_0^t \int_{\mathbb{R}^d} G(s, x, y) \mathrm{d}\xi(s, y),$$

where G was defined in (2.2.21). Note that $V(t, x)$ has the same distribution as $U(t, x)$ in (2.2.22), as they are both Gaussian fields whose mean and correlations coincide by construction. Moreover, the Itô formula quickly shows that $V(t, x)$ converges to

$$U_\infty(x) = \int_0^\infty \int_{\mathbb{R}^d} G(s, x, y) \mathrm{d}\xi(s, y), \tag{2.2.35}$$

in $L^2(\Omega; \mathbb{C})$, since

$$\mathbb{E}|V(t, x) - U_\infty(x)|^2 = \frac{1}{2cH} \int_{\mathbb{R}_k^d} (|k| + ct)^{-2H} |k|^{2H+1} |\widehat{\varphi}(k)|^2 \, \mathrm{d}k.$$

See Apolinário et al. (2023, Corollary 4.11) for additional details.

Finally, using (2.2.31) together with the fact that $\mathbb{E}[U(t, x)\overline{U(t, x')}] = \mathbb{E}[V(t, x)\overline{V(t, x')}]$, the field (2.2.35) is shown to be a zero-mean Gaussian field with the desired correlation structure.

The proof of (2.2.34) may be found in Apolinário et al. (2023, Corollary 4.12). $\square$

Remark 2.2.11 It is easy to show that the inequality (2.2.34) is sharp as $|x - x'| \to 0$ by choosing $\widehat{\varphi}$ to be a (possibly regularized) indicator function over an isotropic region.

Indeed, setting $\ell = x - x'$, one has

$$\mathbb{E}|U_\infty(x) - U_\infty(x')|^2 \geq \int_{\mathbb{R}_k^d} [1 - \cos(2\pi k \cdot \ell)] \, |k|^{-d-2H} \, \vartheta(k) \, \mathrm{d}k.$$

Assuming that $\widehat{\varphi}(k) = \mathbb{1}_{[1,2]}(|k|)$, we have that

$$|k|^{-2H-d} \, \vartheta(k) \geq |k|^{-2H-d} \, \mathbb{1}_{|k|>2} \, (2^{2H+d+1} - 1^{2H+d+1}).$$

As a result, for any $\ell \in \mathbb{R}^d$ with $|\ell| < 1$,

$$\mathbb{E}|U_\infty(x) - U_\infty(x')|^2 \gtrsim_H \int_0^\infty \int_{\mathbb{S}^{d-1}} \left[1 - \cos\left(2\pi|k||\ell|\gamma \cdot \frac{\ell}{|\ell|}\right)\right] \mathbb{1}_{|k|>2} \, |k|^{-2H-1} \, d\mu(\gamma) \, d|k|$$

$$\gtrsim_H |\ell|^{2H} \int_0^\infty \int_{\mathbb{S}^{d-1}} \left[1 - \cos\left(2\pi r \gamma \cdot \frac{\ell}{|\ell|}\right)\right] \mathbb{1}_{r>2} \, r^{-2H-1} \, d\mu(\gamma) \, dr \gtrsim_H |\ell|^{2H}$$

after changing variables to spherical coordinates so that $k \cdot \ell = |k| \, |\ell| \cos \theta$ and performing a dilation in the radial direction $r = |\ell| \, |k|$. Finally, one may repeat this argument with a slightly regularized $\mathbb{1}_{[1,2]}(|k|)$ in order to make $\widehat{\varphi}$ continuous.

2.3 Numerical Method

In this section, we describe a numerical method which allows us to obtain numerical simulations for the model analyzed above. We explain below how the method introduced in the recent article Beck (2024) by the authors of this article can be generalized to deal with a general class of spectra which are not power laws.

To define the numerical method, it is more convenient to consider the formulation of the model in the Fourier domain.

2.3.1 Spatial Discretization: Finite Volume Approximation

To define the numerical scheme, it is convenient to consider a conservative form of the first-order differential operator: one considers

$$
\begin{cases}
\mathrm{d}\widehat{U}(t,k) + c\,\nabla_k \cdot \left(\dfrac{k}{|k|}\widehat{U}(t,k) \right)\mathrm{d}t + c\widetilde{V}(k)\,\widehat{U}(t,k)\mathrm{d}t = \widehat{\varphi}(k)\mathrm{d}\widehat{W}(t,k), & \forall\, t > 0,\, |k| > \kappa > 0, \\
\widehat{U}(t,k) = 0, & \forall\, |k| \le \kappa, \\
\widehat{U}(0,k) = 0,
\end{cases}
$$

$$
\tag{2.3.1}
$$

where the modified potential $\widetilde{V}$ is given by

$$
\widetilde{V}(k) = V(k) - \frac{d-1}{|k|}.
$$

Given a cell $\mathcal{K} \subset \mathbb{R}_k^d$, we denote by $|\mathcal{K}|$ its volume, by $\partial\mathcal{K}$ its boundary and by n the local unit outward normal vector to $\partial\mathcal{K}$. Due to the boundary conditions imposed in the evolution equation, all the cells are assumed to be subsets of $\{k \in \mathbb{R}^d \mid |k| \ge \kappa\}$.

To define a finite volume approximation, it is standard to consider averages of the solution over cells $\mathcal{K}$, i.e. to set

$$
\widehat{U}_\mathcal{K}(t) = \frac{1}{|\mathcal{K}|} \int_\mathcal{K} \widehat{U}(t,k)\mathrm{d}k.
$$

However the Gaussian random field $\widehat{U}$ is not defined pointwise. In this setting it is more convenient to consider

$$
\widehat{U}_\mathcal{K}(t) = \langle \widehat{U}(t), \frac{1}{|\mathcal{K}|}\mathbb{1}_\mathcal{K} \rangle,
$$

however for the first steps of the construction of the scheme let us make formal computations and assume that $\widehat{U}$ has a pointwise meaning.

Note that applying the Stokes formula, one obtains

$$\frac{1}{|\mathcal{K}|}\int_{\mathcal{K}}\nabla_k\cdot\left(\frac{k}{|k|}\widehat{U}(t,k)\right)\mathrm{d}k = \frac{1}{|\mathcal{K}|}\int_{\partial\mathcal{K}}\widehat{U}(t,k)\frac{k\cdot\mathrm{n}}{|k|}\mathrm{d}k.$$

The integration of the evolution equation over a cell $\mathcal{K}$ then gives

$$\mathrm{d}\widehat{U}_{\mathcal{K}}(t) + \frac{c}{|\mathcal{K}|}\int_{\partial\mathcal{K}}\widehat{U}(t,k)\frac{k\cdot\mathrm{n}}{|k|}\mathrm{d}k\mathrm{d}t + \frac{c}{|\mathcal{K}|}\int_{\mathcal{K}}\widehat{U}(t,k)\widetilde{V}(k)\mathrm{d}k\mathrm{d}t = \mathrm{d}\widehat{W}_{\mathcal{K}}^{\varphi}(t)$$

$$(2.3.2)$$

where $\widehat{W}_{\mathcal{K}}^{\varphi}(t)$ is defined as

$$\widehat{W}_{\mathcal{K}}^{\varphi}(t) = \langle\widehat{W}(t),\widehat{\varphi}\frac{1}{|\mathcal{K}|}\mathbb{1}_{\mathcal{K}}\rangle.$$

It is straightforward to check that $\left(\widehat{W}_{\mathcal{K}}^{\varphi}(t)\right)_{t\geq 0}$ is a complex-valued Wiener process, with

$$\mathbb{E}[\widehat{W}_{\mathcal{K}}^{\varphi}(t)\overline{\widehat{W}_{\mathcal{K}}^{\varphi}(t')}] = \frac{(t\wedge t')}{|\mathcal{K}|^2}\langle\widehat{\varphi}\mathbb{1}_{\mathcal{K}},\widehat{\varphi}\mathbb{1}_{\mathcal{K}}\rangle = \frac{(t\wedge t')}{|\mathcal{K}|^2}\int_{\mathcal{K}}|\widehat{\varphi}(k)|^2\mathrm{d}k$$

Moreover, if $\mathcal{K}_1$, $\mathcal{K}_2$ are disjoint cells ($\mathcal{K}_1\cap\mathcal{K}_2 = \emptyset$), then one has

$$\mathbb{E}[\widehat{W}_{\mathcal{K}_1}^{\varphi}(t)\overline{\widehat{W}_{\mathcal{K}_2}^{\varphi}(t')}] = 0.$$

The evolution equation (2.3.2) for $\widehat{U}_{\mathcal{K}}(t)$ above is exact but cannot be implemented, two terms on the right-hand side need to be approximated. This requires to impose a choice for the cells.

As explained in Beck (2024), it is convenient to consider spherical coordinates, i.e. to write $k\neq 0$ as $k = r\theta$ where $r = |k| > 0$ is the radial component of k and $\theta = k/|k|\subset\mathbb{S}^{d-1}$ is the angular component of k. Due to the boundary conditions in the model, one only need to consider a mesh of the set $\{k\in\mathbb{R}^d\mid|k|\geq\kappa\}$. The finite volume mesh is denoted by $\left(\mathcal{K}_{i,a}\right)_{i\geq 1,a\in\mathcal{A}}$, where $\mathcal{A}$ is a finite set, and the cell $\mathcal{K}_{i,a}$ is defined by

$$\mathcal{K}_{i,a} = \{k = |k|\theta\mid\rho_{i-\frac{1}{2}} < |k| < \rho_{i+\frac{1}{2}};\theta\in\Theta_a\},$$

where $\left(\rho_{i-\frac{1}{2}}\right)_{i\geq 1}$ is an increasing sequence such that $\rho_{\frac{1}{2}} = \kappa$ and $\rho_{i-\frac{1}{2}}\underset{i\to+\infty}{\to}+\infty$, and where $\left(\Theta_a\right)_{a\in\mathcal{A}}$ is a mesh of the sphere $\mathcal{S}^{d-1}$. It is assumed that $\left(\Theta_a\right)_{a\in\mathcal{A}}$ is symmetric with respect to the origin, and that it is uniform. For instance, if $d = 2$, this mesh can be defined by setting

$$\Theta_a = \{(\cos\vartheta,\sin\vartheta); a\Delta\vartheta < \vartheta < (a+1)\Delta\vartheta\},$$

for all $a\in\{0,\ldots,2N_\vartheta - 1\}$, with angular resolution $\Delta\vartheta = \pi/N_\vartheta$.

The numerical scheme provides an approximation $\left(\widehat{U}_{i,a}\right)_{i\geq 1,a\in\mathcal{A}}$ of $\left(\widehat{U}_{\mathcal{K}_{i,a}}\right)_{i\geq 1,a\in\mathcal{A}}$. The main observation is that the boundary $\partial\mathcal{K}_{i,a}$ of a cell $\mathcal{K}_{i,a}$ defined as above can be decomposed into three parts, and that $k \cdot n$ vanishes on one of these parts. Then using an upwind approximation, one can approximate the advection contribution as

$$\frac{c}{|\mathcal{K}_{i,a}|} \int_{\partial \mathcal{K}_{i,a}} \widehat{U}(t,k) \frac{k}{|k|} \cdot n \; dk \simeq c \frac{\widehat{U}_{i,a}(t) - \widehat{U}_{i-1,a}(t)}{h_i} + d_i \widehat{U}_{i,a}(t), \qquad (2.3.3)$$

with the convention $\widehat{U}_{0,a}(t) = 0$, where h_i and d_i are defined for all $i \geq 1$ by

$$h_i = \frac{\rho_{i+\frac{1}{2}}^d - \rho_{i-\frac{1}{2}}^d}{\rho_{i-\frac{1}{2}}^{d-1}}, \; d_i = cd \frac{\rho_{i+\frac{1}{2}}^{d-1} - \rho_{i-\frac{1}{2}}^{d-1}}{\rho_{i+\frac{1}{2}}^d - \rho_{i-\frac{1}{2}}^d}. \qquad (2.3.4)$$

We refer to Beck (2024) for the details.

Concerning the last term that remains to be dealt with in (2.3.2) when $\mathcal{K} = \mathcal{K}_{i,a}$, one can approximate

$$\frac{1}{|\mathcal{K}_{i,a}|} \int_{\mathcal{K}_{i,a}} \widehat{U}(t,k) \widetilde{V}(k) dk \approx \frac{\widehat{U}_{i,a}(t)}{|\mathcal{K}_{i,a}|} \int_{\mathcal{K}_{i,a}} \widetilde{V}(k) dk.$$

For all $i \geq 1$, set

$$\rho_i = \frac{\rho_{i+\frac{1}{2}} + \rho_{i-\frac{1}{2}}}{2},$$

and for all $a \in \mathcal{A}$ let $\theta_a \in \Theta_a$ be the center of Θ_a. Recalling the definition of $\widetilde{V}$, one has

$$\begin{aligned}
\frac{1}{|\mathcal{K}_{i,a}|} \int_{\mathcal{K}_{i,a}} \widetilde{V}(k) dk &= \frac{1}{|\mathcal{K}_{i,a}|} \int_{\mathcal{K}_{i,a}} V(k) dk - \frac{1}{|\mathcal{K}_{i,a}|} \int_{\mathcal{K}_{i,a}} \frac{d-1}{|k|} dk \\
&= \frac{1}{|\mathcal{K}_{i,a}|} \int_{\mathcal{K}_{i,a}} V(k) dk - d \frac{\rho_{i+\frac{1}{2}}^{d-1} - \rho_{i-\frac{1}{2}}^{d-1}}{\rho_{i+\frac{1}{2}}^d - \rho_{i-\frac{1}{2}}^d} \\
&\approx V(\rho_i \theta_a) - d_i.
\end{aligned}$$

For all $i \geq 1$ and $a \in \mathcal{A}$, set

$$\widehat{W}_{i,a}^{\varphi}(t) = \widehat{W}_{\mathcal{K}_{i,a}}^{\varphi}(t), \quad \forall t \geq 0.$$

Combining the approximations above, the spatial discretization of the model is given by the system of stochastic differential equations

$$
\begin{cases}
\mathrm{d}\widehat{U}_{i,a}(t) + c\,\dfrac{\widehat{U}_{i,a}(t) - \widehat{U}_{i-1,a}(t)}{h_i}\,\mathrm{d}t + cV(\rho_i\theta_a)\widehat{U}_{i,a}(t)\mathrm{d}t = \mathrm{d}\widehat{W}^{\varphi}_{i,a}(t), & \forall\,t \geq 0,\ i \geq 1,\ a \in \mathcal{A}, \\[2mm]
\widehat{U}_{0,a}(t) = 0, & \forall\,t \geq 0,\ a \in \mathcal{A}, \\[2mm]
\widehat{U}_{i,a}(0) = 0, & \forall\,i \geq 1,\ a \in \mathcal{A}.
\end{cases}
$$

$$(2.3.5)$$

The cells $\left(\mathcal{K}_{i,a}\right)_{i \geq 1,\, a \in \mathcal{A}}$ in the finite volume mesh are pairwise disjoint. Moreover, each cell $\mathcal{K}_{i,a}$ has a symmetric cell with respect to 0 in the mesh, which can be denoted by $\mathcal{K}_{i,-a}$. If $i \neq j$, for all $a, b \in \mathcal{A}$, the Wiener processes $\widehat{W}^{\varphi}_{i,a}$ and $\widehat{W}^{\varphi}_{j,b}$ are independent. Moreover, for any $i \geq 1$, given $a, b \in \mathcal{A}$, if $b \notin \{a, -a\}$, the Wiener processes $\widehat{W}^{\varphi}_{i,a}$ and $\widehat{W}^{\varphi}_{i,b}$ are independent. Finally, $\overline{\widehat{W}^{\varphi}_{i,a}} = \widehat{W}_{i,-a}$ for all $i \geq 1$ and $a \in \mathcal{A}$. The properties above are summarized writing

$$
\begin{cases}
\mathbb{E}[\widehat{W}^{\varphi}_{i,a}(t)\overline{\widehat{W}^{\varphi}_{j,b}(s)}] = \delta_{i,j}\delta_{a,b}\dfrac{(t \wedge s)}{|\mathcal{K}_{i,a}|^2}\displaystyle\int_{\mathcal{K}_{i,a}} |\widehat{\varphi}(k)|^2\mathrm{d}k, & \forall\,t, s \geq 0,\ i, j \geq 1,\ a, b \in \mathcal{A}, \\[3mm]
\overline{\widehat{W}^{\varphi}_{i,a}(t)} = \widehat{W}_{i,-a}(t), & \forall\,t \geq 0,\ i \geq 1,\ \in \mathcal{A}.
\end{cases}
$$

2.3.2   *Temporal Discretization: Splitting Algorithm*

Let us now present the temporal discretization scheme applied to the semi-discrete approximation (2.3.5). The time-step size is denoted by Δt, and for all $n \geq 0$ set $t_n = n\Delta t$. Like in Beck (2024) we propose to use a Lie–Trotter splitting scheme, which on each interval $[t_n, t_{n+1}]$ combines solutions of two systems. First, one considers the system of Ornstein–Uhlenbeck stochastic differential equations

$$
\mathrm{d}\widehat{U}^{\mathrm{ou}}_{i,a}(t) + cV(\rho_i\theta_a)\widehat{U}^{\mathrm{ou}}_{i,a}(t)\mathrm{d}t = \mathrm{d}\widehat{W}^{\varphi}_{i,a}(t), \quad \forall\,t \geq 0,\ i \geq 1,\ a \in \mathcal{A}. \qquad (2.3.6)
$$

Second, one considers the deterministic linear system of differential equations

$$
\begin{cases}
\partial_t \widehat{U}^{\mathrm{ad}}_{i,a}(t) + c\dfrac{\widehat{U}^{\mathrm{ad}}_{i,a}(t) - \widehat{U}^{\mathrm{ad}}_{i-1,a}(t)}{h_i} = 0, & \forall\,t \geq 0,\ i \geq 1,\ a \in \mathcal{A}, \\[3mm]
\widehat{U}^{\mathrm{ad}}_{0,a}(t) = 0, & \forall\,t \geq 0,\ a \in \mathcal{A},
\end{cases} \qquad (2.3.7)
$$

which takes into account only the advection dynamics.

Numerical Approximation for (2.3.6)

Given $i, j \geq 1$ and $a, b \in \mathcal{A}$, the stochastic differential equations (2.3.6) for $\left(\widehat{U}_{i,a}^{\mathrm{ou}}(t)\right)_{t \geq 0}$ and $\left(\widehat{U}_{j,b}^{\mathrm{ou}}(t)\right)_{t \geq 0}$ are driven by independent Wiener processes $\widehat{W}_{i,a}^{\varphi}$ and $\widehat{W}_{j,b}^{\varphi}$, except if one has $i = j$ and $a \in \{-b, b\}$.

Given $i \geq 1$ and $a \in \mathcal{A}$, the solution of (2.3.6) is given by

$$\widehat{U}_{i,a}^{\mathrm{ou}}(t) = e^{-cV(\rho_i \theta_a)t}\, \widehat{U}_{i,a}^{\mathrm{ou}}(0) + \int_0^t e^{-cV(\rho_i \theta_a)(t-s)} \mathrm{d}\widehat{W}_{i,a}^{\varphi}(s).$$

The solution at time t_{n+1} given the solution at time t_n can be expressed as

$$\widehat{U}_{i,a}^{\mathrm{ou}}(t_{n+1}) = e^{-cV(\rho_i \theta_a)\Delta t}\, \widehat{U}_{i,a}^{\mathrm{ou}}(t_n) + \int_{t_n}^{t_{n+1}} e^{-cV(\rho_i \theta_a)(t_{n+1}-s)} \mathrm{d}\widehat{W}_{i,a}^{\varphi}(s).$$

Given $i \geq 1$ and $a \in \mathcal{A}$, the random variables $\int_{t_n}^{t_{n+1}} e^{-cV(\rho_i \theta_a)(t_{n+1}-s)} \mathrm{d}\widehat{W}_{i,a}^{\varphi}(s)$ indexed by $n \geq 0$ are independent complex-valued Gaussian random variables, which can be expressed as

$$\int_{t_n}^{t_{n+1}} e^{-cV(\rho_i \theta_a)(t_{n+1}-s)} \mathrm{d}\widehat{W}_{i,a}^{\varphi}(s) = \varrho_{i,a}\widehat{\gamma}_{i,a}^n$$

where

$$\varrho_i = \sqrt{\frac{1 - e^{-2\Delta t V(\rho_i \theta_a)}}{2V(\rho_i \theta_a)\left|\mathcal{K}_{i,a}\right|^2} \int_{\mathcal{K}_{i,a}} |\widehat{\varphi}(k)|^2 \mathrm{d}k}, \quad \forall\, i \geq 1,\ a \in \mathcal{A}, \tag{2.3.8}$$

and $\left(\widehat{\gamma}_{i,a}^n\right)_{n \geq 0, i \geq 1, a \in \mathcal{A}}$ are Gaussian random variables that satisfy

$$\begin{cases} \mathbb{E}[\widehat{\gamma}_{i,a}^n \overline{\widehat{\gamma}_{j,b}^m}] = \delta_{i,j}\delta_{a,b}\delta_{n,m}, & \forall\, n, m \geq 0,\ i, j \geq 1,\ a, b \in \mathcal{A}, \\ \overline{\widehat{\gamma}_{i,a}^n} = \widehat{\gamma}_{i,-a}^n, & \forall\, n \geq 0,\ i \geq 1,\ \in \mathcal{A}. \end{cases}$$

If one defines

$$\begin{cases} \widehat{U}_{i,a}^{n+1,\mathrm{ou}} = e^{-cV(\rho_i \theta_a)\Delta t}\, \widehat{U}_{i,a}^{n,\mathrm{ou}} + \varrho_{i,a}\widehat{\gamma}_{i,a}^n, & \forall\, n \geq 0,\ i \geq 1,\ a \in \mathcal{A}, \\ \widehat{U}_{i,a}^{0,\mathrm{ou}} = \widehat{U}_{i,a}^{\mathrm{ou}}(0), & \forall\, i \geq 1,\ a \in \mathcal{A}, \end{cases}$$

then for any $n \geq 0$ the Gaussian random variables $\left(\widehat{U}_{i,a}^{n,\mathrm{ou}}\right)_{i \geq 1, a \in \mathcal{A}}$ and $\left(\widehat{U}_{i,a}^{\mathrm{ou}}(t_n)\right)_{i \geq 1, a \in \mathcal{A}}$ are equal in distribution.

Numerical Approximation for (2.3.7)

The system of differential equations (2.3.7) cannot be solved exactly. Applying the standard explicit Euler method, one obtains the numerical scheme

$$\begin{cases} \widehat{U}_{i,a}^{n+1,\mathrm{ad}} - \widehat{U}_{i,a}^{n,\mathrm{ad}} + \dfrac{c\Delta t}{h_i}\left(\widehat{U}_{i,a}^{n,\mathrm{ad}} - \widehat{U}_{i-1,a}^{n,\mathrm{ad}}\right) = 0, & \forall\, n \geq 0,\ i \geq 1,\ a \in \mathcal{A}, \\[2mm] \widehat{U}_{0,a}^{n,\mathrm{ad}} = 0, & \forall\, n \geq 0,\ a \in \mathcal{A}, \end{cases}$$

which can be interpreted as the fully discrete upwind scheme for linear advection equation in dimension 1

$$\begin{cases} \partial_t \widehat{U}_a^{\mathrm{ad}}(t,r) + c\partial_r \widehat{U}_a^{\mathrm{ad}}(t,r) = 0, & \forall\, t \geq 0,\ r \geq \kappa, \\[2mm] \widehat{U}(t,\kappa) = 0, \forall\, t \geq 0, \end{cases}$$

which describes the advection in the radial variable $r = |k|$, for any fixed $a \in \mathcal{A}$, on a mesh with cells of length h_i for $i \geq 1$. It is well-known that ensuring the stability of the above the scheme requires to impose the Courant–Friedrichs–Lewy stability condition

$$\frac{c\Delta t}{h_i} \leq 1, \quad \forall\, i \geq 1.$$

To avoid numerical dissipation and preserve the fundamental properties of the model at the discrete time level, we choose to impose the stronger condition

$$\frac{c\Delta t}{h_i} = 1, \quad \forall\, i \geq 1. \tag{2.3.9}$$

As a result, the sequence $\left(\rho_{i+\frac{1}{2}}\right)_{i \geq 0}$ is chosen such that $h_i = c\Delta t$ is independent of $i \geq 1$, depending on the choice of the time-step size Δt: owing to (2.3.4), this holds if the sequence is defined recursively by

$$\begin{cases} \rho_{i+\frac{1}{2}} = \left(\rho_{i-\frac{1}{2}}^d + c\Delta t \rho_{i-\frac{1}{2}}^{d-1}\right)^{\frac{1}{d}}, & i \geq 1, \\[2mm] \rho_{\frac{1}{2}} = \kappa. \end{cases} \tag{2.3.10}$$

Due to the condition (2.3.9), the solution of the fully discrete upwind scheme above is given by the simple formula

$$\widehat{U}_{i,a}^{n+1,\mathrm{ad}} = \widehat{U}_{i-1,a}^{n,\mathrm{ad}}, \quad \forall\, n \geq 0,\ i \geq 1,\ a \in \mathcal{A}.$$

Splitting Scheme

The fully discrete scheme is constructed using a splitting method. Let $\left(\widehat{U}_{i,a}^{n}\right)_{i\geq 1, a\in\mathcal{A}}$ denote the numerical solution at iteration $n \geq 0$, which is meant to be an approximation of the solution $\left(\widehat{U}_{i,a}(t_n)\right)_{i\geq 1, a\in\mathcal{A}}$ at time t_n to the semi-discrete system (2.3.5). Given the numerical solution $\left(\widehat{U}_{i,a}^{n}\right)_{i\geq 1, a\in\mathcal{A}}$ at iteration n, the numerical solution $\left(\widehat{U}_{i,a}^{n+1}\right)_{i\geq 1, a\in\mathcal{A}}$ at iteration $n + 1$ is defined by integrating first (2.3.6) and second (2.3.7), on the time interval $[t_n, t_{n+1}]$, using the numerical methods described above. We obtain the following numerical scheme: for all $n \geq 0$, $i \geq 1$ and $a \in \mathcal{A}$,

$$\begin{cases} \widehat{U}_{i,a}^{n+\frac{1}{2}} = e^{-cV(\rho_i\theta_a)\Delta t}\,\widehat{U}_{i,a}^{n} + \varrho_{i,a}\widehat{\gamma}_{i,a}^{n}, \\ \widehat{U}_{i,a}^{n+1} = \widehat{U}_{i-1,a}^{n+\frac{1}{2}}, \end{cases} \tag{2.3.11}$$

which is supplemented with the boundary conditions and the initial values

$$\begin{cases} \widehat{U}_{0,a}^{n} = 0, & \forall\, n \geq 0,\ a \in \mathcal{A}, \\ \widehat{U}_{i,a}^{0} = 0, & \forall\, i \geq 1,\ a \in \mathcal{A}. \end{cases} \tag{2.3.12}$$

2.4 Numerical Simulations: Application to the JONSWAP Spectrum

The objective of this section is to illustrate the behavior of the model proposed numerical scheme (2.3.11)–(2.3.12) to generate the JONSWAP spectrum given by (2.1). Note that this spectrum is radial. In addition, $\widehat{\varphi}$ is chosen in the sequel to be radial, i.e. one has $\widehat{\varphi}(k) = \widehat{\varphi}(|k|)$.

Before presenting the numerical simulations, two important aspects are discussed: how to compare the numerical and theoretical spectra, and how to retrieve an approximation in the spatial domain from the finite volume approximation in the Fourier domain.

2.4.1 Wave-Number Spectrum and Physical Space Representation

The results of numerical simulations need to be compared to the theoretical predictions obtained for the continuous model and the expression of the spectrum. At any time $t \geq 0$, the power spectral density (PSD) is defined by

$$\Gamma(t, k) = \int_{\mathbb{R}^d} e^{-2i\pi k\cdot x}\,\mathbb{E}\left[U(t, 0)U(t, x)\right]\,\mathrm{d}x. \tag{2.4.1}$$

When computing the second order moment of the Gaussian random variable $\widehat{U}_{\mathcal{K}}(t)$, which is the average of $\widehat{U}(t, k)$ over a finite volume cell $\mathcal{K}$, one obtains

$$\Gamma_{\mathcal{K}}(t) = \mathbb{E}\left|\widehat{U}_{\mathcal{K}}(t)\right|^2 = \frac{1}{|\mathcal{K}|^2} \int_{\mathcal{K}} \Gamma(t, k)\mathrm{d}k. \qquad (2.4.2)$$

From (2.4.2), to compute the average of the PSD $\Gamma(t, k)$ over a finite volume cell $\mathcal{K}$, one needs to consider $|\mathcal{K}|\,\Gamma_{\mathcal{K}}(t)$, which is therefore the appropriate quantity employed below when comparing the numerical and theoretical spectra. The additional factor $|\mathcal{K}|$ is due to the distributional nature of the field $\widehat{U}$ which is delta correlated and thus of infinite variance. In addition, since the JONSWAP spectrum depends on the radial variable only, we consider the angle averaged version of the spectrum, which is defined by

$$\Gamma_i^{\Theta}(t) = \frac{1}{|\mathcal{A}|} \sum_{a \in \mathcal{A}} \Gamma_{\mathcal{K}_{i,a}}(t). \qquad (2.4.3)$$

It is worth mentioning that the proposed finite volume discretization in the Fourier domain is not associated with a natural method to provide a numerical field defined in the spatial domain, since the finite volume mesh is not a standard Cartesian mesh. To obtain a representation of the numerical solution in the physical space, we propose the field defined by

$$\tilde{U}(t, x) = \sum_{n=1}^{N} \sum_{a \in \mathcal{A}} e^{2i\pi k_{na} \cdot x} \widehat{U}_{\mathcal{K}_{n,a}}(t) \rho_n^{d-1} \Delta \rho_n \Delta \Theta_a, \quad \forall\, x \in \mathbb{R}^d. \qquad (2.4.4)$$

In the expression above, N is the numerical resolution in the radial direction, $\mathcal{A}$ is the set indexing the finite volume discretization of the unit $(d-1)$-dimensional sphere and $\Delta \Theta_a$ is the corresponding differential solid angle at the angular coordinate a. The field constructed this way is statistically homogeneous. In addition, it satisfies a discrete isotropy property: its correlation function is invariant under the discrete set of rotations that preserve the mesh $(\Theta_a)_{a \in \mathcal{A}}$. Numerically, the variable $x \in \mathbb{R}^d$ will be discretized on a Cartesian box of size L with step Δx in every direction. In practice, we choose the mesh in physical space such that $L \simeq \frac{1}{\underline{k}}$ and $\Delta x \leq \rho_N^{-1}$. The discretized version of the continuous field $\tilde{U}$ will be denoted $\tilde{U}_\Delta$.

2.4.2   Numerical Results

We are now going to provide numerical illustration in dimension $d = 2$ of the model for the JONSWAP spectrum

$$S_J(|k|) = \alpha g^{-\frac{1}{2}} |k|^{-5/2} \exp\left(-\frac{5}{4}\left(\frac{k_p}{|k|}\right)^2\right) \gamma^{\exp\left[-\frac{1}{2\sigma^2}\left(\sqrt{\frac{|k|}{k_p}}-1\right)^2\right]}, \tag{2.4.5}$$

given by (2.2) with the dispersion relation $\omega(k) = \sqrt{g|k|}$. The potential function V defined by (2.4) is given by

$$V(|k|) = \frac{1}{2|k|}\left(d + \frac{3}{2} - \frac{5}{2}\left(\frac{k_p}{|k|}\right)^2\right) + \frac{\ln\gamma}{4\sigma^2}\frac{\exp\left(-\frac{1}{2\sigma^2}\left(\sqrt{\frac{|k|}{k_p}}-1\right)^2\right)}{\sqrt{|k|/k_p}}\left(\sqrt{\frac{|k|}{k_p}} - 1\right). \tag{2.4.6}$$

We will also consider Pierson–Moskowitz spectrum that will be denoted by S_{PM} obtained by choosing $\gamma = 1$ in (2.4.5). Owing to (2.2.29), in the large time regime, one has

$$\Gamma(\infty, |k|) = \lim_{t\to+\infty} \Gamma(t, k) = \mathbb{1}_{|k|>\kappa}\, S_J(|k|) \int_\kappa^{|k|} \left|\widehat{\varphi}(s)\right|^2 \frac{1}{S_J(s)}\frac{ds}{c}. \tag{2.4.7}$$

At this point, we have to take care of two points to compare $S_J(|k|)$ and $\Gamma(\infty, |k|)$. First of all, the forcing $\widehat{\varphi}$ is compactly supported in Fourier space. Therefore, the integral on the right hand side of (2.4.7) depends on $|k|$ but is constant for large enough $|k|$. Secondly, it is clear from (2.4.7) that multiplying $S_J(|k|)$ by a multiplicative constant does not modify $\Gamma(\infty, |k|)$. The value of a multiplicative constant in $\Gamma(\infty, |k|)$ can be settled by the choice of the forcing $\widehat{\varphi}$. In the following, the forcing term will be based on a bump function $\widehat{\psi}$ given by

$$\widehat{\psi}(k) = \begin{cases} \exp\left[-\dfrac{1}{(|k| - \rho_1)(\rho_2 - |k|)}\right], & \text{if } \forall \rho_1 < |k| < \rho_2, \\ 0 \text{ else.} \end{cases} \tag{2.4.8}$$

The forcing $\widehat{\varphi}$ is then chosen such that,

$$\widehat{C}_\varphi(|k|) = |\widehat{\varphi}(|k|)|^2 = \mathcal{N}\widehat{\psi}(|k|)^2, \quad \mathcal{N} = \frac{c}{\displaystyle\int_\kappa^{+\infty} S_J(s)^{-1}\widehat{\psi}^2(s)\,ds}. \tag{2.4.9}$$

By doing so, the PSD (2.4.7) satisfies

$$\Gamma(\infty, |k|) = S_J(|k|)I(|k|), \quad \forall\, |k| \geq \kappa,$$

where the auxiliary mappping I satisfies

Table 2.1 Simulations parameters. For γ, the two values correspond to the JONSWAP and Pierson–Moskowitz spectra

N	N_θ	h	κ	c	α	γ	k_p	g	σ^2	ρ_1	ρ_2
2^{13}	2^8	10^{-2}	$1/2$	1	1	$3.3(1)$	5	9.81	0.09	$k_p/2$	$\rho_1 + 1$

$$
I(|k|) = \frac{\displaystyle\int_{\kappa}^{|k|} S_{\mathrm{J}}(s)^{-1}\widehat{\psi}(s)^2 \mathrm{d}s}{\displaystyle\int_{\kappa}^{+\infty} S_{\mathrm{J}}(s)^{-1}\widehat{\psi}(s)^2 \mathrm{d}s} = \begin{cases} 1, & \text{if } |k| \geq \rho_2, \\ 0, & \text{if } |k| \leq \rho_1. \end{cases}
\tag{2.4.10}
$$

Through this mapping, and therefore the forcing support, one controls the regions where the model coincides with the desired spectrum. The simulations presented below are run with $\widehat{C_f}$ given by (2.4.9) and V by (2.4.6). The parameters used in the simulations for the JONSWAP and Pierson–Moskowitz spectra are given in Table 2.1 along with the parameters of the numerical approximation procedure.

Let us now present the results of numerical simulations performed with these parameters.

In Fig. 2.1, we present the results of numerical simulations conducted with the set of parameters given in Table 2.1. The left panel of the figure compares the expected spectrum S_{J} (dotted lines) with the estimation of the angle averaged power spectral densities (solid lines) in the JONSWAP and Pierson–Moskowitz cases. In order to have a statistical estimation of these spectra, we first integrate the dynamics until the time $t^* = \rho_N/c$ is reached. For times larger than t^*, we consider that we are in a statistically steady state. Once in the statistically steady state, the PSD are estimated by averaging $M = 500$ realizations of the squared modulus of the finite volume field every $t^*/5$ units of time. By doing doing so we consider that waiting t^* units of time is sufficient for the instances of the field to be independent—which is not the case but is convenient numerically. The left panel of Fig. 2.1 shows a good agreement between the estimated spectra obtained numerically and the expected ones. In particular, for $\rho \geq \rho_2$ the spectra coincide with the JONSWAP and Pierson–Moskowitz one while it vanishes for $\rho \leq \rho_1$ and is different from S_{J} for $\rho_1 < \rho < \rho_2$. The shape of the spectra obtained by the numerical method for $\rho_1 < \rho < \rho_2$ is settled by the forcing.

On the right panel of Fig. 2.1, we show the physical space representation of a snapshot of the Fourier space field. This field is obtained using (2.4.4) in dimension $d = 2$. The displayed field indeed exhibit statistical homogeneity and isotropy.

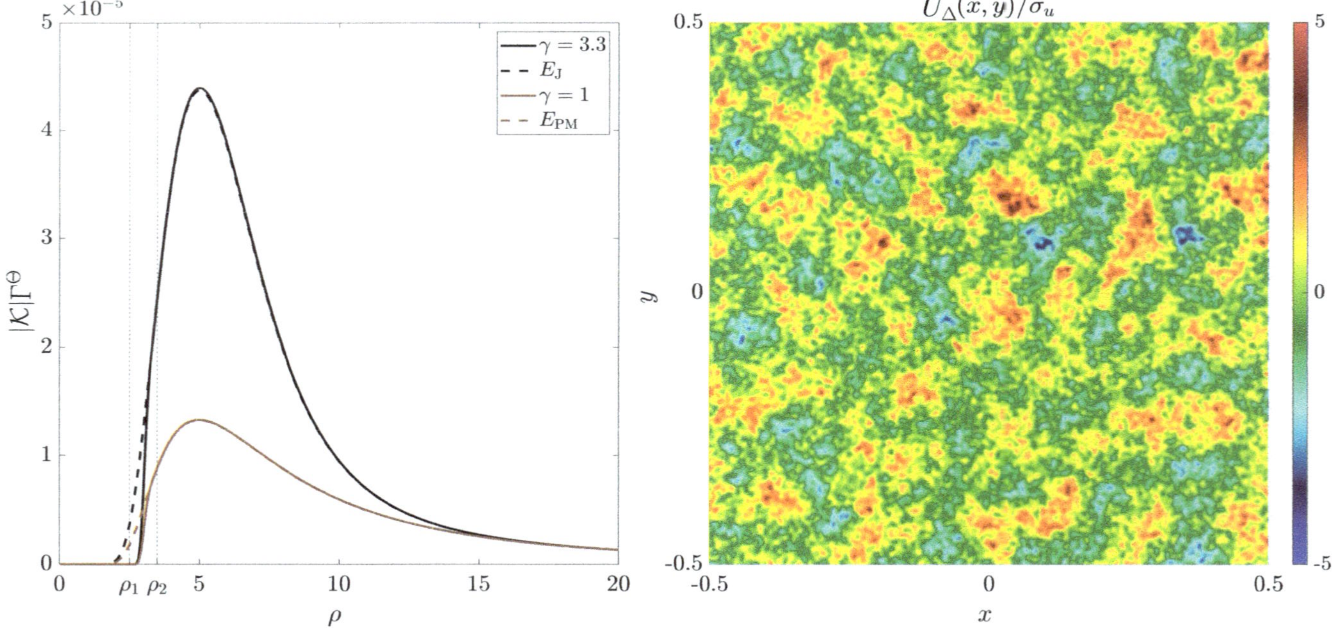

Fig. 2.1 Left, Angle averaged spectra obtained numerically (solid lines) compared to the JONSWAP and Pierson–Mosowitz spectra (2.4.5) (dotted lines). The black lines correspond to the JONSWAP ($\gamma = 3.3$) case while the brown lines are the Pierson–Moskowitz ($\gamma = 1$) case. The vertical dotted lines corresponds to the radii ρ_1 and ρ_2 entering the definition of the forcing term. Right, physical space representation $\widetilde{U}_\Delta$ of the JONSWAP field at a given time in the statistically steady state. σ_u is the expected standard deviation for the solution of the dynamics

Acknowledgements G. B. is supported by the BOURGEONS project, grant ANR-23-CE40-0014-01 of the French National Research Agency (ANR). C.-E. B. and R.G. would like to thank MARGAUx–Fédération Mathématique de Recherche en Région Nouvelle-Aquitaine for funding R. G.'s visit at Université de Pau et des Pays de l'Adour. L.C. is partly supported by the Simons Foundation Award No. 1151711. We thank the referees for suggestions which helped us improve the presentation of this work.

References

Apolinário GB, Beck G, Chevillard L, Gallagher I, Grande R (2023) A linear stochastic model of turbulent cascades and fractional fields. To appear in Ann Sc Norm Super Pisa Cl Sci

Appel JR (1994) An improved model of the ocean surface wave vector spectrum and its effects on radar backscatter. J Geo Res Oceans 99

Batchelor GK (1953) The theory of homogeneous turbulence. Cambridge University Press

Beck G, Bréhier C-E, Chevillard L, Grande R, Ruffenach W (2024) Numerical simulations of a stochastic dynamics leading to cascades and loss of regularity: applications to fluid turbulence and generation of fractional gaussian fields. Phys Rev Res 6:033048

Buckmaster T, Germain P, Hani Z, Shatah J (2021) Onset of the wave turbulence description of the longtime behavior of the nonlinear Schrödinger equation. Inventiones Mathematicae 225:787–855

Collot C, Germain P (2019) On the derivation of the homogeneous kinetic wave equation. arXiv:1912.10368

de Verdìére YC, Saint-Raymond L, Deverdì YC (2019) Attractors for two-dimensional waves with homogeneous Hamiltonians of degree 0. Commun Pure Appl Math 73(2):421–462. https://hal.science/hal-01685278

Deng Y, Hani Z (2023a) Derivation of the wave kinetic equation: full range of scaling laws. arXiv:2301.07063

Deng Y, Hani Z (2023b) Full derivation of the wave kinetic equation. Inventiones Mathematicae 233:543–724

Elfouhaily T, Chapron B, Katsaros K, Vandemark D (1997) A unified directional spectrum for long and short wind-driven waves. J Geo Res 102

Frisch U (1995) Turbulence: the legacy of AN Kolmogorov. Cambridge University Press

Grande R, Hani Z (2024) Rigorous derivation of damped-driven wave turbulence theory. arXiv:2407.10711

Hasselmann K (1962) On the non-linear energy transfer in a gravity-wave spectrum. I. General theory. J Fluid Mech 12:481–500

Hasselmann K (1963) On the non-linear energy transfer in a gravity wave spectrum. II. Conservation theorems; wave-particle analogy; irreversibility. J Fluid Mech 15:273–281

Hasselmann K, Olbers D (1973) Measurements of wind-wave growth and swell decay during the Joint North Sea Wave Project (JONSWAP). Reihe A, Ergänzung zur Deut Hydrogr Z 1–95

Kolmogorov AN (1941) Dissipation of energy in locally isotropic turbulence. Akademiia Nauk SSSR Doklady 32:16

Komen G, Cavaleri L, Donelan M, Hasselmann K, Hasselmann S, Janssen PAEM (1994) Dynamics and modelling of ocean waves. Cambridge University Press

Monin A, Yaglom A (2013) Statistical fluid mechanics, volume II: mechanics of turbulence, vol 2. Dover Publications

Staffilani G, Tran M-B (2021) On the wave turbulence theory for a stochastic KdV type equation. arXiv:2106.09819

Tennekes H, Lumley JL (1972) A first course in turbulence. MIT Press

Tolman HL, the WAVEWATCH III Development Group (2014) User manual and system documentation of wavewatch iii version 4.18. Technical Note, Note 316, pp 282, pp + Appendices

Wilczek M, Daitche A, Friedrich R (2011) On the velocity distribution in homogeneous isotropic turbulence: correlations and deviations from gaussianity. J Fluid Mech 676:191–217

Chapter 3
Improved Blow-Up Criterion in a Variational Framework for Nonlinear SPDEs

Daniel Goodair

Abstract We extend recent existence and uniqueness results for maximal solutions of SPDEs through an improved blow-up criterion. Whilst the maximal time of existence is typically characterised by blow-up in the energy norm of solutions, we show instead that solutions exist until blow-up in the larger spaces of the variational framework. The result is applied through a bootstrap of regularity to show that solutions of 2D and 3D stochastic Navier–Stokes equations retain the higher order regularity of the initial condition on their time of existence.

3.1 Introduction

The variational approach to nonlinear SPDEs with additive and multiplicative noise has long been studied, initially in the works (Gyöngy and Krylov 1980; Krylov and Rozovskii 2007; Pardoux 1975) and more recently (Debussche et al. 2011; Liu and Röckner 2010, 2013; Neelima and Šiška 2020) to name just a few contributions. Motivated by the physical relevance and potential regularising properties of *transport noise*, where the stochastic integral depends on the gradient of the solution, there has been a trend towards unbounded noise in the variational framework as in Agresti and Veraar (2022), Agresti and Veraar (2024), Alonso-Orán et al. (2021), Goodair (2024), Goodair et al. (2023), Röckner et al. (2024). We similarly allow for an unbounded noise, which does not need to be small relative to coercivity.

Our result comes as a strict extension of the author's work (Goodair et al. 2023) with Crisan and Lang. There the authors proved the existence and uniqueness of maximal solutions with maximal time characterised by the blow-up in the energy norm of solutions. Under the exact same assumptions, here we prove that the maximal time can in fact be characterised by blow-up in a weaker norm given by the larger spaces of the framework. The method relies on Proposition 3.4.4, recently proven by

EPFL, Supported by the EPSRC Project 2478902.

D. Goodair (✉)
Imperial College London, London, England
e-mail: daniel.goodair@epfl.ch

B. Chapron et al. (eds.), *Stochastic Transport in Upper Ocean Dynamics IV*, Mathematics of Planet Earth 15, https://doi.org/10.1007/978-3-032-12749-5_3

the author as a development of Glatt-Holtz et al. (2009) Lemma 5.1. The proposition is used to deduce the existence of a limiting process and stopping time under Cauchy and weak equicontinuity properties, as in Glatt-Holtz et al. (2009), with the novelty being that one can characterise the limit stopping time. In our application, the stopping time is understood in terms of the first hitting time in the weaker norm, immediately implying the existence of solutions until blow-up in this norm.

Of the referenced literature we draw particular attention to Agresti and Veraar (2022), Alonso-Orán et al. (2021). In the latter, the authors similarly prove the existence and uniqueness of maximal solutions until blow-up in the largest of the considered spaces. Their framework, however, does not ask for coercivity and in consequence solutions exhibit only pathwise continuity and not the additional square integrability. In the commonplace application of fluid dynamics, this setup is designed for *inviscid* equations whereas ours is for *viscous* equations. The applications and methodology are thus completely different, and we see these results as complementary. The former reference of Agresti and Veraar (2022) is by far the most comprehensive treatment of blow-up criteria in nonlinear SPDEs, where the notion of criticality is at the core of their work and spaces can be selected much more finely. In the vastness of their theory it is not entirely clear to what extent (Agresti and Veraar 2022) could cover our results, though in any case we find value in our work through its simplicity and novel methodology.

We showcase the main result by proving that (local) strong solutions of 2D and 3D stochastic Navier–Stokes equations retain the higher order regularity of the initial condition on their time of existence. The argument is a simple iterated application of the variational result, as for $k \geq 2$ solutions are shown to blow-up in $C\left([0, T]; W^{k,2}\right) \cap L^2\left([0, T]; W^{k+1,2}\right)$ only if they blow-up in $C\left([0, T]; W^{k-1,2}\right) \cap L^2\left([0, T]; W^{k,2}\right)$ and inductively in $C\left([0, T]; W^{1,2}\right) \cap L^2\left([0, T]; W^{2,2}\right)$. To succinctly verify the assumptions we consider only a Lipschitz noise, though more exotic structures such as transport noise can certainly be considered.

The structure of the paper is as follows. We conclude this section with some brief stochastic preliminaries. In Sect. 3.2, we provide the setup, definitions and main result. Section 3.3 is devoted to the proof of the main result. The application to high-order regularity of Stochastic Navier–Stokes is given in Sect. 3.4. The key Proposition 3.4.4 is given as Supplementary Material which concludes the paper.

3.1.1 Stochastic Preliminaries

Let $(\Omega, \mathcal{F}, (\mathcal{F}_t), \mathbb{P})$ be a fixed filtered probability space satisfying the usual conditions of completeness and right continuity. We take $\mathcal{W}$ to be a cylindrical Brownian motion over some Hilbert space $\mathfrak{U}$ with orthonormal basis (e_i). Given a process $F : [0, T] \times \Omega \to \mathscr{L}^2(\mathfrak{U}; \mathcal{H})$ for $\mathscr{L}^2(\mathfrak{U}; \mathcal{H})$ the Hilbert–Schmidt space, progressively measurable and such that $F \in L^2\left(\Omega \times [0, T]; \mathscr{L}^2(\mathfrak{U}; \mathcal{H})\right)$, for any $0 \leq t \leq T$ we define the stochastic integral

$$\int_0^t F_s d\mathcal{W}_s := \sum_{i=1}^{\infty} \int_0^t F_s(e_i) dW_s^i,$$

where the infinite sum is taken in $L^2(\Omega; \mathcal{H})$. We can extend this notion to processes F which are such that $F(\omega) \in L^2\left([0, T]; \mathscr{L}^2(\mathfrak{U}; \mathcal{H})\right)$ for $\mathbb{P} - a.e.\omega$ via the traditional localisation procedure. A complete, direct construction of this integral, a treatment of its properties and the fundamentals of stochastic calculus in infinite dimensions can be found in Goodair and Crisan (2024, Chap. 2).

3.2 Setup and Main Result

3.2.1 Functional Framework

Our object of study is the Itô SPDE

$$\boldsymbol{\Psi}_t = \boldsymbol{\Psi}_0 + \int_0^t \mathcal{A}(s, \boldsymbol{\Psi}_s) ds + \int_0^t \mathcal{G}(s, \boldsymbol{\Psi}_s) d\mathcal{W}_s \tag{3.1}$$

which we pose for a triplet of embedded, separable Hilbert spaces

$$V \hookrightarrow H \hookrightarrow U$$

whereby the embeddings are continuous linear injections. We ask that there is a continuous bilinear form $\langle \cdot, \cdot \rangle_{U \times V} : U \times V \to \mathbb{R}$ such that for $f \in H$ and $\psi \in V$,

$$\langle f, \psi \rangle_{U \times V} - \langle f, \psi \rangle_H.$$

Equation (3.1) is posed on a time interval $[0, T]$ for arbitrary $T \geq 0$. The mappings $\mathcal{A}, \mathcal{G}$ are such that $\mathcal{A} : [0, T] \times V \to U, \mathcal{G} : [0, T] \times V \to \mathscr{L}^2(\mathfrak{U}; H)$ are measurable. Understanding $\mathcal{G}$ as a mapping $\mathcal{G} : [0, T] \times V \times \mathfrak{U} \to H$, we introduce the notation $\mathcal{G}_i(\cdot, \cdot) := \mathcal{G}(\cdot, \cdot, e_i)$. We further impose the existence of a system of elements (a_k) of V with the following properties. Let us define the spaces $V_n := \mathrm{span}\{a_1, \ldots, a_n\}$ and $\mathcal{P}_n$ as the orthogonal projection to V_n in U. It is required that the $(\mathcal{P}_n)$ are uniformly bounded in H, which is to say that there exists a constant c independent of n such that for all $\phi \in H$:

$$\|\mathcal{P}_n f\|_H \leq c \|f\|_H.$$

We also suppose that there exists a real-valued sequence (μ_n) with $\mu_n \to \infty$ such that for any $f \in H$:

$$\|(I - \mathcal{P}_n)f\|_U \le \frac{1}{\mu_n}\|f\|_H$$

where I represents the identity operator in U. Specific bounds on the mappings $\mathcal{A}$ and $\mathcal{G}$ will be imposed in the following subsection. In order to make the assumptions we introduce some more notation here: we shall let $c. : [0, T] \to \mathbb{R}$ denote any bounded function, and for any constant $p \in \mathbb{R}$ we define the functions $K_U : U \to \mathbb{R}$, $K_H : H \to \mathbb{R}$, $K_V : V \to \mathbb{R}$ by

$$K_U(\phi) = 1 + \|\phi\|_U^p, \quad K_H(\phi) = 1 + \|\phi\|_H^p, \quad K_V(\phi) = 1 + \|\phi\|_V^p.$$

We may also consider these mappings as functions of two variables, e.g. $K_U : U \times U \to \mathbb{R}$ by

$$K_U(\phi, \psi) = 1 + \|\phi\|_U^p + \|\psi\|_U^p.$$

Our assumptions will be stated for "the existence of a K such that..." where we really mean "the existence of a p such that, for the corresponding K, ...".

3.2.2 Assumptions

We assume that there exists a $c.$, K and $\gamma > 0$ such that for all $\phi, \psi \in V$, $\phi^n \in V_n$, $f \in H$ and $t \in [0, T]$:

Assumption 3.2.1

$$\|\mathcal{A}(t, \phi)\|_U^2 + \sum_{i=1}^{\infty} \|\mathcal{G}_i(t, \phi)\|_H^2 \le c_t K_U(\phi)\left[1 + \|\phi\|_V^2\right],$$

$$\|\mathcal{A}(t, \phi) - \mathcal{A}(t, \psi)\|_U^2 \le c_t K_V(\phi, \psi)\|\phi - \psi\|_V^2,$$

$$\sum_{i=1}^{\infty} \|\mathcal{G}_i(t, \phi) - \mathcal{G}_i(t, \psi)\|_U^2 \le c_t K_U(\phi, \psi)\|\phi - \psi\|_H^2.$$

Assumption 3.2.2

$$2\langle \mathcal{P}_n\mathcal{A}(t, \phi^n), \phi^n\rangle_H + \sum_{i=1}^{\infty} \|\mathcal{P}_n\mathcal{G}_i(t, \phi^n)\|_H^2 \le c_t K_U(\phi^n)\left[1 + \|\phi^n\|_H^4\right] - \gamma\|\phi^n\|_V^2,$$

$$\sum_{i=1}^{\infty} \langle \mathcal{P}_n\mathcal{G}_i(t, \phi^n), \phi^n\rangle_H^2 \le c_t K_U(\phi^n)\left[1 + \|\phi^n\|_H^6\right].$$

Assumption 3.2.3

$$2\langle \mathcal{A}(t,\phi) - \mathcal{A}(t,\psi), \phi - \psi \rangle_U + \sum_{i=1}^{\infty} \|\mathcal{G}_i(t,\phi) - \mathcal{G}_i(t,\psi)\|_U^2$$

$$\leq c_t K_U(\phi,\psi) \left[1 + \|\phi\|_H^2 + \|\psi\|_H^2 \right] \|\phi - \psi\|_U^2 - \gamma \|\phi - \psi\|_H^2,$$

$$\sum_{i=1}^{\infty} \langle \mathcal{G}_i(t,\phi) - \mathcal{G}_i(t,\psi), \phi - \psi \rangle_U^2 \leq c_t K_U(\phi,\psi) \left[1 + \|\phi\|_H^2 + \|\psi\|_H^2 \right] \|\phi - \psi\|_U^4.$$

Assumption 3.2.4

$$2\langle \mathcal{A}(t,\phi), \phi \rangle_U + \sum_{i=1}^{\infty} \|\mathcal{G}_i(t,\phi)\|_U^2 \leq c_t K_U(\phi) \left[1 + \|\phi\|_H^2 \right],$$

$$\sum_{i=1}^{\infty} \langle \mathcal{G}_i(t,\phi), \phi \rangle_U^2 \leq c_t K_U(\phi) \left[1 + \|\phi\|_H^4 \right].$$

Assumption 3.2.5

$$\langle \mathcal{A}(t,\phi) - \mathcal{A}(t,\psi), f \rangle_U \leq c_t K_U(\phi,\psi)(1 + \|f\|_H) \left[1 + \|\phi\|_V + \|\psi\|_V \right] \|\phi - \psi\|_H.$$

3.2.3 Definitions and Main Result

We state the definitions and main result.

Definition 3.2.6 Let $\boldsymbol{\Psi}_0 : \Omega \to H$ be $\mathcal{F}_0-$ measurable. A pair $(\boldsymbol{\Psi}, \tau)$ where τ is a $\mathbb{P} - a.s.$ positive stopping time and $\boldsymbol{\Psi}$ is a process such that for $\mathbb{P} - a.e.\omega$, $\boldsymbol{\Psi}_.(\omega) \in C([0,T]; H)$ and $\boldsymbol{\Psi}_.(\omega)\mathbb{1}_{.\leq\tau(\omega)} \in L^2([0,T]; V)$ for all $T \geq 0$ and with $\boldsymbol{\Psi}.\mathbb{1}_{.\leq\tau}$ progressively measurable in V, is said to be a local strong solution of Eq. (3.1) if the identity

$$\boldsymbol{\Psi}_t = \boldsymbol{\Psi}_0 + \int_0^{t\wedge\tau} \mathcal{A}(s, \boldsymbol{\Psi}_s)ds + \int_0^{t\wedge\tau} \mathcal{G}(s, \boldsymbol{\Psi}_s)d\mathcal{W}_s$$

holds $\mathbb{P} - a.s.$ in U for all $t \geq 0$.

Definition 3.2.7 A pair $(\boldsymbol{\Psi}, \Theta)$ such that there exists a sequence of stopping times (θ_j) which are $\mathbb{P} - a.s.$ monotone increasing and convergent to Θ, whereby $(\boldsymbol{\Psi}_{.\wedge\theta_j}, \theta_j)$ is a local strong solution of Eq. (3.1) for each j, is said to be a maximal strong solution of Eq. (3.1) if for any other pair $(\boldsymbol{\Phi}, \Gamma)$ with this property then $\Theta \leq \Gamma \mathbb{P} - a.s.$ implies $\Theta = \Gamma \mathbb{P} - a.s.$

Remark We do not require Θ to be finite in this definition, in which case we mean that the sequence (θ_j) is monotone increasing and unbounded for such ω.

Definition 3.2.8 A maximal strong solution $(\boldsymbol{\Psi}, \Theta)$ of Eq. (3.1) is said to be unique if for any other such solution $(\boldsymbol{\Phi}, \Gamma)$, then $\Theta = \Gamma \, \mathbb{P} - a.s.$ and

$$\mathbb{P}\left(\{\omega \in \Omega : \boldsymbol{\Psi}_t(\omega) = \boldsymbol{\Phi}_t(\omega) \quad \forall t \in [0, \Theta)\}\right) = 1.$$

Theorem 3.2.9 *For any given $\mathcal{F}_0$-measurable $\boldsymbol{\Psi}_0 : \Omega \to H$, there exists a unique maximal strong solution $(\boldsymbol{\Psi}, \Theta)$ of Eq. (3.1). Moreover at $\mathbb{P} - a.e.\omega$ for which $\Theta(\omega) < \infty$, we have that*

$$\sup_{r \in [0, \Theta(\omega))} \|\boldsymbol{\Psi}_r(\omega)\|_U^2 + \int_0^{\Theta(\omega)} \|\boldsymbol{\Psi}_r(\omega)\|_H^2 dr = \infty$$

and in consequence for any $\mathbb{P} - a.s.$ positive stopping time τ such that

$$\sup_{r \in [0, \tau(\omega))} \|\boldsymbol{\Psi}_r(\omega)\|_U^2 + \int_0^{\tau(\omega)} \|\boldsymbol{\Psi}_r(\omega)\|_H^2 dr < \infty$$

$\mathbb{P} - a.s.$, $(\boldsymbol{\Psi}_{\cdot \wedge \tau}, \tau)$ *is a local strong solution of Eq. (3.1).*

3.3 Proof of Theorem 3.2.9

This section is devoted to the proof of the main result, Theorem 3.2.9. We recall that the assumptions are identical to those of Goodair et al. (2023, Sect. 3.1). As an extension of Goodair et al. (2023, Theorem 3.15), our first goal of this section is to summarise the method used in Goodair et al. (2023); this is the content of Sect. 3.3.1. Section 3.3.2 then details how our new machinery of Proposition 3.4.4 facilitates the improved result of Theorem 3.2.9, concluding its proof.

3.3.1 A Synopsis of Our Approach

We first consider a bounded initial condition $\boldsymbol{\Psi}_0 \in L^\infty(\Omega; H)$ and the Galerkin equations

$$\boldsymbol{\Psi}_t^n = \boldsymbol{\Psi}_0^n + \int_0^t \mathcal{P}_n \mathcal{A}(s, \boldsymbol{\Psi}_s^n) ds + \int_0^t \mathcal{P}_n \mathcal{G}(s, \boldsymbol{\Psi}_s^n) d\mathcal{W}_s \tag{3.2}$$

for $\boldsymbol{\Psi}_0^n := \mathcal{P}_n \boldsymbol{\Psi}_0$ and $\mathcal{P}_n \mathcal{G}(e_i, s, \cdot) := \mathcal{P}_n \mathcal{G}_i(s, \cdot)$. Central to this work are two norms, for functions $\boldsymbol{\Phi} \in L^\infty([0, T]; U) \cap L^2([0, T]; H)$, $\boldsymbol{\Psi} \in L^\infty([0, T]; H) \cap L^2([0, T]; V)$ defined by

$$\|\boldsymbol{\Phi}\|_{UH,T}^2 := \sup_{r\in[0,T]} \|\boldsymbol{\Phi}_r\|_U^2 + \int_0^T \|\boldsymbol{\Phi}_r\|_H^2 dr$$

$$\|\boldsymbol{\Psi}\|_{HV,T}^2 := \sup_{r\in[0,T]} \|\boldsymbol{\Psi}_r\|_H^2 + \int_0^T \|\boldsymbol{\Psi}_r\|_V^2 dr.$$

The HV norm corresponds to the regularity of strong solutions, so our idea is to show uniform regularity of the Galerkin solutions in the HV norm up until first hitting times in the lower UH norm which sufficiently curbs the nonlinearity. These stopping times are defined for any $M > 1$ and $t \geq 0$ by

$$\tau_n^{M,t} := t \wedge \inf \left\{ s \geq 0 : \|\boldsymbol{\Psi}^n\|_{UH,s}^2 \geq M + \|\boldsymbol{\Psi}_0^n(\omega)\|_U^2 \right\}. \tag{3.3}$$

For any such choices there exists a local strong solution $(\boldsymbol{\Psi}^n, \tau_n^{M,t})$ of Equation (3.2), see Goodair et al. (2023, Lemma 3.18). Relying on Assumption 3.2.2 then the uniform boundedness is proven, Proposition 3.21, stated here.

Proposition 3.3.1 *There exists a constant C dependent on M, t but independent of n such that for the local strong solution $(\boldsymbol{\Psi}^n, \tau_n^{M,t})$ of (3.2),*

$$\mathbb{E}\|\boldsymbol{\Psi}^n\|_{HV,\tau_n^{M,t}}^2 \leq C \left[\mathbb{E}\left(\|\boldsymbol{\Psi}_0^n\|_H^2 \right) + 1 \right]. \tag{3.4}$$

We then look to use the result of Glatt-Holtz et al. (2009, Lemma 5.1), to obtain a limiting process and positive stopping time as a candidate local strong solution of (3.1). It is our extension of this result to Proposition 3.4.4 that is pivotal in the improved Theorem 3.2.9, shown in the next subsection. To apply the Glatt-Holtz and Ziane result, the following were proven as Propositions 3.24 and 3.25.

Proposition 3.3.2 *We have that*

$$\lim_{m\to\infty} \sup_{n\geq m} \left[\mathbb{E}\|\boldsymbol{\Psi}^n - \boldsymbol{\Psi}^m\|_{UH,\tau_m^{M,t}\wedge\tau_n^{M,t}}^2 \right] = 0. \tag{3.5}$$

Proposition 3.3.3 *We have that*

$$\lim_{S\to 0} \sup_{n\in\mathbb{N}} \mathbb{E} \left[\|\boldsymbol{\Psi}^n\|_{UH,\tau_n^{M,t}\wedge S}^2 - \|\boldsymbol{\Psi}_0^n\|_U^2 \right] = 0. \tag{3.6}$$

This allows us to apply the Glatt-Holtz and Ziane result, obtaining Theorem 3.26 of Goodair et al. (2023), stated here.

Proposition 3.3.4 *There exists a stopping time $\tau_\infty^{M,t}$, a subsequence $(\boldsymbol{\Psi}^{n_l})$ and a process $\boldsymbol{\Psi}. = \boldsymbol{\Psi}._{\wedge\tau_\infty^{M,t}}$ whereby $\boldsymbol{\Psi}.\mathbb{1}._{\leq\tau_\infty^{M,t}}$ is progressively measurable in V and such that:*

- $\mathbb{P}\left(\left\{0 < \tau_\infty^{M,t} \leq \tau_{n_l}^{M,t}\right\}\right) = 1;$

- *For $\mathbb{P} - a.e.\omega$, $\boldsymbol{\Psi}^{n_l}(\omega) \to \boldsymbol{\Psi}(\omega)$ in $L^\infty\left([0, \tau_\infty^{M,t}(\omega)]; U\right) \cap L^2\left([0, \tau_\infty^{M,t}(\omega)]; H\right)$, i.e.*

$$\|\boldsymbol{\Psi}^{n_l}(\omega) - \boldsymbol{\Psi}(\omega)\|^2_{UH,\tau_\infty^{M,t}(\omega)} \longrightarrow 0. \tag{3.7}$$

From this point it is reasonably straightforward to show that $(\boldsymbol{\Psi}, \tau_\infty^{M,t})$ is a local strong solution of (3.1), as the uniform boundedness of Proposition 3.3.1 holds for the subsequence on $[0, \tau_\infty^{M,t}]$ allowing $\boldsymbol{\Psi}$ to inherit this regularity. This is done in Sect. 3.7 of Goodair et al. (2023), followed by the uniqueness, maximality and characterisation of the maximal time. Finally one can relieve the boundedness constraint on $\boldsymbol{\Psi}_0$ by partitioning Ω into sets on which an unbounded $\boldsymbol{\Psi}_0$ is bounded, using the unique maximal solution on each set, and piecing these together to obtain a solution for the unbounded $\boldsymbol{\Psi}_0$. This is done in Sect. 3.7.

3.3.2　The Improved Method

Characterisation of the blow-up time in the previous subsection arises only through standard machinery on the energy norm of the solution, which is why the blow-up is given in the HV norm. This machinery begins from the simple existence of *a local strong solution*, making no use of the information that we have on $\tau_\infty^{M,t}$. This stopping time $\tau_\infty^{M,t}$ is in fact constructed as a limit of modified versions of $\tau_{n_l}^{M,t}$, so it is tightly connected with the UH norm and the input parameters M, t. There is a strong intuition saying that for any first hitting time of $\boldsymbol{\Psi}$ in the UH norm, we can choose M and t large enough so that $\tau_\infty^{M,t}$ exceeds it: such a property leads us to the fact that at the maximal time, which must be greater than all $\tau_\infty^{M,t}$, $\boldsymbol{\Psi}$ must blow-up in UH. Proposition 3.4.4 was developed to make this intuition rigorous. To apply it, we must upgrade the *weak equicontinuity at time zero* from Proposition 3.3.3 to a *weak equicontinuity at all times*.

Lemma 3.3.5 *Let θ be a stopping time and (δ_j) a sequence of stopping times which converge to $0\,\mathbb{P} - a.s..$ Then*

$$\lim_{j \to \infty} \sup_{n \in \mathbb{N}} \mathbb{E}\left(\|\boldsymbol{\Psi}^n\|^2_{UH,(\theta+\delta_j)\wedge\tau_n^{M,t}} - \|\boldsymbol{\Psi}^n\|^2_{UH,\theta\wedge\tau_n^{M,t}}\right) = 0.$$

Proof We look at the energy identity satisfied by $\boldsymbol{\Psi}^n$ up until the stopping time $\theta \wedge \tau_n^{M,T}$ and then $(\theta + r) \wedge \tau_n^{M,T}$ for some $r \geq 0$. We have that

$$\|\boldsymbol{\Psi}^n_{\theta\wedge\tau_n^{M,t}}\|^2_U = \|\boldsymbol{\Psi}^n_0\|^2_U + 2\int_0^{\theta\wedge\tau_n^{M,t}} \langle \mathcal{P}_n\mathcal{A}\left(s, \boldsymbol{\Psi}^n_s\right), \boldsymbol{\Psi}^n_s\rangle_U ds + \int_0^{\theta\wedge\tau_n^{M,t}} \sum_{i=1}^\infty \|\mathcal{P}_n\mathcal{G}_i\left(s, \boldsymbol{\Psi}^n_s\right)\|^2_U ds$$

$$+ 2\int_0^{\theta\wedge\tau_n^{M,t}} \langle \mathcal{P}_n\mathcal{G}\left(s, \boldsymbol{\Psi}^n_s\right), \boldsymbol{\Psi}^n_s\rangle_U d\mathcal{W}_s$$

and similarly for $(\theta + r) \wedge \tau_n^{M,t}$, from which the difference of the equalities gives

$$\|\Psi^n_{(\theta+r)\wedge\tau_n^{M,t}}\|^2_U = \|\Psi^n_{\theta\wedge\tau_n^{M,t}}\|^2_U + 2\int_{\theta\wedge\tau_n^{M,t}}^{(\theta+r)\wedge\tau_n^{M,t}} \langle\mathcal{P}_n\mathcal{A}\left(s,\Psi^n_s\right),\Psi^n_s\rangle_U ds$$

$$+ \int_{\theta\wedge\tau_n^{M,t}}^{(\theta+r)\wedge\tau_n^{M,t}} \sum_{i=1}^{\infty}\|\mathcal{P}_n\mathcal{G}_i\left(s,\Psi^n_s\right)\|^2_U ds + 2\int_{\theta\wedge\tau_n^{M,t}}^{(\theta+r)\wedge\tau_n^{M,t}} \langle\mathcal{P}_n\mathcal{G}\left(s,\Psi^n_s\right),\Psi^n_s\rangle_U d\mathcal{W}_s.$$

Using that $\mathcal{P}_n$ is an orthogonal projection in U with Ψ^n in the range of $\mathcal{P}_n$, and invoking Assumption 3.2.4, we reduce to

$$\|\Psi^n_{(\theta+r)\wedge\tau_n^{M,t}}\|^2_U - \|\Psi^n_{\theta\wedge\tau_n^{M,t}}\|^2_U$$

$$\leq c\int_{\theta\wedge\tau_n^{M,t}}^{(\theta+r)\wedge\tau_n^{M,t}}\left(1+\|\Psi^n_s\|^2_H\right)ds + \int_{\theta\wedge\tau_n^{M,t}}^{(\theta+r)\wedge\tau_n^{M,t}}\langle\mathcal{G}\left(s,\Psi^n_s\right),\Psi^n_s\rangle_U d\mathcal{W}_s,$$

where the constant c depends on M, through a bound on the U norm by the stopping time $\tau_n^{M,t}$. We now take the supremum over $r\in[0,\delta_j]$ and expectation, using the Burkholder–Davis–Gundy inequality:

$$\mathbb{E}\left[\sup_{r\in[0,\delta_j]}\|\Psi^n_{(\theta+r)\wedge\tau_n^{M,t}}\|^2_U - \|\Psi^n_{\theta\wedge\tau_n^{M,t}}\|^2_U + \int_{\theta\wedge\tau_n^{M,t}}^{(\theta+\delta_j)\wedge\tau_n^{M,t}}\|\Psi^n_s\|^2_H ds\right]$$

$$\leq c\mathbb{E}\int_{\theta\wedge\tau_n^{M,t}}^{(\theta+\delta_j)\wedge\tau_n^{M,t}}\left(1+\|\Psi^n_s\|^2_H\right)ds + c\mathbb{E}\left(\int_{\theta\wedge\tau_n^{M,t}}^{(\theta+\delta_j)\wedge\tau_n^{M,t}}\sum_{i=1}^{\infty}\langle\mathcal{G}_i\left(s,\Psi^n_s\right),\Psi^n_s\rangle^2_U ds\right)^{\frac{1}{2}}$$

having then added an $\mathbb{E}\int_{\theta\wedge\tau_n^{M,t}}^{(\theta+\delta_j)\wedge\tau_n^{M,t}}\|\Psi^n_s\|^2_H ds$ to both sides. Using the second part of Assumption 3.2.4, again controlling the U norm by a constant, we achieve that

$$\mathbb{E}\left[\sup_{r\in[0,\delta_j]}\|\Psi^n_{(\theta+r)\wedge\tau_n^{M,t}}\|^2_U - \|\Psi^n_{\theta\wedge\tau_n^{M,t}}\|^2_U + \int_{\theta\wedge\tau_n^{M,t}}^{(\theta+\delta_j)\wedge\tau_n^{M,t}}\|\Psi^n_s\|^2_H ds\right]$$

$$\leq c\mathbb{E}\int_{\theta\wedge\tau_n^{M,t}}^{(\theta+\delta_j)\wedge\tau_n^{M,t}}\left(1+\|\Psi^n_s\|^2_H\right)ds + c\mathbb{E}\left(\int_{\theta\wedge\tau_n^{M,t}}^{(\theta+\delta_j)\wedge\tau_n^{M,t}}\left(1+\|\Psi^n_s\|^4_H\right)ds\right)^{\frac{1}{2}}. \tag{3.8}$$

Attentions turn to the last term, for which we use Cauchy–Schwarz and Proposition 3.3.1 to obtain that

$$\mathbb{E}\left(\int_{\theta\wedge\tau_n^{M,t}}^{(\theta+\delta_j)\wedge\tau_n^{M,t}}\left(1+\|\Psi^n_s\|^4_H\right)ds\right)^{\frac{1}{2}} \leq \mathbb{E}\left(\sup_{r\in[0,\tau_n^{M,t}]}\|\Psi^n_s\|^2_H\int_{\theta\wedge\tau_n^{M,t}}^{(\theta+\delta_j)\wedge\tau_n^{M,t}}\left(1+\|\Psi^n_s\|^2_H\right)ds\right)^{\frac{1}{2}}$$

$$\leq\left[\mathbb{E}\left(\sup_{r\in[0,\tau_n^{M,t}]}\|\Psi^n_s\|^2_H\right)\right]^{\frac{1}{2}}\left[\mathbb{E}\int_{\theta\wedge\tau_n^{M,t}}^{(\theta+\delta_j)\wedge\tau_n^{M,t}}\left(1+\|\Psi^n_s\|^2_H\right)ds\right]^{\frac{1}{2}}$$

$$\leq c\left[\mathbb{E}\int_{\theta\wedge\tau_n^{M,t}}^{(\theta+\delta_j)\wedge\tau_n^{M,t}}\left(1+\|\Psi^n_s\|^2_H\right)ds\right]^{\frac{1}{2}}$$

where c is dependent on the boundedness of the initial condition $\boldsymbol{\Psi}_0$ from Proposition 3.3.1. For both this control and the remaining term of (3.8), we show that

$$\mathbb{E} \int_{\theta \wedge \tau_n^{M,t}}^{(\theta+\delta_j) \wedge \tau_n^{M,t}} \|\boldsymbol{\Psi}_s^n\|_H^2 ds$$

$$\leq \mathbb{E}\left(\sup_{r \in [0,\tau_n^{M,t}]} \|\boldsymbol{\Psi}_r^n\|_H \int_{\theta \wedge \tau_n^{M,t}}^{(\theta+\delta_j) \wedge \tau_n^{M,t}} \|\boldsymbol{\Psi}_s^n\|_H ds \right)$$

$$\leq \left[\mathbb{E}\left(\sup_{r \in [0,\tau_n^{M,t}]} \|\boldsymbol{\Psi}_r^n\|_H^2 \right) \right]^{\frac{1}{2}} \left[\mathbb{E}\left(\int_{\theta \wedge \tau_n^{M,t}}^{(\theta+\delta_j) \wedge \tau_n^{M,t}} \|\boldsymbol{\Psi}_s^n\|_H ds \right)^2 \right]^{\frac{1}{2}}$$

$$\leq c \left[\mathbb{E}\left(\int_{\theta \wedge \tau_n^{M,t}}^{(\theta+\delta_j) \wedge \tau_n^{M,t}} \|\boldsymbol{\Psi}_s^n\|_H ds \right)^2 \right]^{\frac{1}{2}}$$

$$\leq c \left[\mathbb{E}\left(\left(\int_{\theta \wedge \tau_n^{M,t}}^{(\theta+\delta_j) \wedge \tau_n^{M,t}} (1 + \|\boldsymbol{\Psi}_s^n\|_H^2) \, ds \right) \left(\int_{\theta \wedge \tau_n^{M,t}}^{(\theta+\delta_j) \wedge \tau_n^{M,t}} \|\boldsymbol{\Psi}_s^n\|_H ds \right) \right) \right]^{\frac{1}{2}}$$

$$\leq c \left[\mathbb{E}\left(\int_{\theta \wedge \tau_n^{M,t}}^{(\theta+\delta_j) \wedge \tau_n^{M,t}} \|\boldsymbol{\Psi}_s^n\|_H ds \right) \right]^{\frac{1}{2}}$$

$$\leq c \left[\left[\mathbb{E}\left(\sup_{r \in [0,\tau_n^{M,t}]} \|\boldsymbol{\Psi}_r^n\|_H^2 \right) \right]^{\frac{1}{2}} \left[\mathbb{E}\left(\int_{\theta \wedge \tau_n^{M,t}}^{(\theta+\delta_j) \wedge \tau_n^{M,t}} 1 ds \right)^2 \right]^{\frac{1}{2}} \right]^{\frac{1}{2}}$$

$$\leq c \left[\mathbb{E}(\delta_j^2) \right]^{\frac{1}{4}}$$

having utilised that $\int_0^{\tau_n^{M,t}} \|\boldsymbol{\Psi}_s^n\|_H^2 ds \leq c$ again from the definition of the first hitting time. Noting that δ_j is $\mathbb{P} - a.s.$ monotone decreasing (as $j \to \infty$) and convergent to 0, the monotone convergence theorem thus justifies that

$$c \left[\mathbb{E}(\delta_j^2) \right]^{\frac{1}{4}} = o_j$$

where o_j represents a constant independent of n which goes to zero as $\delta_j \to 0$. Revisiting (3.8), we have now justified that

$$\mathbb{E}\left[\sup_{r \in [0,\delta_j]} \|\boldsymbol{\Psi}_{(\theta+r) \wedge \tau_n^{M,t}}^n\|_U^2 - \|\boldsymbol{\Psi}_{\theta \wedge \tau_n^{M,t}}^n\|_U^2 + \int_{\theta \wedge \tau_n^{M,t}}^{(\theta+\delta_j) \wedge \tau_n^{M,t}} \|\boldsymbol{\Psi}_s^n\|_H^2 ds \right] \leq o_j. \quad (3.9)$$

It remains to relate the expression on the left-hand side with what we are interested in, which is $\|\boldsymbol{\Psi}^n\|_{UH,(\theta+\delta_j) \wedge \tau_n^{M,t}}^2 - \|\boldsymbol{\Psi}^n\|_{UH,\theta \wedge \tau_n^{M,t}}^2$. We have that

$$\|\boldsymbol{\Psi}^n\|^2_{UH,(\theta+\delta_j)\wedge\tau_n^{M,t}} - \|\boldsymbol{\Psi}^n\|^2_{UH,\theta\wedge\tau_n^{M,t}}$$

$$= \sup_{s\in[0,(\theta+\delta_j)\wedge\tau_n^{M,t}]}\|\boldsymbol{\Psi}^n_s\|^2_U - \sup_{s\in[0,\theta\wedge\tau_n^{M,t}]}\|\boldsymbol{\Psi}^n_s\|^2_U + \int_{\theta\wedge\tau_n^{M,t}}^{(\theta+\delta_j)\wedge\tau_n^{M,t}}\|\boldsymbol{\Psi}^n_s\|^2_H ds$$

and claim

$$\sup_{s\in[0,(\theta+\delta_j)\wedge\tau_n^{M,t}]}\|\boldsymbol{\Psi}^n_s\|^2_U - \sup_{s\in[0,\theta\wedge\tau_n^{M,t}]}\|\boldsymbol{\Psi}^n_s\|^2_U \le \sup_{r\in[0,\delta_j]}\|\boldsymbol{\Psi}^n_{(\theta+r)\wedge\tau_n^{M,t}}\|^2_U - \|\boldsymbol{\Psi}^n_{\theta\wedge\tau_n^{M,t}}\|^2_U. \qquad (3.10)$$

Indeed, we have that

$$\sup_{s\in[0,(\theta+\delta_j)\wedge\tau_n^{M,t}]}\|\boldsymbol{\Psi}^n_s\|^2_U \le \sup_{s\in[0,\theta\wedge\tau_n^{M,t}]}\|\boldsymbol{\Psi}^n_s\|^2_U + \sup_{s\in[\theta\wedge\tau_n^{M,t},(\theta+\delta_j)\wedge\tau_n^{M,t}]}\|\boldsymbol{\Psi}^n_s\|^2_U - \|\boldsymbol{\Psi}^n_{\theta\wedge\tau_n^{M,t}}\|^2_U$$

as the left-hand side must equal either $\sup_{s\in[0,\theta\wedge\tau_n^{M,t}]}\|\boldsymbol{\Psi}^n_s\|^2_U$ or $\sup_{s\in[\theta\wedge\tau_n^{M,t},(\theta+\delta_j)\wedge\tau_n^{M,t}]}\|\boldsymbol{\Psi}^n_s\|^2_U$, both of which are greater than the subtracted term $\|\boldsymbol{\Psi}^n_{\theta\wedge\tau_n^{M,t}}\|^2_U$. Appreciating that

$$\sup_{s\in[\theta\wedge\tau_n^{M,t},(\theta+\delta_j)\wedge\tau_n^{M,t}]}\|\boldsymbol{\Psi}^n_s\|^2_U = \sup_{r\in[0,\delta_j]}\|\boldsymbol{\Psi}^n_{(\theta+r)\wedge\tau_n^{M,t}}\|^2_U$$

then yields the claim (3.10), which in combination with (3.9) grants that

$$\mathbb{E}\left[\|\boldsymbol{\Psi}^n\|^2_{UH,(\theta+\delta_j)\wedge\tau_n^{M,t}} - \|\boldsymbol{\Psi}^n\|^2_{UH,\theta\wedge\tau_n^{M,t}}\right] \le o_j.$$

This proves the result. □

In combination with Proposition 3.3.2 we are entitled to apply Proposition 3.4.4 for the spaces $X_s := L^\infty([0,s];U) \cap L^2([0,s];H)$ with $\|\cdot\|_{UH,s}$ norm. We obtain, therefore, that for any $t \ge 0$ and any given $R > 0$ we can choose M such that the $\tau_\infty^{M,t}$ of Proposition 3.3.4, for which $(\boldsymbol{\Psi},\tau_\infty^{M,t})$ is a local strong solution, satisfies $\tau^{R,t} \le \tau_\infty^{M,t}\,\mathbb{P} - a.s.$ with

$$\tau^{R,t} := t \wedge \inf\left\{s \ge 0 : \|\boldsymbol{\Psi}\|^2_{UH,s} \ge R\right\}.$$

We now inspect how this affects the maximal time Θ, and argue that at $\mathbb{P} - a.e.\omega$ for which $\Theta(\omega) < \infty$,

$$\|\boldsymbol{\Psi}(\omega)\|^2_{UH,\Theta(\omega)} := \sup_{r\in[0,\Theta(\omega))}\|\boldsymbol{\Psi}_r(\omega)\|^2_U + \int_0^{\Theta(\omega)}\|\boldsymbol{\Psi}_r(\omega)\|^2_H dr = \infty. \qquad (3.11)$$

The maximality of Θ ensures that for every R and t, $\tau^{R,t} \le \Theta\,\mathbb{P} - a.s.$ as it must exceed any stopping time which is the lifetime of a local strong (see, for example, Goodair et al. (2023, Corollary 3.35)). Suppose for a contradiction that there exists a set of positive probability on which $\Theta < \infty$ and $\|\boldsymbol{\Psi}\|^2_{UH,\Theta} < \infty$. Classically there must exist a set of positive probability $\mathscr{A}$ and values t, R such that for all $\omega \in \mathscr{A}$, $\Theta < t$ and $\|\Psi\|^2_{UH,\Theta} < R$. This implies that for $\omega \in \mathscr{A}$, $\Theta(\omega) < \tau^{R,t}(\omega)$ which provides the contradiction. The property (3.11) is thus proven.

This proves the first assertion of Theorem 3.2.9 in the case that $\Psi_0 \in L^\infty(\Omega; H)$. Extension to the unbounded case is exactly as in Sect. 3.3.1, c.f. Goodair et al. (2023, Sect. 3.7), which preserves this blow-up at the maximal time. To completely prove Theorem 3.2.9 we now only need to justify the second assertion. To this end we take a positive stopping time τ such that $\|\Psi\|^2_{UH,\tau} < \infty \mathbb{P} - a.s.$ We must show that $\Psi_{\cdot \wedge \tau} \in C([0, T]; H)$ and $\Psi_{\cdot} \mathbb{1}_{\cdot \leq \tau} \in L^2([0, T]; V)$ for all $T \geq 0 \mathbb{P} - a.s.$, that $\Psi_{\cdot \wedge \tau} \mathbb{1}_{\cdot \leq \tau}$ progressively measurable in V, and that the identity

$$\Psi_{t \wedge \tau} = \Psi_0 + \int_0^{t \wedge \tau} \mathcal{A}(s, \Psi_s)ds + \int_0^{t \wedge \tau} \mathcal{G}(s, \Psi_s)d\mathcal{W}_s$$

holds $\mathbb{P} - a.s.$ in U for all $t \geq 0$. By definition of Θ there exists a sequence of stopping times $(\theta_j)\mathbb{P} - a.s.$ monotone increasing and convergent to Θ, whereby $(\Psi_{\cdot \wedge \theta_j}, \theta_j)$ is a local strong solution of the equation (3.1) for each j. We consider the possibility that τ is infinite on some measurable subset $\mathscr{B} \subset \Omega$, implying that $\Theta = \infty$ on $\mathscr{B}$ by the blow-up, and also note that on $\mathscr{B}^C$ where $\tau < \infty$ then again due to blow-up we must have that $\tau < \Theta$. In any case, $(\theta_j \wedge \tau)$ is monotone convergent to $\tau \mathbb{P} - a.s.$, and $(\Psi_{\cdot \wedge \theta_j \wedge \tau}, \theta_j \wedge \tau)$ is a local strong solution of Eq. (3.1). For the progressive measurability, for each fixed $T > 0$ we understand $\Psi_{\cdot \wedge \tau} \mathbb{1}_{\cdot \leq \tau}$ as the $\mathbb{P} \times \lambda - a.e.$ limit of the $\mathcal{F}_T \times \mathcal{B}([0, T])-$measurable $(\Psi_{\cdot \wedge \theta_j \wedge \tau} \mathbb{1}_{\cdot \leq \theta_j \wedge \tau})$ on $\Omega \times [0, T]$. Such a limit preserves the measurability on the product sigma algebra, justifying the required progressive measurability. The remaining properties are pathwise hence even clearer, as for $\mathbb{P} - a.e. \omega$ in $\mathscr{B}$ and for any given T there exists a j such that $\theta_j(\omega) > T$ and $\Psi_{\cdot \wedge \theta_j}$ has the required regularity. Similarly on $\mathscr{B}^C$ for $\mathbb{P} - a.e. \omega$ there exists a j such that $\tau(\omega) < \theta_j(\omega)$, so $\Psi_{\cdot \wedge \tau(\omega)}(\omega) = \Psi_{\cdot \wedge \theta_j(\omega) \wedge \tau(\omega)}(\omega)$ which has the necessary properties as $\Psi_{\cdot \wedge \theta_j \wedge \tau}$ is a local strong solution. This concludes the proof.

3.4 Application: High-Order Regularity for Stochastic Navier–Stokes

As an application of this improved blow-up criterion, we demonstrate high-order regularity of a stochastic Navier–Stokes equation:

$$u_t = u_0 - \int_0^t \mathcal{L}_{u_s} u_s\, ds + \nu \int_0^t \Delta u_s\, ds + \int_0^t \mathcal{G}(u_s)d\mathcal{W}_s - \nabla \rho_t \tag{3.12}$$

where u represents the fluid velocity, $\nu > 0$ is the viscosity, ρ is the pressure and $\mathcal{L}$ is the nonlinear term defined by $\mathcal{L}_f g = \sum_{j=1}^N f^j \partial_j g$ with Laplacian $\Delta f = \sum_{j=1}^N \partial_j^2 f$.

We pose the equation over the torus $\mathbb{T}^N$ in $N = 2$ or 3 dimensions. On the noise $\mathcal{G}$ we assume that for each $i, j \in \mathbb{N}$, there exists constants $c_{i,j}$ such that $\mathcal{G}_i$ is $c_{i,j}-$Lipschitz on $W^{j,2}(\mathbb{T}^N; \mathbb{R}^N)$ and $\sum_{i=1}^{\infty} c_{i,j}^2 < \infty$. Of course more exciting noise structures could be considered in this framework, but we choose the Lipschitz case for a simple demonstration. We require the divergence-free property of solutions, which is to say that $\sum_{k=1}^{N} \partial_k u^k = 0$. To facilitate the analysis, we introduce some additional function spaces. Recall that any function $f \in L^2(\mathbb{T}^N; \mathbb{R}^N)$ admits the representation

$$f(x) = \sum_{k \in \mathbb{Z}^N} f_k e^{ik \cdot x} \tag{3.13}$$

whereby each $f_k \in \mathbb{C}^N$ is such that $f_k = \overline{f_{-k}}$ and the infinite sum is defined as a limit in $L^2(\mathbb{T}^N; \mathbb{R}^N)$, see, e.g. Robinson et al. (2016, Sect. 1.5 for details).

Definition 3.4.1 We define L^2_σ as the subset of $L^2(\mathbb{T}^N; \mathbb{R}^N)$ of zero-mean functions f whereby for all $k \in \mathbb{Z}^N$, $k \cdot f_k = 0$ with f_k as in (3.13). For general $m \in \mathbb{N}$ we introduce $W^{m,2}_\sigma$ as the intersection of $W^{m,2}(\mathbb{T}^N; \mathbb{R}^N)$, respectively, with L^2_σ.

Note that the dimensionality N is not explicitly included in the spaces, but will be made clear from context. We define the Leray Projector $\mathcal{P}$ as the orthogonal projection in $L^2(\mathbb{T}^N; \mathbb{R}^N)$ onto L^2_σ. For $m \in \mathbb{N}$ the inner product $\langle f, g \rangle_m := \langle (-\mathcal{P}\Delta)^{m/2} f, (-\mathcal{P}\Delta)^{m/2} g \rangle$ is equivalent to the usual $W^{m,2}(\mathbb{T}^N; \mathbb{R}^N)$ inner product on $W^{m,2}_\sigma$ and we consider $W^{m,2}_\sigma$ as a Hilbert Space equipped with this inner product. Further details can be found in Robinson et al. (2016, Exercises 2.12, 2.13 and the discussion in Sect. 2.3). Following the typical study of incompressible Navier–Stokes, we work with the projected equation

$$u_t = u_0 - \int_0^t \mathcal{P}\mathcal{L}_{u_s} u_s \, ds + \nu \int_0^t \mathcal{P}\Delta u_s \, ds + \int_0^t \mathcal{P}\mathcal{G}(u_s) d\mathcal{W}_s \tag{3.14}$$

which is now in the form of (3.1). The existence of a unique local strong solution to (3.14) in 3D, and a unique global strong solution in 2D, is by this point standard: see, for instance, Glatt-Holtz et al. (2009), Goodair (2024), Goodair and Crisan (2023). We state the result here.

Proposition 3.4.2 *Let $u_0 : \Omega \to W^{1,2}_\sigma$ be $\mathcal{F}_0-$measurable. Then there exists a pair (u, τ) where τ is a $\mathbb{P} - a.s.$ positive stopping time and u is a process such that for $\mathbb{P} - a.e.\omega$, $u.(\omega) \in C\left([0, T]; W^{1,2}_\sigma\right)$ and $u.(\omega)\mathbb{1}_{\cdot \leq \tau(\omega)} \in L^2\left([0, T]; W^{2,2}_\sigma\right)$ for all $T \geq 0$ and with $u.\mathbb{1}_{\cdot \leq \tau}$ progressively measurable in $W^{2,2}_\sigma$, satisfying*

$$u_t = u_0 - \int_0^{t \wedge \tau} \mathcal{P}\mathcal{L}_{u_s} u_s \, ds + \nu \int_0^{t \wedge \tau} \mathcal{P}\Delta u_s \, ds + \int_0^{t \wedge \tau} \mathcal{P}\mathcal{G}(u_s) d\mathcal{W}_s$$

$\mathbb{P} - a.s.$ in L^2_σ for all $t \geq 0$. Moreover if (v, γ) was any other such local strong solution then

$$\mathbb{P}\left(\{\omega \in \Omega : u_t(\omega) = v_t(\omega) \ \ \forall t \in [0, \tau \wedge \gamma]\}\right) = 1.$$

If $N = 2$ then for any given $T > 0$ one can choose $\tau := T$.

The problem that we consider is, if $u_0 : \Omega \to W_\sigma^{k,2}$ is $\mathcal{F}_0-$measurable for some $k \in \mathbb{N}$, then does u belong pathwise to $C\left([0, \tau]; W_\sigma^{k,2}\right) \cap L^2\left([0, \tau]; W_\sigma^{k+1,2}\right)$? The result is affirmative and proven through iterated applications of Theorem 3.2.9. In fact the framework and assumptions of Sects. 3.2.1 and 3.2.2 were completely verified for the spaces

$$V := W_\sigma^{3,2}, \qquad H := W_\sigma^{2,2}, \qquad U := W_\sigma^{1,2} \tag{3.15}$$

under a transport noise in Goodair and Crisan (2023, Sect. 3), whilst much more easily holding for the Lipschitz noise. The maximal solution that we obtain must agree with u on its lifetime of existence by uniqueness. From Proposition 3.4.2 it is certainly true that $\sup_{r\in[0,\tau)} \|u_r\|_{W_\sigma^{1,2}}^2 + \int_0^\tau \|u_r\|_{W_\sigma^{2,2}}^2 dr < \infty \, \mathbb{P} - a.s.$, so from Theorem 3.2.9 with the spaces established in (3.15) we verify that for $\mathbb{P} - a.e.\omega$, $u.(\omega) \in C\left([0, T]; W_\sigma^{2,2}\right)^1$ and $u.(\omega)\mathbb{1}_{\cdot \leq \tau(\omega)} \in L^2\left([0, T]; W_\sigma^{3,2}\right)$ for all $T \geq 0$. In particular, $\sup_{r\in[0,\tau)} \|u_r\|_{W_\sigma^{2,2}}^2 + \int_0^\tau \|u_r\|_{W_\sigma^{3,2}}^2 dr < \infty$. The inductive method is now apparent, where we consider spaces

$$V := W_\sigma^{j+1,2}, \qquad H := W_\sigma^{j,2}, \qquad U := W_\sigma^{j-1,2}.$$

For the Lipschitz noise a verification of the assumptions in these higher spaces provides little additional difficulty to the case of (3.15), so we omit the complete details here and content ourselves with applying Theorem 3.2.9 in any such case. Repeating this procedure for $j = 3, 4, \ldots, k$, we show the following to see that if u belongs pathwise to $C\left([0, T]; W_\sigma^{k-2,2}\right) \cap L^2\left([0, T]; W_\sigma^{k-1,2}\right)$ then indeed it does to $C\left([0, T]; W_\sigma^{k-1,2}\right) \cap L^2\left([0, T]; W_\sigma^{k,2}\right)$ and as above to $C\left([0, T]; W_\sigma^{k,2}\right) \cap L^2\left([0, T]; W_\sigma^{k+1,2}\right)$. Iterating this procedure down to the base case of (3.15), we then only need the regularity given to us in Proposition 3.4.2. We have shown the following.

Theorem 3.4.3 *For any given $k \in \mathbb{N}$ let $u_0 : \Omega \to W_\sigma^{k,2}$ be $\mathcal{F}_0-$measurable. Then any local strong solution (u, τ) of (3.14) as specified in Proposition 3.4.2 is such that for $\mathbb{P} - a.e.\omega, u.(\omega) \in C\left([0, T]; W_\sigma^{k,2}\right)$ and $u.(\omega)\mathbb{1}_{\cdot \leq \tau(\omega)} \in L^2\left([0, T]; W_\sigma^{k+1,2}\right)$ for all $T \geq 0$. If $N = 2$ then for $\mathbb{P} - a.e.\omega, u.(\omega) \in C\left([0, T]; W_\sigma^{k,2}\right) \cap L^2\left([0, T]; W_\sigma^{k+1,2}\right)$ for all $T \geq 0$.*

It should be noted that the assumptions cannot be verified for the spaces $V := W_\sigma^{2,2}, H := W_\sigma^{1,2}, U := L_\sigma^2$ as the algebra property for H is lost. Furthermore we do not obtain strong solutions in 3D on the lifespan of weak solutions.

---

[1] Recall that $u. = u._{\wedge\tau}$.

Appendix

Proposition 3.4.4 *Fix $T > 0$. For $t \in [0, T]$ let X_t denote a Banach space with norm $\| \cdot \|_{X,t}$ such that for all $s > t$, $X_s \hookrightarrow X_t$ and $\| \cdot \|_{X,t} \leq \| \cdot \|_{X,s}$. Suppose that (Ψ^n) is a sequence of processes $\Psi^n : \Omega \mapsto X_T$, $\|\Psi^n\|_{X,\cdot}$ is adapted and $\mathbb{P} - a.s.$ continuous, $\Psi^n \in L^2(\Omega; X_T)$, and such that $\sup_n \|\Psi^n\|_{X,0} \in L^\infty(\Omega; \mathbb{R})$. For any given $M > 1$ define the stopping times*

$$\tau_n^{M,T} := T \wedge \inf \left\{ s \geq 0 : \|\Psi^n\|_{X,s}^2 \geq M + \|\Psi^n\|_{X,0}^2 \right\}. \tag{3.16}$$

Furthermore suppose

$$\lim_{m \to \infty} \sup_{n \geq m} \mathbb{E}\left[\|\Psi^n - \Psi^m\|_{X, \tau_m^{M,t} \wedge \tau_n^{M,t}}^2 \right] = 0 \tag{3.17}$$

and that for any stopping time γ and sequence of stopping times (δ_j) which converge to $0 \, \mathbb{P} - a.s.$,

$$\lim_{j \to \infty} \sup_{n \in \mathbb{N}} \mathbb{E}\left(\|\Psi^n\|_{X, (\gamma + \delta_j) \wedge \tau_n^{M,T}}^2 - \|\Psi^n\|_{X, \gamma \wedge \tau_n^{M,T}}^2 \right) = 0. \tag{3.18}$$

Then there exists a stopping time $\tau_\infty^{M,T}$, a process $\Psi : \Omega \mapsto X_{\tau_\infty^{M,T}}$ whereby $\|\Psi\|_{X, \cdot \wedge \tau_\infty^{M,T}}$ is adapted and $\mathbb{P} - a.s.$ continuous, and a subsequence indexed by (m_j) such that

- $\tau_\infty^{M,T} \leq \tau_{m_j}^{M,T} \mathbb{P} - a.s.$,
- $\lim_{j \to \infty} \|\Psi - \Psi^{m_j}\|_{X, \tau_\infty^{M,T}} = 0 \, \mathbb{P} - a.s.$

Moreover for any $R > 0$ we can choose M to be such that the stopping time

$$\tau^{R,T} := T \wedge \inf \left\{ s \geq 0 : \|\Psi\|_{X, s \wedge \tau_\infty^{M,T}}^2 \geq R \right\} \tag{3.19}$$

satisfies $\tau^{R,T} \leq \tau_\infty^{M,T} \mathbb{P} - a.s..$ Thus $\tau^{R,T}$ is simply $T \wedge \inf \left\{ s \geq 0 : \|\Psi\|_{X,s}^2 \geq R \right\}$.

Proof See Goodair (2024, Proposition 6.1). $\square$

References

Agresti A, Veraar M (2022) Nonlinear parabolic stochastic evolution equations in critical spaces part II: blow-up criteria and instantaneous regularization. J Evol Equ 22(2):56

Agresti A, Veraar M (2024) The critical variational setting for stochastic evolution equations. Probab Theory Relat Fields 188(3):957–1015

Alonso-Orán D, Rohde C, Tang H (2021) A local-in-time theory for singular SDEs with applications to fluid models with transport noise. J Nonlinear Sci 31(6):98

Debussche A, Glatt-Holtz N, Temam R (2011) Local martingale and pathwise solutions for an abstract fluids model. Phys D 240(14):1123–1144

Glatt-Holtz N, Ziane M et al (2009) Strong pathwise solutions of the stochastic Navier-Stokes system. Adv Differ Equ 14(5/6):567–600

Goodair D (2024) Weak and strong solutions to nonlinear SPDEs with unbounded noise. Nonlinear Differ Equ Appl 31(6):106

Goodair D, Crisan D (2023) On the 3D Navier-Stokes equations with stochastic lie transport. In: Stochastic transport in upper ocean dynamics II: STUOD 2022 Workshop, London, UK, September 26–29, vol 11. Springer Nature, p 53

Goodair D, Crisan D (2024) Stochastic calculus in infinite dimensions and SPDEs. Springer Nature

Goodair D, Crisan D, Lang O (2023) Existence and uniqueness of maximal solutions to SPDEs with applications to viscous fluid equations. In: Stochastics and partial differential equations: analysis and computations, pp 1–64

Gyöngy I, Krylov NV (1980) On stochastic equations with respect to semimartingales I. Stoch Int J Probab Stoch Process 4(1):1–21

Krylov NV, Rozovskii BL (2007) Stochastic evolution equations. In: Stochastic differential equations: theory and applications: a volume in honor of Professor Boris L Rozovskii. World Scientific, pp 1–69

Liu W, Röckner M (2010) SPDE in Hilbert space with locally monotone coefficients. J Funct Anal 259(11):2902–2922

Liu W, Röckner M (2013) Local and global well-posedness of SPDE with generalized coercivity conditions. J Differ Equ 254(2):725–755

Neelima, Šiška D (2020) Coercivity condition for higher moment a priori estimates for nonlinear SPDEs and existence of a solution under local monotonicity. Stochastics 92(5):684–715

Pardoux E (1975) Equations aux dérivées partielles stochastiques monotones. University these

Robinson JC, Rodrigo JL, Sadowski W (2016) The three-dimensional Navier–Stokes equations: classical theory, vol 157. Cambridge University Press

Röckner M, Shang S, Zhang T (2024) Well-posedness of stochastic partial differential equations with fully local monotone coefficients. Mathematische Annalen 1–51

Chapter 4
A Thermal Green–Naghdi Model with Time-Dependent Bathymetry and Complete Coriolis Force

Darryl D. Holm and Oliver D. Street

Abstract This paper extends the theoretical Euler–Poincaré framework for modelling ocean mixed layer dynamics. Through a symmetry-broken Lie group invariant variational principle, we derive a generalised Green–Naghdi equation with time-dependent bathymetry, a complete Coriolis force, and inhomogeneity of the thermal buoyancy. The nature of the model derived here lends it a potential future application to wave dynamics generated by changes to the bathymetry.

4.1 Introduction

The Green–Naghdi model is a two-dimensional geophysical fluid model for the dynamics of a shallow fluid with free upper boundary. The lower boundary of the domain is usually taken to be the (fixed) bathymetry below the fluid. In reduced barotropic-mode models (also called 1.5 layer models) for oceanic mixed layer dynamics, the upper boundary of the domain is a free surface and the lower boundary is taken to be the thermocline. In these models, the mixed layer of the ocean comprises the transitional state of fluid dynamics coupling the wind forcing at the ocean free surface and the oscillations of the thermocline below, riding the passive stably stratified depths of the deep ocean. With the assumption of a dynamically passive deep lower layer, the thermocline moves exactly out of phase with the upper free surface. In keeping with the dynamically passive lower layer assumption, most of this variation results from the deformation of the bottom boundary (thermocline). Thus, this presents the modelling challenge of identifying the influence of introducing time dependence into the 'bathymetry' on the properties and geometric structure of the Green–Naghdi model.

D. D. Holm
Department of Mathematics, Imperial College London, London, England

O. D. Street (✉)
Grantham Institute, Imperial College London, London, England
e-mail: o.street18@imperial.ac.uk

© The Author(s) 2026

B. Chapron et al. (eds.), *Stochastic Transport in Upper Ocean Dynamics IV*, Mathematics of Planet Earth 15, https://doi.org/10.1007/978-3-032-12749-5_4

These large movements of the thermocline give rise to non-hydrostatic pressure terms which, in vertically integrated shallow water models such as the Green–Naghdi equations, show up as dispersive terms involving higher-order derivatives of the velocity (Green and Naghdi 1976; Pogutse et al. 1987; Holm 1988; Nadiga et al. 1996). Such nonlinearly dispersive higher-order terms are known to matter dramatically in shallow water flow over an obstacle, Nadiga et al. (1996). Nonetheless, these terms are usually neglected in the hydrostatic approximations, although their nonlinear dispersive effects on the low-frequency variability may sometimes be significant. It should be noted that Green–Naghdi, in particular, is not advised in the multi-layer configuration, because the multi-layer Green–Naghdi equations are ill-posed (Cotter et al. 2010).

Traditional approximated models of geophysical flow involve assumptions due to the aspect ratio of the domain and flow. When approximating a full rotating fluid model in spherical geometry, it is standard to assume that the locally horizontal components of the rotation vector are sufficiently small to be ignored. This approximation can be poor when horizontal length scales are sufficiently well resolved, or when the flow has significant dynamics in the vertical direction. A correction to this can be made by reintroducing horizontal components into the rotation vector (Dellar and Salmon 2005; Stewart and Dellar 2010), thereby considering the complete Coriolis force in an approximate two-dimensional model.

In this work, following Holm and Luesink (2021), we will also be introducing thermal effects into the Green–Naghdi model. Modelling thermal fluids has a long and diverse history. There is naturally a particular interest within the geophysical sciences, and we here bring attention to the development of two-dimensional thermal models of fluid flow in a shallow domain. Such models were notably developed by Ripa (1993, 1995), who contributed to the thermal shallow water equation (sometimes referred to as the 'Ripa model', or IL^0 in Ripa's notation), later considered further in Dellar (2003); Warneford and Dellar (2013). The thermal shallow water equation was extended in Beron-Vera (2021a) to include polynomial (of degree α) stratification in the vertical direction (the $IL^{0,\alpha}$ model) and, in a separate work Beron-Vera (2021b), to the multi-layer setting. See Cotter et al. (2010) for a discussion of the Green–Naghdi model in a multi-layer context. Recent interest in thermal ocean modelling has involved the consideration of stochastic parameterisation and data assimilation methodologies (Holm et al. 2021), the variational structure of the thermal vertical slice model attributed to Eady (Holm et al. 2024a), and the study of wave–current interaction in a thermal fluid (Holm et al. 2024b).

In this work, we formulate the theoretical framework for comparing the solution behaviour of vertically integrated shallow water models, using traditional Coriolis force and fixed bathymetry, with the solution behaviour of a new extended Green–Naghdi model. The new model presented here is a variant of the thermal rotating Green–Naghdi equations, using a complete Coriolis force and featuring interactions between the fluid and a moving bathymetry representing the variability of the thermocline at the bottom boundary of the oceanic mixed layer. In formulating the theoretical model for this comparison, we expect to enable measurements of the sensitivity of surface wave propagation to its nonlinear coupling to the dynamics of the bathymetry, as well as the effect of thermal buoyancy, nonlinear dispersion, and the complete Coriolis force. In addition, this work creates the theoretical methodology needed to include contemporary methodologies of stochastic uncertainty parameterisation which preserve the model's behaviour (Holm 2015; Street and Takao 2024; Holm and Hu 2021; Luesink 2021).

Overview of the present work. This paper extends the theoretical basis for simulations of mixed layer dynamics in 1.5 layer models to determine the impacts of non-hydrostatic and thermal effects as well as non-traditional Coriolis forces. This is accomplished by deriving a version of the classic Green–Naghdi model which we extend to describe the mixed layer flow with long-wave dynamics of the upper free surface coupled to the dynamics of the thermocline layer at the bottom of the mixed layer, all riding on inert depths below.[1]

We expect that coupling the thermocline dynamics to horizontal gradients of the buoyancy in the mixed layer may have observable effects on the solution for the surface wave elevation. If so, these effects might be useful as diagnostics for the data produced by the NASA and ESA Surface Water and Ocean Topography (SWOT) satellite mission.

4.1.1 Some Applications of the Model and a Description of the Domain

Historically, the Green–Naghdi (GN) model is a two-dimensional model for a thin three-dimensional fluid with a free surface. That is, through a vertical averaging procedure, it provides a description of two-dimensional flow and wave dynamics within a 'thin' fluid layer. The setup of the domain is therefore as follows. In the horizontal directions, we have coordinates $\boldsymbol{x} = (x, y) \in \mathcal{D} \subseteq \mathbb{R}^2$ and in the vertical direction we have a coordinate $z \in [-H(x, y), \eta(x, y, t)]$, where $z = -H$ is the bottom boundary and $z = \eta$ is the free upper surface. Thus, the three-dimensional domain is $\mathcal{D} \times [-H, \eta] \subset \mathbb{R}^3$. In this paper, we modify this setup such that the bottom boundary is permitted to have some exogenous time dependence. This setup is interesting for several physical reasons, including tsunami/wave generation and the ocean mixed layer. The derivation of the Green–Naghdi model for the mixed layer assumes a dynamically passive, hydrostatic lower layer. This assumption leads to the following relations for partitioning the thickness of the mix-layer, denoted as $\eta(\boldsymbol{x}, t)$ (Gill 1982; Nadiga et al. 1996)

---

[1] This paper is based on unpublished notes from the mid-1990s during early work in the modelling phase of the COSIM (Computational Ocean and Sea Ice Model) project at Los Alamos National Laboratory. Some of the bibliography from that time is retained in this section as historical references whose presence might be helpful in tracing their influence in the development of modern methods of ocean mixed layer modelling during the past three decades.

$$H(\boldsymbol{x}, t) = \rho_r h(\boldsymbol{x}, t) - (1 + \rho_r) B,$$
$$\eta(\boldsymbol{x}, t) = (1 - \rho_r) h(\boldsymbol{x}, t) + (1 + \rho_r) B. \tag{4.1.1}$$

The symbol B denotes the (constant) mean depth of the upper layer, $H(\boldsymbol{x}, t)$ is the depth of the thermocline, and $\eta(\boldsymbol{x}, t)$ is the height of the free surface, using the undisturbed free surface as the reference level, at $\eta = 0$. The model parameter ρ_r is the ratio of the density of the upper layer to that of the lower layer. Here, we assume $\rho_r = 1 - \epsilon$ with $\epsilon \ll 1$, representing the small density difference of say 0.5% across the thermocline. It is clear from the relation $h = \eta + H$ in (4.1.1) that most of the barotropic variation in the surface layer depth $h(\boldsymbol{x}, t)$ is driven by the movement of the thermocline $H(\boldsymbol{x}, t)$; since the thermocline thickness H in (4.1.1) is more heavily weighted than the wave elevation, $\eta(\boldsymbol{x}, t)$.

4.1.2 The Historical Green–Naghdi Equations

The historical Green–Naghdi equations are given by Green and Naghdi (1976)

$$\frac{\partial \boldsymbol{u}}{\partial t} = -(\boldsymbol{u} \cdot \nabla)\boldsymbol{u} - g\nabla(h - H) + \frac{1}{h}\nabla\left(h^2 \frac{dA}{dt}\right) - \frac{dB}{dt}\nabla H,$$
$$\frac{\partial h}{\partial t} = -\nabla \cdot (h\boldsymbol{u}). \tag{4.1.2}$$

Here, $\boldsymbol{u}$ is the *two-dimensional* velocity field, $h = \eta(\boldsymbol{x}, t) + H(\boldsymbol{x})$ is the local depth of the water, $z = \eta(\boldsymbol{x}, t)$ is the free surface height, and $z = -H(\boldsymbol{x})$ is the *fixed* bottom boundary for the standard GN equations. Moreover, ∇ denotes the two-dimensional gradient in the (x, y) coordinates and $d/dt = \partial_t + \boldsymbol{u} \cdot \nabla$ is the material derivative. The quantities A and B in (4.1.2) are given by

$$A = \frac{1}{3}h\nabla \cdot \boldsymbol{u} + \frac{1}{2}\boldsymbol{u} \cdot \nabla H,$$
$$B = \frac{1}{2}h\nabla \cdot \boldsymbol{u} + \boldsymbol{u} \cdot \nabla H. \tag{4.1.3}$$

The historical GN equations in (4.1.2) provide a depth-averaged (barotropic) description of shallow water motion with a free surface under gravity, g, in *nondominant* asymptotics. That is, the GN equations are derived using shallow water scaling asymptotics for a domain of small aspect ratio, but no restriction is placed on the Froude number, or nonlinearity parameter of the flow, cf. Camassa and Holm (1992).

The historical GN equations in (4.1.2) have been derived by making a solution Ansatz of columnar motion in which the horizontal fluid velocity is independent of the vertical coordinate. One then imposes three-dimensional incompressibility and certain symmetry requirements. The GN equations are derived in Green and Naghdi (1976) by requiring the incompressible columnar motion to satisfy conservation of energy and invariance under rigid body translations. Equations were rediscovered in Pogutse et al. (1987) by inserting the columnar motion solution Ansatz, $\partial \boldsymbol{u}/\partial z = 0$, and incompressibility into the Euler equations, then averaging over depth. These equations have also been derived from a variational principle in Miles and Salmon (1985) by inserting the columnar motion Ansatz into the variational principle for the Euler equations for an inviscid incompressible fluid, and explicitly performing the vertical integrations before varying. The Hamiltonian structure of these equations has been discussed in Holm (1988).

The approach of Miles and Salmon (1985) in obtaining the historical GN equations by restricting to columnar motion in Hamilton's principle for an incompressible fluid with a free surface affords a convenient starting point for extending the GN equations to include effects of a rotating frame and weak buoyancy stratification. Our derivation by Hamilton's principle, following the derivation in Holm and Luesink (2021); Luesink (2021) inspired by Miles and Salmon, also provides a systematic derivation of the Kelvin circulation theorem and potential vorticity conservation laws for the extended GN equations.

4.2 An Euler–Poincaré Derivation of the Equations of Motion

In this section, we will augment the historical Green–Naghdi equations with the complete Coriolis force, thermal effects, and a time-varying lower boundary. We will derive the model by performing approximations in Hamilton's principle. For further details about these approximations, we note that the nondimensional form of a stochastic perturbation of the historical Green–Naghdi equations with thermal effects was given in (Holm and Luesink 2021). Moreover, in Stewart and Dellar (2010), the non-traditional Coriolis terms are discussed with reference to the Rossby number and aspect ratio.

We begin with the Lagrangian for the rotating Euler–Boussinesq equations in our three-dimensional domain

$$\ell_{EB} = \int_{\mathcal{D}\times[-H,\eta]} Dd^3x \left[\underbrace{\frac{1}{2}\left(|\boldsymbol{u}|^2 + w^2\right)}_{\text{Kinetic energy}} + \underbrace{\boldsymbol{R}(\boldsymbol{x})\cdot\boldsymbol{u} + S(\boldsymbol{x})w}_{\text{Coriolis terms}} - \underbrace{\rho gz}_{\substack{\text{Potential}\\\text{energy}}} - \underbrace{p(D^{-1} - 1)}_{\substack{\text{Incompressibility}\\\text{in 3D}}} \right],$$

$$(4.2.1)$$

where $Dd^3x \in \text{Vol}(M)$ is the advected mass density and ρ is an advected scalar thermal buoyancy. Denoting the three-dimensional gradient by ∇_3, the vector quantity

$$2\boldsymbol{\Omega} = \nabla_3 \times \boldsymbol{R}(\boldsymbol{x}) + \nabla_3 S(\boldsymbol{x}) \times \widehat{\boldsymbol{z}} \qquad (4.2.2)$$

is twice the rotation vector, $\boldsymbol{\Omega}$, which we note has a horizontal component when S has a nonvanishing gradient. The domain is bounded below by bathymetry $z = -H(\boldsymbol{x}, t)$

and the free surface $z = \eta(\boldsymbol{x}, t)$, and we introduce the quantity $h = \eta + H$, which is the layer thickness. In a previous work Holm and Luesink (2021), a similar variational method was used to derive thermal ocean models where, instead of the full thermal density ρ, the authors considered a stratified fluid by using a buoyancy variable b related to ρ through $\rho = \rho_0(1 + b)$, where ρ_0 is some reference density. In this paper, we will instead consider the variable ρ to be our advected variable. The effect of this change on the calculations to follow is minimal, as can be verified by comparing this paper with the Appendix of Holm and Luesink (2021).

Since the domain has moving boundaries, care must be taken over the boundary conditions. On a moving surface defined by $F(x, y, z, t) = 0$, a particle remains on this surface if its Lagrangian differential is zero. That is, if $\partial_t F + \boldsymbol{u} \cdot \nabla F + w \partial_z F = 0$. The standard kinematic boundary for free surfaces is recovered by setting $F = z - \eta(x, y, t)$, that is:

$$\partial_t \eta + \boldsymbol{u}|_{z=\eta} \cdot \nabla \eta = w|_{z=\eta}. \tag{4.2.3}$$

Similarly, in this case the bottom boundary, $z = -H$, is permitted to move according to some exogenous dynamical processes. In this case, setting $F = z + H$ gives the condition that a particle on the bottom boundary remains on the bottom boundary, i.e.:

$$\partial_t H + \boldsymbol{u}|_{z=-H} \cdot \nabla H = -w|_{z=-H}. \tag{4.2.4}$$

Since the three-dimensional flow governed by the Euler–Boussinesq model is incompressible, we can express the vertical derivative of the vertical velocity in terms of the two-dimensional gradient, ∇, as

$$\partial_z w = -\nabla \cdot \boldsymbol{u}. \tag{4.2.5}$$

To determine the vertical velocity at some point (x, y, z), we integrate this constraint from $z = -H$ to z to give

$$w = -\int_{-H}^{z} \nabla \cdot \boldsymbol{u} \, dz' + w|_{z=-H}$$

$$= -\nabla \cdot \int_{-H}^{z} \boldsymbol{u} \, dz' + \boldsymbol{u}|_{z=-H} \cdot \nabla H + w|_{z=-H} = -\nabla \cdot \int_{-H}^{z} \boldsymbol{u} \, dz' - \partial_t H, \tag{4.2.6}$$

where we have used a multidimensional version of Leibniz's integration rule and the boundary condition (4.2.4). To arrive at the Green–Naghdi model, we must introduce a columnar motion approximation. Put simply, this can be achieved by replacing the horizontal velocity field $\boldsymbol{u}$ by its vertical average[2] $\overline{\boldsymbol{u}}$. Since, the vertically averaged horizontal velocity depends only on x, y, t, we have that

---

[2] We will, in general, denote vertically averaged variables using the 'overbar' notation.

$$w = -\nabla \cdot (\overline{\boldsymbol{u}}(z + H)) - \partial_t H$$

$$= -(z + H)\nabla \cdot \overline{\boldsymbol{u}} - \overline{\boldsymbol{u}} \cdot \nabla H - \partial_t H = -(z + H)\nabla \cdot \overline{\boldsymbol{u}} - \frac{dH}{dt}, \qquad (4.2.7)$$

where we have introduced the notation

$$\frac{d}{dt} := \partial_t + \overline{\boldsymbol{u}} \cdot \nabla \qquad (4.2.8)$$

for the material derivative of a scalar quantity along a Lagrangian trajectory of the vertically averaged velocity field $\overline{\boldsymbol{u}}$. The Lagrangian can be therefore approximated as

$$\begin{aligned}
\ell_{GN} &= \int_{\mathcal{D}} dx\,dy \int_{-H(\boldsymbol{x},t)}^{\eta(\boldsymbol{x},t)} dz \left[\frac{|\overline{\boldsymbol{u}}|^2}{2} + \frac{1}{2}\left((z+H)(\nabla \cdot \overline{\boldsymbol{u}}) + \frac{dH}{dt} \right)^2 \right. \\
&\qquad\qquad \left. + \boldsymbol{R}(\boldsymbol{x}) \cdot \overline{\boldsymbol{u}} - S\left((z+H)(\nabla \cdot \overline{\boldsymbol{u}}) + \frac{dH}{dt} \right) - \overline{\rho} g z \right] \\
&= \int_{\mathcal{D}} dx\,dy \left[\frac{h}{2}|\overline{\boldsymbol{u}}|^2 + h\boldsymbol{R}(\boldsymbol{x}) \cdot \overline{\boldsymbol{u}} + \frac{1}{6}h^3(\nabla \cdot \overline{\boldsymbol{u}})^2 + \frac{1}{2}h^2(\nabla \cdot \overline{\boldsymbol{u}})\frac{dH}{dt} + \frac{h}{2}\left(\frac{dH}{dt} \right)^2 \right. \\
&\qquad\qquad \left. - \frac{1}{2}h^2\,S(\boldsymbol{x})(\nabla \cdot \overline{\boldsymbol{u}}) - hS(\boldsymbol{x})\frac{dH}{dt} - \frac{1}{2}\overline{\rho}(\eta^2 - H^2) \right] \\
&= \int_{\mathcal{D}} dx\,dy\, h \left[\frac{|\overline{\boldsymbol{u}}|^2}{2} + \boldsymbol{R}(\boldsymbol{x}) \cdot \overline{\boldsymbol{u}} + \frac{1}{6}h^2(\nabla \cdot \overline{\boldsymbol{u}})^2 + \frac{1}{2}h(\nabla \cdot \overline{\boldsymbol{u}})\left(\frac{dH}{dt} - S(\boldsymbol{x}) \right) \right. \\
&\qquad\qquad \left. + \frac{1}{2}\frac{dH}{dt}\left(\frac{dH}{dt} - 2S(\boldsymbol{x}) \right) - \frac{1}{2}\overline{\rho}g(h - 2H) \right],
\end{aligned} \qquad (4.2.9)$$

and this is the Lagrangian for our thermal rotating Green–Naghdi model with variable bathymetry. The equation of motion then follows from the Euler–Poincaré theorem and takes the form:

$$\frac{d}{dt}\left(\frac{1}{h}\frac{\delta\ell_{GN}}{\delta\overline{\boldsymbol{u}}} \right) + \frac{1}{h}\frac{\delta\ell_{GN}}{\delta\overline{u}^j}\nabla\overline{u}^j = \nabla\frac{\delta\ell_{GN}}{\delta h} - \frac{1}{h}\frac{\delta\ell_{GN}}{\delta\overline{\rho}}\nabla\overline{\rho}, \qquad (4.2.10)$$

where the right-hand side of this equation results from the so-called 'diamond terms' inherent to the semidirect product theory of continuum dynamics with advected quantities (Holm et al. 1998). Here, this is the symmetry-broken Euler–Poincaré equation where the variables naturally live within a semidirect product between the diffeomorphisms on $\mathrm{Diff}(\mathcal{D})$ (or its associated algebra of vector fields) and the space of advected quantities $\Omega^0(\mathcal{D}) \oplus \Omega^2(\mathcal{D})$. The left-hand side of the above equation is a coordinate expression of

$$(\partial_t + \mathcal{L}_{\overline{u}})\left(\frac{1}{h}\frac{\delta\ell_{GN}}{\delta\overline{u}} \right), \qquad (4.2.11)$$

where $\mathcal{L}_{\overline{u}}$ denotes the Lie derivative by the vector field $\overline{u}$.

The variational derivatives of the Lagrangian are computed as

$$\frac{\delta \ell_{GN}}{\delta h} = \frac{1}{2}|\overline{u}|^2 + R(x) \cdot \overline{u} + \frac{1}{2}h^2(\nabla \cdot \overline{u})^2$$

$$+ h(\nabla \cdot \overline{u})\left(\frac{dH}{dt} - S(x)\right) + \frac{1}{2}\frac{dH}{dt}\left(\frac{dH}{dt} - 2S(x)\right) - \overline{\rho}g(h - H),$$

$$\frac{\delta \ell_{GN}}{\delta \overline{\rho}} = -\frac{1}{2}gh(h - 2H),$$

$$\frac{1}{h}\frac{\delta \ell_{GN}}{\delta \overline{u}} = \overline{u} + R(x) - \frac{1}{3h}\nabla\left(h^3(\nabla \cdot \overline{u})\right) + \frac{1}{2}h(\nabla \cdot \overline{u})\nabla H - \frac{1}{2h}\nabla\left(h^2\frac{dH}{dt}\right) + \frac{1}{2h}\nabla\left(h^2 S(x)\right)$$

$$+ \frac{dH}{dt}\nabla H - S(x)\nabla H$$

$$= \overline{u} + R(x) - \frac{1}{h}\nabla(h^2 F) + G\nabla H,$$

$$(4.2.12)$$

where F and G are given by (cf. Eq. (4.1.3))

$$F = \frac{1}{3}h\nabla \cdot \overline{u} + \frac{1}{2}\left(\frac{dH}{dt} - S(x)\right),$$
$$G = \frac{1}{2}h\nabla \cdot \overline{u} + \left(\frac{dH}{dt} - S(x)\right).$$

$$(4.2.13)$$

To make sense of the left-hand side of the Euler–Poincaré equation (4.2.10), we will make use of two equivalent forms of the Lie derivative of a vector field acting on a one-form, in two-dimensional domains, which are

$$\mathcal{L}_{\overline{u}}(A \cdot dx) = \left(\overline{u} \cdot \nabla A + A_j \nabla \overline{u}^j\right) \cdot dx = \left((\nabla^{\perp} \cdot A)\overline{u}^{\perp} + \nabla(\overline{u} \cdot A)\right) \cdot dx,$$

where we have introduced the perpendicular gradient, using the notation $(x, y)^{\perp} = (-y, x)$. Thus, after substituting in the variational derivatives, we have that the left-hand side of the Euler–Poincaré momentum equation is given by

$$(\partial_t + \mathcal{L}_{\overline{u}})\left(\frac{1}{h}\frac{\delta \ell_{GN}}{\delta \overline{u}}\right) = \left(\partial_t\overline{u} + \overline{u} \cdot \nabla\overline{u} + \overline{u}_j\nabla\overline{u}^j + (\nabla^{\perp} \cdot R)\overline{u}^{\perp} + \nabla(\overline{u} \cdot R)\right) \cdot dx$$

$$+ (\partial_t + \mathcal{L}_{\overline{u}})\left(\left(-\frac{1}{h}\nabla\left(h^2 F\right) + G\nabla H\right) \cdot dx\right),$$

$$(4.2.14)$$

where dx is the coordinate basis for one-forms. In the following calculations, we will make use of the fact that $h\,d^2x$ is advected, i.e.:

$$0 = (\partial_t + \mathcal{L}_{\overline{u}})(h\,d^2x) = (\partial_t h + \nabla \cdot (h\overline{u}))d^2x = \left(\frac{dh}{dt} + h(\nabla \cdot \overline{u})\right)d^2x. \quad (4.2.15)$$

In the expression for $\frac{1}{h}\frac{\delta \ell_{GN}}{\delta \overline{u}}$, since we have divided through by the density, when h appears it does so as a *function* without its coordinate basis d^2x. Thus, we have that expanding out Eq. (4.2.14) produces

$$(\partial_t + \mathcal{L}_{\bar{u}}) \left(\frac{1}{h} \frac{\delta \ell_{GN}}{\delta \bar{u}} \right) = \left(\partial_t \bar{u} + \bar{u} \cdot \nabla \bar{u} + \bar{u}_j \nabla \bar{u}^j + (\nabla^{\perp} \cdot R)\bar{u}^{\perp} + \nabla(\bar{u} \cdot R) - \frac{1}{h} \nabla \left(h^2 \frac{dF}{dt} \right) \right.$$

$$\left. + \frac{dG}{dt} \nabla H + \frac{1}{h^2} \frac{dh}{dt} \nabla(h^2 F) - \frac{1}{h} \nabla \left(2h \frac{dh}{dt} F \right) + G \nabla \frac{dH}{dt} \right) \cdot dx,$$

$$(4.2.16)$$

where we have used that $\partial_t + \mathcal{L}_{\bar{u}}$ obeys the standard rules of calculus and is d/dt when it acts on the functions F, G, h, H. Making use of the explicit formulae for F and G, as well as the fact that $h\, d^2 x$ is advected as a volume form, Eq. (4.2.14) can now be expressed as

$$(\partial_t + \mathcal{L}_{\bar{u}}) \left(\frac{1}{h} \frac{\delta \ell_{GN}}{\delta \bar{u}} \right) = \left(\partial_t \bar{u} + \bar{u} \cdot \nabla \bar{u} + \bar{u}_j \nabla \bar{u}^j + (\nabla^{\perp} \cdot R)\bar{u}^{\perp} + \nabla(\bar{u} \cdot R) - \frac{1}{h} \nabla \left(h^2 \frac{dF}{dt} \right) + \frac{dG}{dt} \nabla H \right.$$

$$- \frac{\nabla \cdot \bar{u}}{h} \nabla \left(h^2 \left(\frac{1}{3} h \nabla \cdot \bar{u} + \frac{1}{2} \left(\frac{dH}{dt} - S \right) \right) \right)$$

$$+ \frac{1}{h} \nabla \left(2h(h \nabla \cdot \bar{u}) \left(\frac{1}{3} h \nabla \cdot \bar{u} + \frac{1}{2} \left(\frac{dH}{dt} - S \right) \right) \right)$$

$$\left. + \left(\frac{1}{2} h \nabla \cdot \bar{u} + \left(\frac{dH}{dt} - S \right) \right) \nabla \frac{dH}{dt} \right) \cdot dx$$

$$= \left(\partial_t \bar{u} + \bar{u} \cdot \nabla \bar{u} + \bar{u}_j \nabla \bar{u}^j + (\nabla^{\perp} \cdot R)\bar{u}^{\perp} + \nabla(\bar{u} \cdot R) - \frac{1}{h} \nabla \left(h^2 \frac{dF}{dt} \right) + \frac{dG}{dt} \nabla H \right.$$

$$- \frac{1}{3}(h \nabla \cdot \bar{u}) \nabla (h \nabla \cdot \bar{u}) - \frac{2}{3}(h \nabla \cdot \bar{u})^2 \frac{\nabla h}{h} - \frac{h \nabla \cdot \bar{u}}{2} \nabla \left(\frac{dH}{dt} - S \right)$$

$$- (\nabla \cdot \bar{u}) \left(\frac{dH}{dt} - S \right) \nabla h + \frac{2}{3} \nabla \left((h \nabla \cdot \bar{u})^2 \right) + \frac{2}{3}(h \nabla \cdot \bar{u})^2 \frac{\nabla h}{h}$$

$$(h \nabla \cdot \bar{u}) \left(\frac{dH}{dt} - S \right) \frac{\nabla h}{h} + \nabla \left((h \nabla \cdot \bar{u}) \left(\frac{dH}{dt} - S \right) \right) + \frac{1}{2}(h \nabla \cdot \bar{u}) \nabla \frac{dH}{dt}$$

$$\left. + \left(\frac{dH}{dt} - S \right) \nabla \frac{dH}{dt} \right) \cdot dx,$$

where we have applied the product and chain rules to go from the first line to the second. Cancelling terms and grouping terms which can be written as a gradient, we have that

$$(\partial_t + \mathcal{L}_{\bar{u}}) \left(\frac{1}{h} \frac{\delta \ell_{GN}}{\delta \bar{u}} \right) = \left(\partial_t \bar{u} + \bar{u} \cdot \nabla \bar{u} + \bar{u}_j \nabla \bar{u}^j + (\nabla^{\perp} \cdot R)\bar{u}^{\perp} + \nabla(\bar{u} \cdot R) - \frac{1}{h} \nabla \left(h^2 \frac{dF}{dt} \right) + \frac{dG}{dt} \nabla H \right.$$

$$+ \frac{1}{2} \nabla \left((h \nabla \cdot \bar{u})^2 \right) + \nabla \left((h \nabla \cdot \bar{u}) \left(\frac{dH}{dt} - S \right) \right) + \frac{1}{2} \nabla \left(\frac{dH}{dt} \right)^2$$

$$\left. - S \nabla \frac{dH}{dt} + \frac{h \nabla \cdot \bar{u}}{2} \nabla S \right) \cdot dx$$

$$= \left(\partial_t \bar{u} + \bar{u} \cdot \nabla \bar{u} + \bar{u}_j \nabla \bar{u}^j + (\nabla^{\perp} \cdot R)\bar{u}^{\perp} + \nabla(\bar{u} \cdot R) - \frac{1}{h} \nabla \left(h^2 \frac{dF}{dt} \right) + \frac{dG}{dt} \nabla H \right.$$

$$+ \nabla \left(\frac{1}{2}(h \nabla \cdot \bar{u})^2 + (h \nabla \cdot \bar{u}) \left(\frac{dH}{dt} - S \right) + \frac{1}{2} \frac{dH}{dt} \left(\frac{dH}{dt} - 2S \right) \right)$$

$$\left. + \left(\frac{dH}{dt} + \frac{h \nabla \cdot \bar{u}}{2} \right) \nabla S \right) \cdot dx.$$

The right-hand side of the Euler–Poincaré equation (4.2.10) is

$$\nabla \frac{\delta \ell_{GN}}{\delta h} - \frac{1}{h} \frac{\delta \ell_{GN}}{\delta \overline{\rho}} \nabla \overline{\rho} = \nabla \left(\frac{1}{2} |\overline{u}|^2 + R(x) \cdot \overline{u} + \frac{1}{2} h^2 (\nabla \cdot \overline{u})^2 + h(\nabla \cdot \overline{u}) \left(\frac{dH}{dt} - S \right) \right.$$

$$+ \frac{1}{2} \frac{dH}{dt} \left(\frac{dH}{dt} - 2S \right) - \overline{\rho} g(h - H) \right) + \frac{1}{2} g(h - 2H) \nabla \overline{\rho}$$

$$= \nabla \left(\frac{|\overline{u}|^2}{2} + R \cdot \overline{u} \right) - \frac{1}{2} g h \nabla \overline{\rho} - g \overline{\rho} \nabla (h - H)$$

$$+ \nabla \left(\frac{1}{2} (h \nabla \overline{u})^2 + (h \nabla \cdot \overline{u}) \left(\frac{dH}{dt} - S \right) + \frac{1}{2} \frac{dH}{dt} \left(\frac{dH}{dt} - 2S \right) \right).$$

$$(4.2.17)$$

Combining these calculations finally gives the following extended Green–Naghdi equations:

$$\frac{d\overline{u}}{dt} + (\nabla^{\perp} \cdot R) \overline{u}^{\perp} = -g \overline{\rho} \nabla (h - H) - \frac{1}{2} g h \nabla \overline{\rho} + \frac{1}{h} \nabla \left(h^2 \frac{dF}{dt} \right) - \frac{dG}{dt} \nabla H - \left(\frac{dH}{dt} + \frac{h \nabla \cdot \overline{u}}{2} \right) \nabla S,$$

$$(4.2.18)$$

$$\frac{d\overline{\rho}}{dt} = 0, \tag{4.2.19}$$

$$\frac{dh}{dt} = -h \nabla \cdot \overline{u}. \tag{4.2.20}$$

4.3 Some Features of the Model

The equations presented in the previous section have a potential vorticity formulation; however, since the model includes thermal effects, this potential vorticity is not conserved. The most direct way to see this is through Kelvin's circulation theorem, which follows from the Euler–Poincaré formulation of the model as follows. To express Kelvin's circulation theorem, we first must introduce the notion of closed continuous loops within the fluid, $c : C^1 \to \mathcal{D}$, the space of which is denoted by $\mathcal{C}(\mathcal{D})$. The group $\mathrm{Diff}(\mathcal{D})$ acts on $\mathcal{C}(\mathcal{D})$ from the left by composition, that is, $\phi c := \phi \circ c : C^1 \to \mathcal{D}$ is a group action for $\phi \in \mathrm{Diff}(M)$ and $c \in \mathcal{C}(\mathcal{D})$. For an initial loop of fluid $c_0 \in \mathcal{C}(\mathcal{D})$, if the flow map of the fluid $\phi_t \in \mathrm{Diff}(\mathcal{D})$ acts on this initial loop then we obtain a closed loop moving with the flow $c_t = \phi_t c_0$.

Proposition 1 *For a closed loop, $c_t \in \mathcal{C}(\mathcal{D})$, enclosing a region of fluid, $S_t \subset \mathcal{D}$, and moving with the flow of the Green–Naghdi Equations (4.2.18)–(4.2.20), we have the following Kelvin circulation relation:*

$$\frac{d}{dt} \oint_{c_t} \left(\overline{u} + R(x) - \frac{1}{h} \nabla (h^2 F) + G \nabla H \right) \cdot dx = \frac{g}{2} \int_{S_t} \nabla^{\perp} (h - 2H) \cdot \nabla \overline{\rho} \, dS.$$

$$(4.3.1)$$

Proof The proof of this follows from the transport theorem for the contour c_t moving with the fluid velocity $\overline{\boldsymbol{u}}$, and the form of the momentum equation presented in Eq. (4.2.10). Indeed,

$$
\frac{d}{dt} \oint_{c_t} \left(\frac{1}{h} \frac{\delta \ell_{GN}}{\delta \overline{\boldsymbol{u}}} \right) \cdot d\boldsymbol{x} = \oint_{c_t} \left[\frac{d}{dt} \left(\frac{1}{h} \frac{\delta \ell_{GN}}{\delta \overline{\boldsymbol{u}}} \right) + \frac{1}{h} \frac{\delta \ell_{GN}}{\delta \overline{u}^j} \nabla \overline{u}^j \right] \cdot d\boldsymbol{x}
$$

$$
= \oint_{c_t} \left[\nabla \frac{\delta \ell_{GN}}{\delta h} - \frac{1}{h} \frac{\delta \ell_{GN}}{\delta \overline{\rho}} \nabla \overline{\rho} \right] \cdot d\boldsymbol{x} = \oint_{c_t} \frac{1}{2} g(h - 2H) \nabla \overline{\rho} \cdot d\boldsymbol{x},
$$

$$\tag{4.3.2}$$

and Green's theorem then gives the required result. $\qquad\square$

Notice that, for this model, we have that circulation is created whenever the gradient of the bathymetry or surface elevation does not align with the gradient of the thermal buoyancy. The Kelvin circulation result (4.3.1) permits us to understand the potential vorticity dynamics. In particular, we define the potential vorticity by

$$
q = \frac{1}{h} \nabla^{\perp} \cdot \left(\overline{\boldsymbol{u}} + \boldsymbol{R}(\boldsymbol{x}) - \frac{1}{h} \nabla(h^2 F) + G \nabla H \right),
\tag{4.3.3}
$$

and the dynamics of q are then governed by

$$
\partial_t q + \overline{\boldsymbol{u}} \cdot \nabla q = \frac{g}{2} \frac{1}{h} \nabla^{\perp}(h - 2H) \cdot \nabla \overline{\rho}.
\tag{4.3.4}
$$

In the definition of q given above, it has been expressed in terms of the perpendicular two-dimensional gradient to ensure that the equations are expressed in terms of a self-consistent two-dimensional vector calculus convention. Reintroducing the third coordinate in the vertical direction, we note that this is equivalent to $q = \frac{1}{h} \hat{\boldsymbol{z}} \cdot \nabla_3 \times \frac{1}{h} \frac{\delta \ell_{GN}}{\delta \overline{\boldsymbol{u}}}$.

Remark 4.1 (*The influence of thermal effects*) The presence of thermal inhomogeneity in the model has resulted in forcing terms on the right-hand side of the Kelvin circulation statement (4.3.1) and on the potential vorticity Equation (4.3.4). When the advected quantity $\overline{\rho}$ is not present (i.e. when it is constant), these terms vanish and the potential vorticity is purely advected. Thus, in the case without thermal effects, the equations have infinitely many conserved quantities (Casimirs) which are the arbitrary functions of q integrated against the measure $h\,dx\,dy$.

The equations presented in this paper, complete with thermal effects, also possess infinitely many conserved integral quantities. Corresponding to suitably well-behaved functions Φ, Ψ of the thermal buoyancy $\overline{\rho}$, we have the following Casimir conserved quantity

$$
C_{\Phi,\Psi} = \int_{\mathcal{D}} (\Phi(\overline{\rho}) + q \Psi(\overline{\rho}))\, h\, dx dy.
\tag{4.3.5}
$$

There are infinitely many such quantities, inherited from the dimension of the space of suitable functions of $\overline{\rho}$. We here note that, in the absence of thermal effects, the Casimirs for the standard Green–Naghdi (or shallow water) equations are dramatically different. That is, potential vorticity is an advected variable when $\overline{\rho}$ is a constant and, as is known for standard rotating shallow water equations (Dellar and Salmon 2005), the Casimir invariants would be simply arbitrary functions of potential vorticity integrated against the depth weighted measure. In the symmetry-broken thermal fluid model considered here, the form of the Casimirs is given in (4.3.5) and is standard for such semidirect product systems (see, e.g. Dellar (2003); Holm et al. (2025)) and was also presented for the Green–Naghdi equations in Luesink (2021). These Casimirs have the same structure as those for the thermal shallow water model (also known as the IL^0 equations) (Beron-Vera 2021a)

4.4 Comments and Outlook

In this paper, we introduce an extension to the standard Green–Naghdi model for vertically averaged geophysical flows. In particular, we introduce a complete Coriolis force (Dellar and Salmon 2005), thermal effects, and a time-dependent bathymetry. This time-dependent bathymetry makes the model suitable to be used as the topmost layer of a multi-layer ocean model, in which the bottom boundary is taken to be the thermocline or a free boundary between layers with different modelling assumptions. This model has been derived using the Euler–Poincaré approach, and is therefore intrinsically geometric in nature. This has permitted the formulation of Kelvin's circulation theorem, a potential vorticity formulation and infinitely many 'Casimir' conserved quantities. In a future work, it may be of interest to introduce the nondimensional form of the equations and illuminate the relationship between the model derived here and a family of related models. Such a family is well known for the standard case for which the bottom boundary is a fixed function of space.

This work has application towards other geophysical problems for which a changing bottom boundary is relevant. In particular, this work is of relevance to wave generation by landslides or earthquakes (Firdaus and Behrens 2024), a problem for which accurately accounting for the time dependence in the bathymetry is crucial for accurate forecasting. The equations presented here are deterministic in nature, and follow from classical geometric approaches to model derivation. It is possible, following, e.g. Holm (2015); Street and Takao (2024); Holm and Hu (2021); Luesink (2021), to include stochastic terms through the geometric structure of the model for the purposes of data assimilation and uncertainty quantification. There is particular interest in interpreting data in this field due to its application to tsunami forecasting and, in particular, the assimilation of experimental wave data (see, e.g. Whittaker et al. (2015)) into a stochastic Green–Naghdi model is an interesting proposition. Testing this proposal is beyond the scope of the present paper.

Acknowledgements We would like to thank our friends R. Hu, C. Cotter, J. Woodfield, and E. Luesink for their helpful discussions on geophysical fluids during the course of this work and beyond. Oliver D. Street acknowledges funding for a research fellowship from Quadrature Climate Foundation. Darryl D. Holm was partially supported during the present work by European Research Council (ERC) Synergy grant Stochastic Transport in Upper Ocean Dynamics (STUOD)—DLV-856408, as well as partially supported by Office of Naval Research (ONR) grant award N00014-22-1-2082, Stochastic Parameterization of Ocean Turbulence for Observational Networks.

References

Beron-Vera FJ (2021a) Extended shallow-water theories with thermodynamics and geometry. Phys Fluids 33(10):106605

Beron-Vera FJ (2021b) Multilayer shallow-water model with stratification and shear. Revista Mexicana de Física 67(3 May–Jun):351–364

Camassa R, Holm DD (1992) Dispersive barotropic equations for stratified mesoscale ocean dynamics. Physica D 60(1):1–15

Cotter CJ, Holm DD, Percival JR (2010) The square root depth wave equations. Proc Royal Soc A Math Phys Eng Sci 466:3621–3633

Dellar PJ (2003) Common Hamiltonian structure of the shallow water equations with horizontal temperature gradients and magnetic fields. Phys Fluids 15(2):292–297

Dellar PJ, Salmon R (2005) Shallow water equations with a complete Coriolis force and topography. Phys Fluids 17:106601–19

Gill AE (1982) Atmosphere–ocean dynamics. Elsevier

Green AE, Naghdi PM (1976) A derivation of equations for wave propagation in water of variable depth. J Fluid Mech 78(2):237–246

Holm DD (1988) Hamiltonian structure for two-dimensional hydrodynamics with nonlinear dispersion. Phys Fluids 31(8):2371–2373

Holm DD (2015) Variational principles for stochastic fluid dynamics. Proce Royal Soc A: Math Phys Eng Sci 471(2176):20140963

Holm DD, Hu R (2021) Stochastic effects of waves on currents in the ocean mixed layer. J Math Phys 62(7):073102

Holm DD, Ruiao H, Street OD (2024a) Geometric mechanics of the vertical slice model. Geom Mech 01(02):77–121

Holm DD, Hu R, Street OD (2024b) On the interactions between mean flows and inertial gravity waves in the WKB approximation. In: Chapron B, Crisan D, Holm D, Mémin E, Radomska A (eds) Stochastic transport in upper ocean dynamics II. Springer Nature Switzerland, pp 111–141

Holm DD, Hu R, Street OD (2025) Plasma dynamics in thin domains

Holm DD, Luesink E (2021) Stochastic wave–current interaction in thermal shallow water dynamics. J Nonlinear Sci 31:29

Holm DD, Luesink E, Pan W (2021) Stochastic mesoscale circulation dynamics in the thermal ocean. Phys Fluids 33(4):046603

Holm DD, Marsden JE, Ratiu TS (1998) The Euler–Poincaré equations and semidirect products with applications to continuum theories. Adv Math 137:1–81

Firdaus K, Behrens J (2024) Non-hydrostatic model for simulating moving bottom-generated waves: a shallow water extension with quadratic vertical pressure profile

Luesink E (2021) Stochastic geometric mechanics of thermal ocean dynamics. PhD thesis, Imperial College London

Miles J, Salmon R (1985) Weakly dispersive nonlinear gravity waves. J Fluid Mech 157:519–531

Nadiga BT, Margolin LG, Smolarkiewicz PK (1996) Different approximations of shallow fluid flow over an obstacle. Phys Fluids 8(8):2066–2077

Pogutse OP, Bazdenkov SV, Morozov NN (1987) Dispersive effects in two-dimensional hydrodynamics. Dokl Akad Nauk SSSR 293:818–822

Ripa P (1993) Conservation laws for primitive equations models with inhomogeneous layers. Geophys Astrophys Fluid Dyn 70(1–4):85–111

Ripa P (1995) On improving a one-layer ocean model with thermodynamics. J Fluid Mech 303:169–201

Stewart AL, Dellar PJ (2010) Multilayer shallow water equations with complete Coriolis force. Part 1. Derivation on a non-traditional beta-plane. J Fluid Mech 651:387–413

Street OD, Takao S (2024) Semimartingale driven mechanics and reduction by symmetry for stochastic and dissipative dynamical systems

Warneford ES, Dellar PJ (2013) The quasi-geostrophic theory of the thermal shallow water equations. J Fluid Mech 723:374–403

Whittaker C, Nokes R, Davidson M (2015) Tsunami forcing by a low Froude number landslide. Environ Fluid Mech 15(6):1215–1239

Chapter 5
Ensemble Data Assimilation in Time-Evolving Reproducing Kernel Hilbert Spaces

Maël Jaouen, Benjamin Dufée, Étienne Mémin, and Gilles Tissot

Abstract Traditional data assimilation methods like ensemble Kalman filters (EnKF) face intrinsic challenges with system nonlinearity and high-dimensional dynamics. We propose a novel approach based on reproducing kernel Hilbert spaces (RKHS), embedding dynamical system observables into time-evolving RKHSs. This framework leads to reconstruction of trajectories with time-invariant linear combinations of ensemble members, which enables aggregating observations over time and assimilate them during a same analysis step. The aim of the present paper is to construct an ensemble data assimilation method formulated in this time-evolving RKHS. We assess the approach using a multilayer quasi-geostrophic ocean model with synthetic sea surface height (SSH) observations derived from satellite tracks. Results show significantly improved reconstruction accuracy and robustness compared to a classical ensemble filter. This work summarizes a more detailed study (Jaouen et al. 2025).

5.1 Introduction

Data assimilation in ocean modelling integrates sparse, spatiotemporal observations into circulation models to improve ocean state estimation. A major challenge is assimilating sea surface height (SSH) data from satellite swaths to accurately capture mesoscale dynamics, including eddies, jets, and large fronts, which are crucial for global ocean circulation modeling. Various data assimilation methods (Carrassi et al. 2018; Evensen 2009; Reich and Cotter 2015) have been developed to address this challenge and are widely applied in oceanographic studies.

Variational methods minimize a cost function that strikes a balance between observation and state estimation errors. Among them, four-dimensional variational assim-

M. Jaouen (✉) · É. Mémin · G. Tissot
IRMAR - UMR CNRS 6625, Univ Rennes, Inria, Rennes, France
e-mail: mael.jaouen@inria.fr

B. Dufée
Department of Data Science, Univ Bretagne Sud, IUT Vannes, Vannes, France

© The Author(s) 2026

B. Chapron et al. (eds.), *Stochastic Transport in Upper Ocean Dynamics IV*, Mathematics of Planet Earth 15, https://doi.org/10.1007/978-3-032-12749-5_5

ilation (4DVar) integrates both spatial and temporal observations. It relies on optimization techniques such as adjoint methods (Le Dimet and Talagrand 1986), which efficiently calculate the gradients of the cost function by propagating the dynamics forward, then running the adjoint system backward in time. Although effective, these methods can be computationally expensive, difficult to parallelize and require the construction of an accurate tangent-linear operator.

To overcome these difficulties, neural networks such as 4DVarNet (Fablet et al. 2021) offer an alternative by learning jointly the model dynamics and the estimated state. Recent applications of 4DVarNet have been applied to the reconstruction of sea surface currents from SSH and SST data (Fablet et al. 2024). More generally, machine learning is increasingly integrated with data assimilation to improve forecasts (Cheng et al. 2023), improve interpolation of sparse observations (Ouala et al. 2018; Manucharyan et al. 2021), and refine prediction of future state (Brajard et al. 2020; Farchi et al. 2021).

However, these approaches raise concerns about assessing the physical reliability of predictions without an explicit dynamical model. Neural networks also face generalization challenges, potentially leading to poor performance when encountering previously unseen physical phenomena. In the present study, we use the knowledge provided by a dynamical model, but forbid iterative runs.

To manage the computational demands of high-dimensional models and operational constraints, ensemble Kalman filters (EnKFs) are commonly used. They approximate the state distribution with an ensemble of realizations and sequentially assimilate observations, assuming Gaussian errors. While EnKFs provide an optimal estimator in a linear Gaussian context, their accuracy in highly nonlinear systems is less assured, as they do not converge to the true filtering distribution (Le Gland et al. 2011). Nevertheless, EnKFs have been successfully applied in ocean modeling, particularly for SSH and sea surface temperature (SST) assimilation (Leeuwenburgh et al. 2005; Wan et al. 2010). A key limitation, however, is their sequential nature, as they update the model only at observation times within a given window. To address this, smoothing techniques like 4DEnVar (Desroziers et al. 2014; Liu et al. 2008; Lorenc 2003) incorporate information from the entire assimilation window using ensemble approximations of the tangent-linear operator, offering a computationally efficient alternative (Yang et al. 2015). Other methods reconstruct trajectories using time-invariant weight vectors, though this assumption is postulated from the start (Hunt et al. 2004, 2007) or results from an approximation (Sakov et al. 2010).

Reproducing kernel Hilbert space (RKHS)-based methods provide an alternative for addressing system nonlinearity and relaxing Kalman filter assumptions. Mauran et al. (2023) reformulated the Ensemble Transform Kalman Filter (ETKF) (Bishop et al. 2001; Hunt et al. 2007) using feature maps derived from ensemble state variables, while Gottwald and Reich (2021) proposed an EnKF that simultaneously estimates state variables and learns the system dynamics via random feature maps. However, these methods require solving for weight vectors that scale with state space size, leading to a computational bottleneck in high-dimensional settings.

The Koopman operator (Koopman 1931) is an infinite-dimensional linear operator that describes the evolution of observables in a dynamical system. By leveraging linear operator theory, it provides a powerful framework for analyzing nonlinear dynamics. Since Mezić's pioneering work (Mezić 2005), various techniques have been developed to approximate its spectral decomposition (Eisner et al. 2015; Mezić 2021; Brunton et al. 2022). Such techniques have seen extensive use in multiple fields including fluid mechanics (Rowley et al. 2009; Schmid 2010; Tu et al. 2014). More recently, Koopman theory has been combined with kernel methods to improve data-driven spectral estimation based on long time series (Williams et al. 2015; Das and Giannakis 2020) or ensemble of numerical runs (Dufée et al. 2024). This spectral analysis has been used to study ocean circulation and climate based on satellite data (Hogg et al. 2020; Navarra et al. 2021; Zhen et al. 2022). In particular, it has been used in relation to the spectral representation of time-series autocovariance (Ghil et al. 2002; Kondrashov et al. 2020), which was rigorously demonstrated for ergodic systems with a finite invariant measure in (Zhen et al. 2022).

The work in Navarra et al. (2021), built upon Klus et al. (2020), integrates the Koopman spectral decomposition within an RKHS framework. However, it assumes that the RKHS remains invariant under the Koopman operator—a highly restrictive condition, as it would require defining time-varying RKHS kernels based on the system dynamics.

The approach we adopt addresses this constraint by constructing a family of dynamically evolving RKHS (Dufée et al. 2024). Due to this intrinsic non-stationarity, the RKHS is constructed based on an ensemble of simulations rather than long time series—rendering it suitable for ensemble data assimilation methods. More precisely, this approach involves an evolution law for the RKHS feature maps, defining an isometry between kernel evaluations across different times. As a result of this isometry, trajectories can be reconstructed as time-invariant linear combinations of ensemble members, allowing observations from multiple times to be aggregated into a single assimilation step. It also follows that this framework enables the analyzed ensemble to be restarted at the initial time without iterative forward–backward assimilation.

In addition to its ability to handle isochrone linear combinations of the system trajectories, this technique operates on vectors of function evaluations that match the ensemble dimension, which makes it computationally very efficient, as the ensemble size is much smaller than the state space dimension. Such efficiency enables us to handle high-dimensional systems.

In this study, we apply this approach to a multilayer quasi-geostrophic model in the North Atlantic, by using synthetic observations derived from real satellite tracks within the framework of an observing system simulation experiment (OSSE). The results demonstrate that our method enhances both reconstruction accuracy and robustness.

This short paper corresponds to a summary of an extended paper (Jaouen et al. 2025), which provides an in-depth exploration of the method and its theoretical underpinnings.

5.2 Theoretical Framework

This section presents the theoretical background on RKHS theory, the Koopman operator, and the framework presented in Dufée et al. (2024), which constructs a family of RKHSs by evolving an initial kernel over time through the dynamics of the system.

5.2.1 Background on RKHS

A reproducing kernel Hilbert space (RKHS), $\mathcal{H}$, is a Hilbert space of functions $f : E \to \mathbb{C}$, where E is a non-empty set, equipped with an inner product $\langle \cdot, \cdot \rangle_{\mathcal{H}}$. A function $k : E \times E \to \mathbb{C}$ is the reproducing kernel associated to $\mathcal{H}$, if it verifies

- membership property: $\forall x \in E, k\left(\cdot, x\right) \in \mathcal{H}$,
- reproducing property: $\forall x \in E, \forall f \in \mathcal{H}, \langle f, k\left(\cdot, x\right)\rangle_{\mathcal{H}} = f\left(x\right)$.

The RKHS is uniquely defined by k (Aronszajn 1950). The feature map $k(\cdot, x)$ span forms a dense subset of $\mathcal{H}$. The integral operator $\mathcal{L}_k : L^2(E, \nu) \to L^2(E, \nu)$, defined as

$$\left(\mathcal{L}_k f\right)\left(x\right) = \int_E k\left(x, y\right) f\left(y\right) \nu\left(dy\right), \tag{5.1}$$

acts on square-integrable functions, where ν is a measure on E. The inner product in $L^2(E, \nu)$ is given by

$$\langle f, g \rangle_{L^2(E,\nu)} = \int_E f\left(y\right) \overline{g\left(y\right)} \nu\left(dy\right), \tag{5.2}$$

which leads to

$$\langle f, k\left(\cdot, x\right)\rangle_{L^2(E,\nu)} = \left(\mathcal{L}_k f\right)\left(x\right). \tag{5.3}$$

Assuming that the kernel k is continuous, symmetric, positive, and semi-definite, Mercer's theorem (Mercer and Forsyth 1909) guarantees that the eigenfunctions of the integral operator $\mathcal{L}_k$ define a basis of the RKHS $\mathcal{H}$. Consequently, the kernel k fully characterizes the structure of the RKHS $\mathcal{H}$. In addition, the operator $\mathcal{L}_k$ provides a way to compute the RKHS inner product

$$\langle f, g \rangle_{\mathcal{H}} = \langle \mathcal{L}_k^{-\frac{1}{2}}(f), \mathcal{L}_k^{-\frac{1}{2}}(g)\rangle_{L^2(E,\nu)}. \tag{5.4}$$

5.2.2 Dynamical System and Koopman Operator

We consider a continuous dynamical system represented by a flow Φ_t acting on functions $X \in L^2(\mathbb{R}^+ \times \Omega_x, \mathbb{R}^d)$ within an invariant phase space manifold, where Ω_x is the spatial domain. The system state at time t is given by $X_t = \Phi_t(X_0)$, with the initial condition being X_0. In differential form,

$$\frac{\partial X_t}{\partial t} = \mathcal{M}(X_t), \quad X(0) = X_0, \tag{5.5}$$

where X_t belongs to a compact subset of $L^2(\Omega_x, \mathbb{R}^d)$, and $\mathcal{M}$ is a differentiable operator with continuous derivatives. The observables of this system are square-integrable functions mapping states to complex-valued vectors. We assume the system to be invertible, and the finite measure ν to be preserved by the dynamical system.

The Koopman operator U_t acts on observables (Koopman 1931), propagating them over time as

$$U_t f(X) = f(\Phi_t(X)). \tag{5.6}$$

This operator is linear. However, as it operates on functions, it is of infinite dimension, and can rarely be thought as an infinite matrix, in the sense that it can admit a continuous spectrum in addition to punctual eigenvalues. Embedding the Koopman infinitesimal generator in an RKHS enables formally compactifying this operator to extract an approximation of its spectral representation.

5.2.3 Embedding the Dynamical System in a RKHS

Following Dufée et al. (2024), we define an initial condition ensemble Ω_0 and the ensemble at time $t \in \mathbb{R}^+$, $\Omega_t = \Phi_t(\Omega_0)$. We note that Ω_t corresponds to the function set at time t, which is not to be confused with the spatial domain Ω_x. Given an initial kernel $k_0 : \Omega_0 \times \Omega_0$, we construct the kernel at time t as

$$\forall X_t, Y_t \in \Omega_t, \ k_t(X_t, Y_t) = k_0\left(\Phi_t^{-1}(X_t), \Phi_t^{-1}(Y_t)\right). \tag{5.7}$$

A family of RKHSs $\mathcal{W} = (\mathcal{H}_t)_{t \geq 0}$ can be constructed by considering the RKHSs along time associated to the feature maps k_t. As seen in Sect. 5.2.1, $\mathcal{H}_t$ is completely defined by k_t. An isometry denoted $\mathcal{U}_t$ can be defined between $\mathcal{H}_0$ and $\mathcal{H}_t$:

$$\mathcal{U}_t k_0(\cdot, X_0) = k_t(\cdot, \Phi_t(X_0)). \tag{5.8}$$

This operator closely relates to the adjoint of the Koopman operator, known as the Perron–Frobenius operator. The kernel k_t evaluated on a pair of state X_t, Y_t from

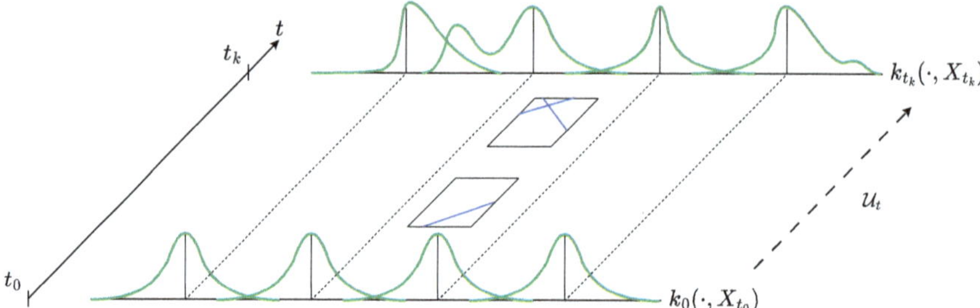

Fig. 5.1 Feature maps transported over time from t_0 to t_k by the dynamical system, with the observations collected between t_0 and t_k represented at the center of the figure in the form of satellite swath. The operator $\mathcal{U}_t$ transports the feature maps between t_0 and t_k

Ω_t can be seen as the inner product in the RKHS $\mathcal{H}_t$ between the associated feature maps:

$$k_t\left(\boldsymbol{X}_t, \boldsymbol{Y}_t\right) = \langle k_t\left(\cdot, \boldsymbol{Y}_t\right), k_t\left(\cdot, \boldsymbol{X}_t\right)\rangle_{\mathcal{H}_t}. \tag{5.9}$$

The observables of this dynamical system are assumed to be sufficiently smooth to belong to the RKHS $\mathcal{H}_t$, these observables can then be expressed as linear combinations of the RKHS feature maps $k_t(\cdot, \boldsymbol{X}_t)$ for $\boldsymbol{X}_t \in \mathcal{H}_t$. Figure 5.1 illustrates the construction of the feature maps $k_t\left(\cdot, \boldsymbol{X}_t\right)$ as the initial feature maps transported in time by $\mathcal{U}_t$. Let us note that although the $\mathcal{H}_t$ are infinite-dimensional, in practice, we have to consider a finite ensemble denoted as $\left\{\boldsymbol{X}^{(i)}\right\}_{i=1}^{p}$. As the ensemble size increases, the finite-dimensional approximation of the operator $\mathcal{L}_k$ admits a richer set of eigenfunctions, effectively expanding the representational capacity of the RKHS $\mathcal{H}_t$.

One very important property for data assimilation ensemble methods concerns the superposition principle, which is commonly assumed inappropriately—or considered as an approximation—for nonlinear systems. This principle holds rigorously within the RKHS framework. Given a smooth observable g_x belonging to $\mathcal{H}_t$, the reproducing property enables to write

$$g_x\left(\boldsymbol{X}_t\right) = \langle g_x\left(\cdot\right), k_t\left(\cdot, \boldsymbol{X}_t\right)\rangle_{\mathcal{H}_t}. \tag{5.10}$$

We approximate $k_t\left(\cdot, \boldsymbol{X}_t\right)$ by $k_t^*\left(\cdot, \boldsymbol{X}_t\right)$, which correspond to the solution of a least-square minimization problem. According to the representer theorem (Kimeldorf and Wahba 1971), this solution belongs to the finite-dimensional subspace spanned by the kernel functions $\left\{k(\cdot, \boldsymbol{X}^{(i)})\right\}_{i=1}^{p}$,

$$k_t^*\left(\cdot, \boldsymbol{X}_t\right) = \sum_{i=1}^{p} \beta_i k_t\left(\cdot, \boldsymbol{X}_t^{(i)}\right), \tag{5.11}$$

we obtain

$$g_x^*(X_t) = \sum_{i=1}^{p} \beta_i g_x\left(X_t^{(i)}\right). \tag{5.12}$$

This equation being true for every $x \in \Omega_x$, it leads to

$$X_t^* = \sum_{i=1}^{p} \beta_i X_t^{(i)}. \tag{5.13}$$

Thus, new trajectories can be constructed as linear combinations of existing ones. This approximation becomes exact for an infinite number of ensemble members p.

In this section, we have introduced a framework that combines the Koopman operator and RKHSs. The RKHS $\mathcal{H}_t$ provides a functional space for the observables of the dynamical system at time t. The reproducing property in this space justifies the reconstruction of the system state at time t as a linear combination of an ensemble of states. The Koopman operator propagates the observables along time. A mapping $\mathcal{U}_t : \mathcal{H}_0 \rightarrow \mathcal{H}_t$, closely related to the adjoint of the Koopman operator, is defined. This mapping propagates observables embedded in the initial RKHS $\mathcal{H}_0$ forward in time and establishes an isometry between $\mathcal{H}_0$ and $\mathcal{H}_t$. Consequently, the kernel evaluations—and then the reconstruction coefficients—remain constant over time. This enables the reconstruction of trajectories as linear combination of an ensemble of trajectories. This property leads to appealing data assimilation techniques, which will be presented in the next section after a brief overview of classical ensemble filters.

5.3 Ensemble Data Assimilation Techniques

5.3.1 Classical Methods

Data assimilation aims to estimate the state of a dynamical system from partial observations. The system state $X \in \mathbb{R}^n$ and the observations $Y \in \mathbb{R}^m$ are related by

$$Y = H(X) + \epsilon, \tag{5.14}$$

where $H : \mathbb{R}^m \mapsto \mathbb{R}^n$ is the observation operator, generally with $n \gg m$, and $\epsilon \in \mathbb{R}^m$ represents an additive Gaussian noise with zero mean and covariance $R \in \mathbb{R}^{m \times m}$.

The goal of assimilation methods is to solve $H(X) \simeq Y$, seeking the most probable state given prior knowledge. For simplicity, we assume that H is linear, represented as $\mathbb{H} \in \mathbb{R}^{m \times n}$, so we can write $H(X) = \mathbb{H}X$. To ensure a well-posed problem, we seek the solution closest to a prior forecast X^f, minimizing the cost function:

$$J(X) = \frac{1}{2} \|\mathbb{H}X - Y\|_{R^{-1}}^2 + \frac{1}{2} \|X - X^f\|_{P^{-1}}^2, \tag{5.15}$$

where $P \in \mathbb{R}^{n \times n}$ is the state covariance matrix. Standard optimization techniques yield the optimal analyzed state:

$$X^a = X^f - G\left(\mathbb{H}X^f - Y\right), \tag{5.16}$$

where the Kalman gain is given by

$$G = P\mathbb{H}^T\left(R + \mathbb{H}P\mathbb{H}^T\right)^{-1}. \tag{5.17}$$

5.3.1.1 Ensemble Square Root Filter (ESRF)

Ensemble Kalman filters approximate the system distribution using p ensemble members together with a Gaussian prior, where the ensemble mean and covariance serve as empirical estimators of the system's mean and covariance.

The ensemble is represented by a matrix $X \in \mathbb{R}^{n \times p}$, where each column $X^{(i)}$ corresponds to the i-th ensemble member. The Ensemble Square Root Filter (ESRF) (Whitaker and Hamill 2002) expresses the state covariance through its square root, known as the anomaly matrix:

$$A = [\left(X^{(1)} - \overline{X}\right), \left(X^{(2)} - \overline{X}\right), \ldots, \left(X^{(p)} - \overline{X}\right)], \tag{5.18}$$

where $\overline{X}$ is the ensemble mean. The empirical covariance is given by

$$P = \frac{1}{p-1}AA^T. \tag{5.19}$$

The covariance update is expressed as

$$P^a = (I - G\mathbb{H})\,P^f, \tag{5.20}$$

where G is the Kalman gain. Exponents $\bullet^f$ and $\bullet^a$ stand for *forecast* and *analyzed*, respectively.

Defining the square root matrix:

$$S = \left(I + \frac{1}{p-1}\left(\mathbb{H}A^f\right)^T R^{-1}\left(\mathbb{H}A^f\right)\right)^{-1/2}, \tag{5.21}$$

the analyzed anomalies are updated as

$$A^a = A^f S. \tag{5.22}$$

The mean update follows:

$$\overline{X^a} = \overline{X^f} - G\left(\overline{\mathbb{H}X^f} - Y\right), \tag{5.23}$$

and finally the analyzed ensemble members are constructed as

$$X^a = \overline{X^a} + A^f S. \tag{5.24}$$

5.3.2 Ensemble Filter on RKHS Family

The framework proposed by Dufée et al. (2024), described in Sect. 5.2, operates directly on functions defined in an RKHS and evaluated on the ensemble, and enable trajectory estimation from time-varying satellite swath observations. This RKHS-based ensemble filter (RKHS EnF) assumes that the observation operator is well defined on the RKHS family $\mathcal{W}$. The method is referred to as a "filter" rather than a "smoother" because it reconstructs system trajectories by balancing the background trajectory with observational data, without incorporating future observations to estimate present states. It is a filter on trajectories, similarly as asynchronous ensemble Kalman filters (Sakov et al. 2010). However, as it assimilates observations taken at different times within the same assimilation step and reconstructs in particular the initial conditions of this piece of trajectory (earlier than the observations), it shares similarities with smoothing approaches. The "filter" terminology is still retained to stress the strongly structuring role of the Koopman operator on these pieces of trajectories. The kernel is taken as real valued without loss of generality. The RKHS EnF is formulated using a 4DVar-like cost function with a regularization term on the weight anomaly norm in the RKHS:

$$J_{\text{RKHS}}\left(\beta\right) = \frac{1}{2} \sum_{j=0}^{J} \left\| \sum_{i=1}^{p} \beta_i \mathbb{H}_{t_j} X_{t_j}^{(i)} - Y_{t_j} \right\|_{R_{t_j}^{-1}}^2 + \frac{1}{2} \left\| \beta - \overline{\beta^f} \right\|_K^2, \tag{5.25}$$

where the kernel matrix $K \in \mathbb{R}^{p \times p}$ is defined as

$$K_{i,j} = k_0\left(X_0^{(i)}, X_0^{(j)}\right), \tag{5.26}$$

for $1 \le i, j \le p$. The cost function can be rewritten as

$$J_{\text{RKHS}}\left(\beta\right) = \frac{1}{2}\left(VK\beta - Y\right)^T \tilde{R}^{-1}\left(VK\beta - Y\right) + \frac{1}{2}\left(\beta - \overline{\beta^f}\right)^T K\left(\beta - \overline{\beta^f}\right), \tag{5.27}$$

with $Y = \left(Y_{t_j}\right)_{j \in [\![1, J]\!]}$ and $\tilde{R} = I_J \otimes \left(R_{t_j}\right)_{j \in [\![0, J]\!]}$, where $\otimes$ represents the Kronecker product. V is defined such that $\mathbb{H}_t \overline{X_t} = V_t K \beta$ and $V = \left(V_{t_j}\right)_{j \in [\![1, J]\!]}$. V is a $m_{tot} \times p$ matrix, with m_{tot} the number of observations along time. As $\mathbb{H}_{t_j} X_{t_j}^f =$

$V_{t_j} K \beta^f$, and β^f is the identity matrix, we have in practice $V_{t_j} K = \mathbb{H}_{t_j} X_{t_j}{}^f$: the observation operator applied to the ensemble members.

5.3.2.1 Kernel Matrix Update

The inverse kernel matrix K^{-1} acts as the weight covariance matrix P_w^f. We define its square root decomposition:

$$P_w^f = K^{-1/2} \left(K^{-1/2}\right)^T. \tag{5.28}$$

The gain matrix is expressed as

$$G_{\text{RKHS}} = P_w^f \left(VK\right)^T \left(\widetilde{R} + (VK) P_w^f (VK)^T\right)^{-1}, \tag{5.29}$$

and the covariance is updated via

$$P_w^a = \left(I - G_{\text{RKHS}} \left(VK\right)\right) P_w^f. \tag{5.30}$$

5.3.2.2 Weights Update

Unlike the ESRF, RKHS EnF does not provide a direct transformation matrix S to update the anomaly weight matrix. We denote this $p \times p$ matrix as $(\beta)'$, where each column represents the anomaly weight vector of an ensemble member. Since the anomaly weight matrix forms a zero-mean ensemble, a whitening transformation can be applied using $K^{1/2}$, followed by a coloring step using $\left(P_w^a\right)^{1/2}$, giving

$$\left(\beta^a\right)' = \left(\beta^f\right)' K^{1/2} \left(P_w^a\right)^{1/2}, \tag{5.31}$$

thus the analyzed weight ensembles have as covariance P_w^a. The mean update follows:

$$\overline{\beta^a} = \overline{\beta^f} - G_{\text{RKHS}} \left(VK\overline{\beta^f} - Y\right), \tag{5.32}$$

and the analyzed weight ensemble is

$$\beta^a = \overline{\beta^a} + \left(\beta^f\right)' K^{1/2} \left(P_w^a\right)^{1/2}. \tag{5.33}$$

5.3.3 Ensemble Update Strategy

Embedding the dynamics within the RKHS framework results in ensemble weights that are constant over time. This property allows for reconstructing analyzed ensemble members at any time within the assimilation window as a linear combination of the forecast ensemble. Since the weights are computed based on the initial kernel, it is consistent to define the analyzed members at the start of the assimilation window. However, due to system nonlinearity and the finite ensemble size, the forward-propagated analyzed members may differ from those reconstructed at the end of the assimilation window. In theory, with an infinite ensemble size, both reconstructions would be identical. In practice, numerical dissipation and the limited ensemble size introduce slight discrepancies. This ability to reconstruct the ensemble at the start of the assimilation window is used to generate the analyzed ensemble trajectories. This approach, named the tiled RKHS ensemble filter, is detailed in Algorithm 1. Contrary to smoothers, the reconstruction errors presented in Sect. 5.5 are computed before running back the analyzed members on the assimilation window. This run is used only to set the ensemble before the next assimilation. In our method, the ensemble is propagated in time only during the forecast step of the assimilation, not during the analysis. Classical smoothing methods run the system during analysis in order to obtain the analyzed ensemble and to evaluate errors.

Algorithm 1 Tiled RKHS EnF

Require: Ensemble X_0, initial time t_0, end time t_{end}, time step dt, assimilation interval T_a.

$X_t^f \leftarrow X_0$

$t \leftarrow t_0$

$t_a \leftarrow T_a$ the assimilation time horizon.

Save X_t for future reconstruction.

Construct the kernel matrix K.

Initialize VK and Y as empty.

while $t < t_{end}$ **do**

    $t \leftarrow t + dt$

    Propagates the ensemble members in time to update X_t^f.

    Append to VK the observations of the ensemble members.

    Append to Y the reference observations.

    if It is time to assimilate *i.e.* t is a multiple of t_a **then**

        $\beta^a \leftarrow \text{assimilate}(VK, Y, K)$ $\triangleright$ Equations (5.32) and (5.33).

        Empty VK and Y.

        Apply the weights β^a to $X_{t-t_a}^f$ saved at $t - t_a$, to obtain $X_{t-t_a}^a$.

        Propagates $X_{t-t_a}^a$ on the time interval $[t - t_a, t]$, to obtain X_t^a.

        $X_t^f \leftarrow X_t^a$

        Save X_t^f for future reconstruction.

        Update the kernel matrix K.

    end if

end while

5.4 Experimental Framework

This section outlines the experimental setup used to evaluate the proposed data assimilation method.

5.4.1 Dynamical Model

The assimilation methods are tested using a North Atlantic Ocean model covering the region between 9°N and 48°N latitude and -98°E and -3.84°E longitude with realistic coastlines. The domain spans $4329\,\text{km} \times 8658\,\text{km}$, assuming a flat ocean bottom and closed boundaries, including normally open ones at the equator and northern limits. The Gulf Stream forms along the North American east coast and is the dominant flow structure in the considered domain. Coasts are represented by treating regions shallower than $-100\,\text{m}$ as land. The model incorporates realistic wind forcing.

A multilayer quasi-geostrophic (QG) model, derived from a rotating shallow-water system (Thiry et al. 2024), is employed. The shallow-water component follows Roullet and Gaillard (2022), while the QG projection uses the elliptic solver from Thiry et al. (2023). The model consists of three layers with thicknesses $H = (400, 1100, 2600)^T$ m, simulating three prognostic variables: u_k, v_k are the horizontal velocity components and h_k is the height anomaly in the layer $k \in \{1, \ldots, 3\}$. Their evolution is governed by

$$
\begin{aligned}
\partial_t u_k &= (\omega_k + f)\, v_k - \partial_x \left(p_k + E_k \right) \\
\partial_t v_k &= -(\omega_k + f)\, u_k - \partial_y \left(p_k + E_k \right) \\
\partial_t h_k &= -H_k \left(\partial_x u_k + \partial_y v_k \right) - \partial_x \left(u_k h_k \right) - \partial_y \left(v_k h_k \right),
\end{aligned}
\tag{5.34}
$$

where $\omega_k = \partial_x v_k - \partial_y u_k$ is the relative vorticity, p_k is the pressure vector, $E_k = \left(u_k^2 + v_k^2 \right)/2$ is the kinetic energy, $f = f_0 + \beta y$ is the Coriolis parameter, with $f_0 = 9.375 \times 10^{-5}\,\text{s}^{-1}$ and $\beta = 1.754 \times 10^{-11}\,\text{m}^{-1}\text{s}^{-1}$. Layer pressure follows:

$$
p_k = \rho_1 \sum_{j=1}^{k} \left(g'_j \sum_{l=j}^{3} h_l \right),
\tag{5.35}
$$

where $g_1 = g = 9.81\,\text{m} \cdot \text{s}^{-2}$ and $g'_i = g \left(\rho_i - \rho_{i-1} \right)/\rho_1$ for $i = 2, 3$, ρ being the density in the different layers.

The QG equations result from projecting this shallow-water system using an elliptic operator. Details and numerical codes are in Thiry et al. (2024). Simulations run on a $512 \times 1024 \times 3$ grid, yielding a spatial resolution of $8.46 \times 8.46\,\text{km}^2$ square. The time step is 2000 s. This model is used both to generate the reference simulation, from which observations are extracted, and to propagate the ensemble members over time.

5.4.2 Ensemble Generation

The initial ensemble X_0 is generated using spectral decomposition of random local fluctuations (Bauer et al. 2020; Brecht et al. 2021). For a field q, defined as either u or v, respectively, on a $n = n_x \times n_y$ grid, let us define a local spatial window $W_{i,j}$ of size $n_w \times n_w$ (with odd n_w) around each grid point $x_{i,j}$. Within each window, $n_o > 1$ samples are drawn from an uniform law, forming n_l-dimensional vectors, denoted $w_k^{(i,j)}$, with $1 \leq k \leq n_o$. The sampled values are centered as

$$\left(w_k^{(i,j)} \right)' = w_k^{(i,j)} - \overline{w}^{(i,j)} \quad \text{with} \quad \overline{w}^{(i,j)} = \frac{1}{n_o} \sum_{\ell=1}^{n_o} w_\ell^{(i,j)}. \tag{5.36}$$

These vectors gathered in column form a matrix $Z \in \mathbb{R}^{n \times n_o}(\mathbb{R})$. A singular value decomposition (SVD) is then applied to this matrix, leading to $Z = U \Sigma V^T$. This decomposition defines perturbation modes U_i from the columns of U and singular values $\lambda_j = \Sigma_{jj}$. Perturbations $\{q_0^{'(i)}, i = 1 \ldots, p\}$ are generated as

$$q^{'(i)} = \sum_{j=1}^{n_o} \lambda_j U_j \, \eta_j^{(i)}, \tag{5.37}$$

where η is a multivariate Gaussian random vector with n_o components and $\eta^{(i)}$ is its i th realization. These perturbations are added to the initial velocity field in order to generate the ensemble.

For experiments, 76 members were generated using $n_w = 5$ and $n_o = 21$. One member serves as the reference, while the remaining 75 form the training ensemble.

5.4.3 Simulating Observations

A synthetic observing system simulation experiment (OSSE) is conducted using sea surface height (SSH) as the observed variable. The SSH represents the sum of height anomalies h across vertical layers. To generate synthetic data, spatial masks are applied to the reference simulation, mirroring realistic patterns from the Copernicus along-track SSH database (DOI: https://doi.org/10.48670/moi-00147.) Observations were initialized on January 1, 2022. Figure 5.2 displays observation masks for two durations: a single simulation step (33 min 20 s) and a more extended (41 h 40 min) period. Over the longer period, altimeter swaths provide a reasonable coarse-grid

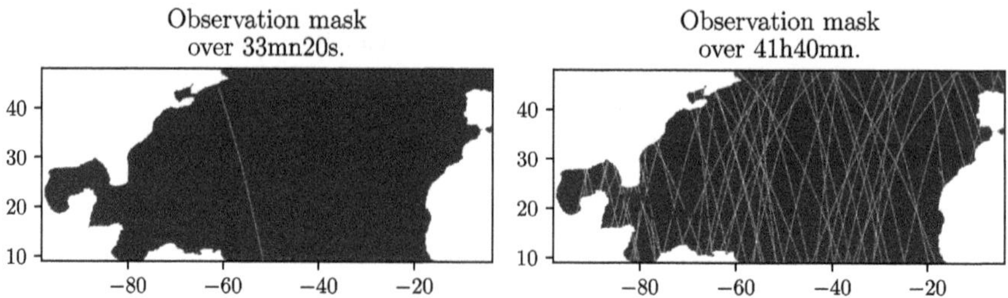

Fig. 5.2 Observation masks for the periods of 33 min 20 s (left) and 41 h 40 min (right), both from January 1, 2022

coverage of the basin. Traditional optimal interpolation (OI) methods reconstruct dense observation fields by assuming simultaneous acquisition, whereas our approach treats observations as a time series, avoiding spatial interpolation.

Two experiments were conducted, with additive white Gaussian noise perturbing observations. The noise standard deviations were set at 1% and 10% of the reference SSH root mean square (RMS) at the initial time, corresponding to $\sigma_o = 0.18$ cm and $\sigma_o = 1.8$ cm, respectively. The observation error covariance matrix $\widetilde{R}$ used in the filters is diagonal, with identical error variance σ_o^2 for all observations.

5.4.4 Filter Configuration

5.4.4.1 Assimilation Frequency

The RKHS EnF assimilation frequency can be adjusted to optimize reconstruction accuracy. A short assimilation window may limit the benefit of the RKHS framework, whereas an excessively long one can compromise the invertibility assumption of the dynamical system, leading to inaccuracies in ensemble trajectory combinations. However, tiling the filter over longer windows enhances small-scale variability and helps suppress non-physical structures.

For this study, the assimilation window is set to 41 h 40 min (75 model time steps) for the RKHS EnF. The ESRF is updated at each step (every 33 min 20 s).

5.4.4.2 Kernel Selection

A centered Gaussian kernel was used, defined as

$$\forall X, Y \in \Omega_0, \ k_g\left(X, Y\right) = \exp\left(-\frac{\|u_X - u_Y\|_{\Omega_x}^2 + \|v_X - v_Y\|_{\Omega_x}^2}{l_G^2}\right), \tag{5.38}$$

where u_X and v_X (resp. u_Y and v_Y) denote the velocity components of the ensemble member X (resp. Y), and $\| \cdot \|_{\Omega_x}$ represents the L^2 norm over domain Ω_x. The parameter l_G is chosen is order to ensure a consistent ratio μ_p/μ_1 between the smallest and largest eigenvalues of the kernel matrix K_G. This is achieved for each assimilation window by assembling iteratively the kernel matrix K_G by varying the parameter l_G until convergence to the target ratio. This ratio is fixed at $\mu_p/\mu_1 = 10^{-2}$. Preliminary tests have shown good performances with this value. Additionally, tests conducted on various dynamical models (not presented in this study) indicate that this parameter is robust to both the choice of the dynamical system and the level of noise in the observations.

Normalization and centering is performed by defining

$$K = \alpha \, (p - 1) \, (I - \mathbb{1}/p) \, K_G \, (I - \mathbb{1}/p) , \tag{5.39}$$

where α is set to 0.75 to balance the impact of K in the regularization term of the cost function. Since the rank of $(I - \mathbb{1}/p)$ is $p - 1$, K is not invertible, requiring the use of its Moore–Penrose pseudo-inverse:

$$K^{\dagger} = \frac{1}{\alpha \, (p - 1)} \, (I - \mathbb{1}/p) \, K_G^{-1} \, (I - \mathbb{1}/p) . \tag{5.40}$$

Invertibility of K_G is ensured by the control of its eigenvalues via l_G. The pseudo-inverse and its square root are computed through SVD for gain computation (5.29) and weight anomaly updates (5.33).

5.4.5 Performance Evaluation

Reconstruction quality is assessed using relative vorticity fields, which provide a more sensitive diagnostic than velocity or water height due to their sharper gradients. The root mean square error (RMSE) is used to quantify deviations between the analyzed state estimation and the reference state. Denoting the mean of vorticity along the ensemble as $\overline{\omega}$ and the reference as ω_{ref}, RMSE is computed as

$$e_{\mathrm{RMSE}} = \sqrt{\frac{1}{n} \, \|\overline{\omega} - \omega_{\mathrm{ref}}\|^2_{L^2(\Omega_x)}} . \tag{5.41}$$

To enhance interpretability, a normalized RMSE (NRMSE) is derived:

$$e_{\mathrm{NRMSE}} = \frac{e_{\mathrm{RMSE}}}{\|\omega_{\mathrm{ref}}\|_{L^2(\Omega_x)}} . \tag{5.42}$$

The NRMSE provides a clearer metric of ensemble accuracy, offering an assessment of deviation from the reference state. The standard deviation of the ensemble is defined as follows:

$$\sigma_e = \sqrt{\frac{1}{p} \sum_{i=1}^{p} \left\| \boldsymbol{\omega}^{(i)} - \overline{\boldsymbol{\omega}} \right\|_{\Omega_x}^2} \; . \tag{5.43}$$

If the ensemble standard deviation is too small, the reference state may fall outside its range, making future adjustments challenging and increasing the risk of divergence of the filter. Conversely, an overly large spread suggests poor representation of the system by individual members. A broad distribution can also affect the RMSE, as excessive dispersion may smooth out the ensemble mean. Achieving an optimal ensemble spread is therefore critical for robust predictions, with the RMSE and standard deviation ideally being of the same order of magnitude, classically referred to as bias-variance trade-off.

5.5   Results

This section evaluates the performance of the proposed data assimilation methods over a 100-day period. The problem is particularly challenging due to the high-dimensional state space ($\sim 10^6$) and the limited ensemble size (75 members). In such scenarios, ensemble methods typically require localization to prevent filter divergence. However, this study focuses on the efficiency and robustness of the RKHS embedding, particularly its ability to solve a 4DVar problem at low computational cost via a tiled simulation strategy. While localization could enhance these methods, its integration is beyond the scope of this work and remains an avenue for future exploration.

As an initial comparison, Fig. 5.3 presents the NRMSE for both the classical ESRF and the RKHS EnF. The results are averaged across five reference datasets, each with a distinct initial ensemble generated according to Sect. 5.4.2. The shaded regions around the mean NRMSE curves indicate the standard deviation across these experiments. A wide shaded area shows high sensitivity to the initial conditions, whereas a narrower band implies greater robustness. With 1% Gaussian noise in observations, the classical ESRF shows a steady increase in error and significant variability, highlighting its lack of robustness. On the other hand, the EnF RKHS maintains a consistently low error over time, demonstrating greater long-term reliability. It also proves to be robust with respect to initial conditions. For higher observation noise (10% of the reference RMS height field), both methods show increased error. However, the RKHS EnF consistently outperforms the ESRF, achieving lower NRMSE and reduced variance across different experiments, confirming its resilience to noise. As previously noted, performance could likely be improved with localization. For the RKHS EnF, this could involve applying localized observables, leading to spatially varying weights rather than global ones. These experiments underscore the

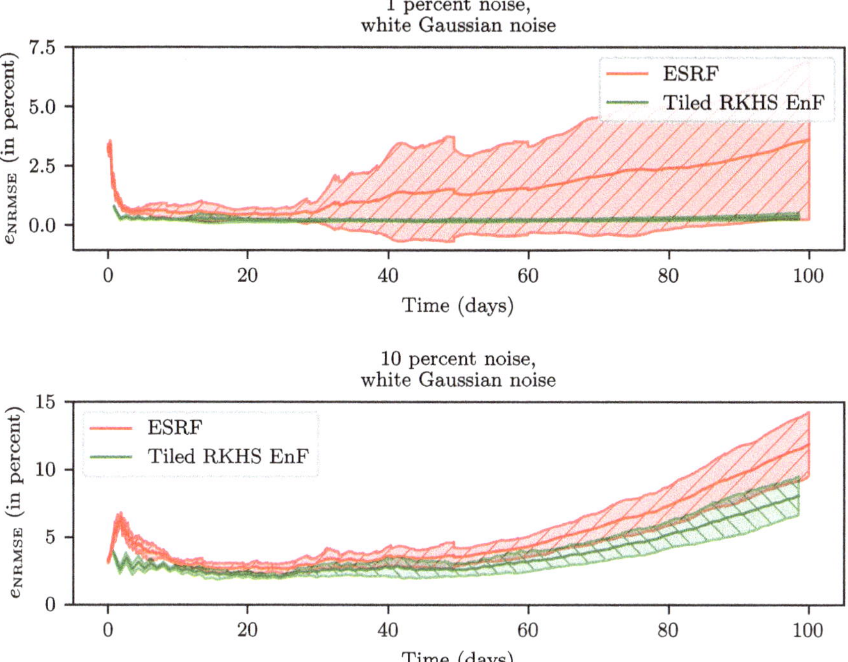

Fig. 5.3 Comparison of the normalized root mean square error (NRMSE) between the ESRF (red) and the tiled RKHS EnF (green). The first row corresponds to experiments with 1% standard deviation noise, while the second row represents experiments with 10% standard deviation noise

advantages of the RKHS EnF. It consistently yields better results, exhibiting less sensitivity to initial conditions, as reflected in its tighter standard deviation across noise levels. The method leverages the theoretical benefits of embedding the dynamical system in an evolving RKHS, allowing 4DVar assimilation to be formulated as a single-iteration filtering process, akin to sequential Bayesian filters.

Figure 5.4 presents the upper layer of the reference field (i.e., the simulation used to generate the observations). It shows the RMSE for the upper layer for both methods after 100 days for an experiment where observations were perturbed by 10% noise. Both methods successfully reconstruct a vorticity map closely matching the reference field, with minimal qualitative differences. But, the ESRF exhibits higher RMSE, whereas the tiled RKHS EnF achieves lower values, indicating improved accuracy. A similar trend is observed in other layers, though they are not shown here.

To assess the influence of the kernel and the tiling strategy on the assimilation performances, we conduct an additional experiment using the empirical covariance kernel (which is equivalent to replacing $\boldsymbol{K}_G$ by the identity matrix). As such, the kernel eigenvalues associated with the p evaluations are all equal to 1 by construction. They do not decrease as required at the continuous level for a valid RKHS reproducing kernel. It is then sensitive to the number of members since adding a new

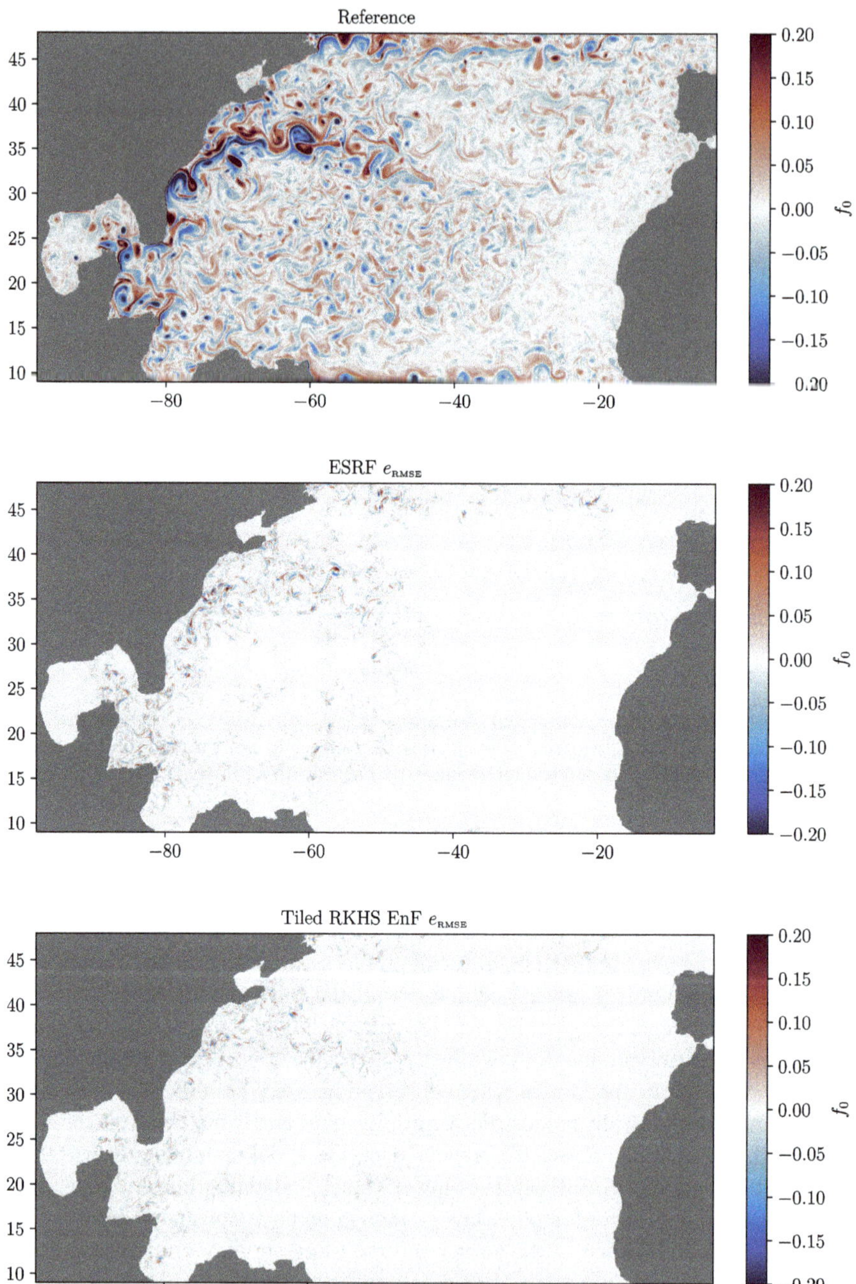

Fig. 5.4 Reference (the simulation from which the observations are extracted) upper layer relative vorticity field (top row) after 100 days, RMSE field after 100 days for the ESRF (second row), tiled RKHS EnF (bottom row), both corresponding to an experiment with 10% standard deviation noise

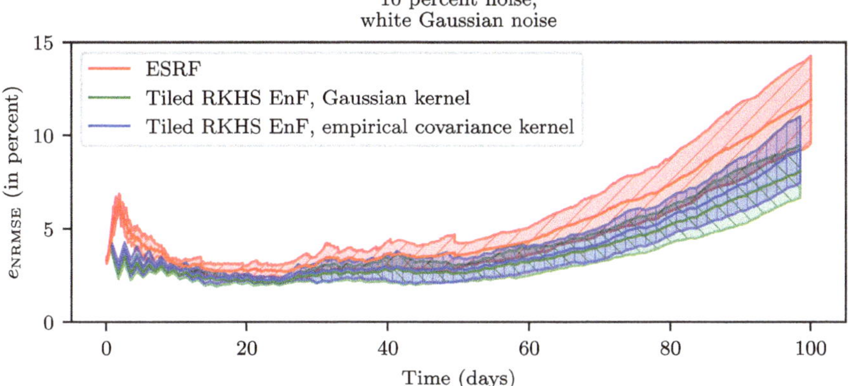

Fig. 5.5 Comparison of the normalized root mean square error (NRMSE) between the ESRF (red), the tiled RKHS EnF with a Gaussian kernel (green), and the tiled RKHS EnF with the empirical covariance kernel (blue) for the experiments with 10% standard deviation noise

one affects drastically the evaluated part of the Mercer spectrum. The reconstructions are then subject to spurious artifacts. As discussed in Sect. 3.3 of Jaouen et al. (2025), assimilating data at each time step using the identity kernel without tiling is equivalent to the ESRF. Data assimilation was performed under 10% additive white Gaussian noise, with the tiling strategy applied. The results of this experiment, shown in Fig. 5.5, indicate that the tiled assimilation with the identity kernel outperforms the classical ESRF. This highlights the benefit of aggregating observations over a 41h40min time window and employing the tiling strategy during the analysis step. Furthermore, the RKHS EnF with a Gaussian kernel leads to an additional improvement in assimilation performance. These results confirm that defining the filtering problem in an RKHS enhances the effectiveness of the method. In particular, the use of a suitable kernel can contribute to a more favorable balance between RMSE and ensemble spread as observed in Mauran et al. (2023).

Figure 5.6 presents the standard deviation of the ensemble in the upper layer for the experiment with 10% standard deviation noise. The ESRF exhibits a reduced spread, suggesting ensemble collapse. In contrast, the tiled RKHS EnF maintains a higher standard deviation. Given its lower RMSE, this suggests greater robustness and improved performance compared to the ESRF.

5.6 Conclusion and Perspectives

This work presents a novel approach for assimilating partial observations of nonlinear dynamical systems using a family $\mathcal{W}$ of reproducing kernel Hilbert spaces (RKHS). Each RKHS in this family is associated with kernels k_t, which is defined by transporting an initial kernel k_0 along the system dynamics. As introduced by

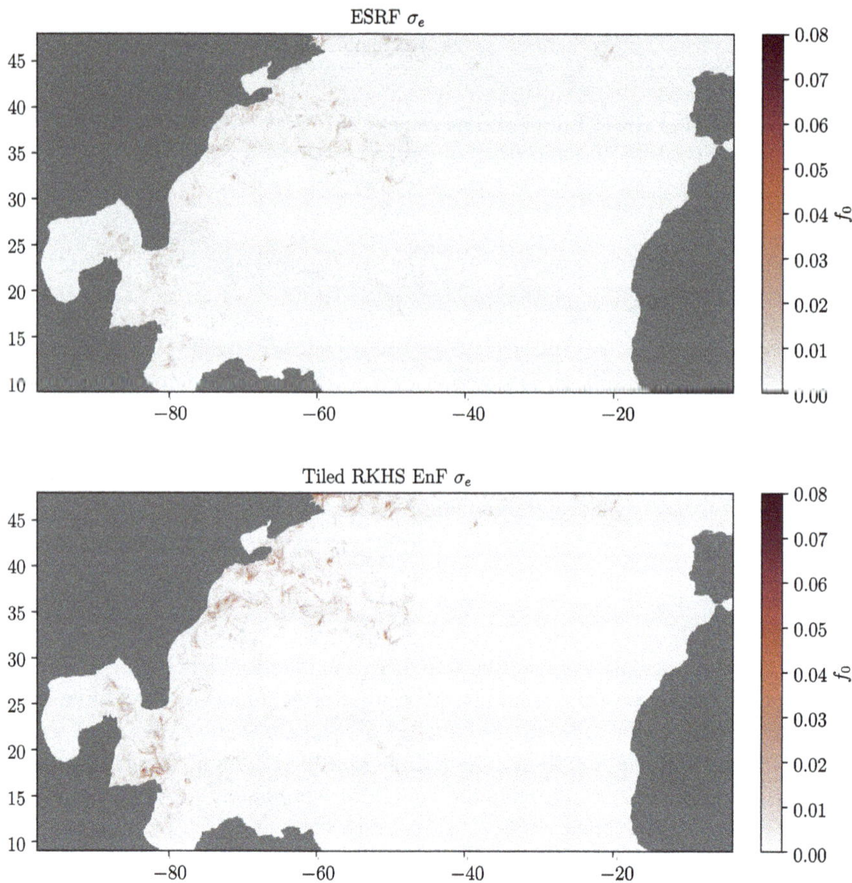

Fig. 5.6 Upper layer relative vorticity standard deviation field after 100 days for the ESRF (top row), tiled RKHS EnF (bottom row), both corresponding to the experiment with 10% standard deviation noise

Dufée et al. (2024), this framework establishes an isometry between the RKHSs of $\mathcal{W}$, allowing the construction of trajectories as linear combinations of an ensemble of trajectories, fully justifying the principle of superposition.

An ensemble filter is designed within this evolving RKHS framework. Its theoretical properties naturally allow for the assimilation of time-series observations, accumulating information over time before updating the entire trajectory. This leads to more robust state estimates. Moreover, the tiled implementation of the filter enables efficient propagation of the analyzed ensemble.

The proposed methods have been tested on a quasi-geostrophic model of the North Atlantic Ocean and compared against the classical ensemble square root filter (ESRF). The results demonstrate clear advantages of the RKHS EnF in terms of stability and performance.

Further improvements could be explored by optimizing kernel selection. One promising direction, following Akian et al. (2022), involves adjusting kernel eigenvalues to minimize accuracy loss when data points are removed. Additionally, considering alternative kernels beyond the Gaussian, such as the Matérn kernel, which offers greater flexibility (Stein 1999), may further enhance performance.

A key challenge in ensemble-based filters remains the limited number of ensemble members due to computational constraints. Localization techniques, as presented by Sakov and Bertino (2010), are commonly used to address this issue by either localizing observations or the covariance matrix. The definition of localization methods in the RKHS would constitute an efficient way for considering the adaptation of localization methods that propagate over time, as suggested by Desroziers et al. (2016), Bocquet (2016), since reproducing property circumvents the commutation issue between the localization tensor and the tangent-linear dynamical system operator. Additionally, applying a tiled filter in combination with local analysis may help mitigate imbalances that arise from localized reconstructions.

References

Akian J-L, Bonnet L, Owhadi H, Savin E (2022) Learning "bestâŁž kernels from data in Gaussian process regression with application to aerodynamics. J Comput Phys 470:111595

Aronszajn N (1950) Theory of reproducing kernels. Trans Am Math Soc 68:337–404

Bauer W, Chandramouli P, Chapron B, Li L, Mémin E (2020) Deciphering the role of small-scale inhomogeneity on geophysical flow structuration: a stochastic approach. J Phys Oceanogr 50:983–1003

Bishop CH, Etherton BJ, Majumdar SJ (2001) Adaptive sampling with the ensemble transform Kalman filter. Part I: Theoretical aspects. Monthly Weather Rev 129:420–436

Bocquet M (2016) Localization and the iterative ensemble Kalman smoother. Q J R Meteorol Soc 142:1075–1089

Brajard J, Carrassi A, Bocquet M, Bertino L (2020) Combining data assimilation and machine learning to emulate a dynamical model from sparse and noisy observations: a case study with the Lorenz 96 model. J Comput Sci 44:101171

Brecht R, Li L, Bauer W, Mémin E (2021) Rotating shallow water flow under location uncertainty with a structure-preserving discretization. J Adv Model Earth Syst 13

Brunton SL, Budišić M, Kaiser E, Kutz JN (2022) Modern Koopman theory for dynamical systems. SIAM Rev 64:229–340

Carrassi A, Bocquet M, Bertino L, Evensen G (2018) Data assimilation in the geosciences: an overview of methods, issues, and perspectives. WIREs Climate Change 9

Cheng S, Quilodrán-Casas C, Ouala S, Farchi A, Liu C, Tandeo P, Fablet R, Lucor D, Iooss B, Brajard J, Xiao D, Janjic T, Ding W, Guo Y, Carrassi A, Bocquet M, Arcucci R (2023) Machine learning with data assimilation and uncertainty quantification for dynamical systems: a review. IEEE/CAA J Autom Sin 10:1361–1387

Das S, Giannakis D (2020) Koopman spectra in reproducing Kernel Hilbert spaces. Appl Comput Harmon Anal 49:573–607

Desroziers G, Camino J-T, Berre L (2014) 4DEnVar: link with 4D state formulation of variational assimilation and different possible implementations. Q J R Meteorol Soc 140:2097–2110

Desroziers G, Arbogast E, Berre L (2016) Improving spatial localization in 4DEnVar. Q J R Meteorol Soc 142:3171–3185

Dufée B, Hug B, Mémin E, Tissot G (2024) Ensemble forecasts in reproducing Kernel Hilbert space family. Physica D 459:134044

Eisner T, Farkas B, Haase M, Nagel R (2015) Operator theoretic aspects of ergodic theory. Graduate texts in mathematics. Springer

Evensen G (2009) Data assimilation, the ensemble Kalman filter. Springer, Berlin, Heidelberg

Fablet R, Chapron B, Drumetz L, Mémin E, Pannekoucke O, Rousseau F (2021) Learning variational data assimilation models and solvers. J Adv Model Earth Syst 13

Fablet R, Chapron B, Le Sommer J, Sévellec F (2024) Inversion of sea surface currents from satellite-derived SST-SSH synergies with 4DVarNets. J Adv Model Earth Syst 16:e2023MS003609. E2023MS003609 2023MS003609

Farchi A, Laloyaux P, Bonavita M, Bocquet M (2021) Using machine learning to correct model error in data assimilation and forecast applications. Q J R Meteorol Soc 147:3067–3084

Ghil M, Allen MR, Dettinger MD, Ide K, Kondrashov D, Mann ME, Robertson AW, Saunders A, Tian Y, Varadi F, Yiou P (2002) Advanced spectral methods for climatic time series. Rev Geophys 40:3-1-3–41

Gottwald GA, Reich S (2021) Supervised learning from noisy observations: combining machine-learning techniques with data assimilation. Physica D 423:132911

Hogg J, Fonoberova M, Mezić I (2020) Exponentially decaying modes and long-term prediction of sea ice concentration using Koopman mode decomposition. Sci Rep 10:16313

Hunt BR, Kalnay E, Kostelich E, Ott E, Patil D, Sauer T, Szunyogh I, Yorke J, Zimin A (2004) Four-dimensional ensemble Kalman filtering. Tellus A Dyn Meteorol Oceanogr 56:273–277

Hunt BR, Kostelich EJ, Szunyogh I (2007) Efficient data assimilation for spatiotemporal chaos: a local ensemble transform Kalman filter. Phys D Nonlinear Phenom 230:112–126. Data Assimilation

Jaouen M, Dufée B, Mémin E, Tissot G (2025) Ensemble filter on dynamically driven RKHS: application to a multilayer quasi-geostrophic ocean model. https://inria.hal.science/hal-04915412

Kimeldorf G, Wahba G (1971) Some results on Tchebycheffian spline functions. J Math Anal Appl 33:82–95

Klus S, Schuster I, Muandet K (2020) Eigen decompositions of transfer operators in reproducing Kernel Hilbert spaces. J Nonlinear Sci 30:283–315

Kondrashov D, Ryzhov E, Berloff P (2020) Data-adaptive harmonic analysis of oceanic waves and turbulent flows. Chaos 30(6):061105

Koopman BO (1931) Hamiltonian systems and transformation in Hilbert space. Proc Natl Acad Sci 17:315–318

Le Dimet F-X, Talagrand O (1986) Variational algorithms for analysis and assimilation of meteorological observations: theoretical aspects. Tellus A 38A:97–110

Leeuwenburgh O, Evensen G, Bertino L (2005) The impact of ensemble filter definition on the assimilation of temperature profiles in the tropical pacific. Q J R Meteorol Soc 131:3291–3300

Le Gland F, Monbet V, Tran V (2011) Large sample asymptotics for the ensemble Kalman filter. In: Crisan D, Rozovskii B (eds) Handbook on nonlinear filtering. Oxford University Press

Liu C, Xiao Q, Wang B (2008) An ensemble-based four-dimensional variational data assimilation scheme. Part I: Technical formulation and preliminary test. Monthly Weather Rev 136:3363–3373

Lorenc AC (2003) The potential of the ensemble Kalman filter for NWP-a comparison with 4D-Var. Q J R Meteorol Soc 129:3183–3203

Manucharyan GE, Siegelman L, Klein P (2021) A deep learning approach to spatiotemporal sea surface height interpolation and estimation of deep currents in geostrophic ocean turbulence. J Adv Model Earth Syst 13:e2019MS001965. E2019MS001965 2019MS001965

Mauran S, Mouysset S, Simon E, Bertino L (2023) A kernel extension of the ensemble transform Kalman filter. In: Mikyška J, de Mulatier C, Paszynski M, Krzhizhanovskaya VV, Dongarra JJ, Sloot PM (eds) Computational Science - ICCS 2023. Springer Nature Switzerland, Cham, pp 438–452

Mercer J, Forsyth AR (1909) Xvi. Functions of positive and negative type, and their connection the theory of integral equations. Philos Trans Royal Soc Lond Ser A Containing Papers of a Mathematical or Physical Character 209:415–446

Mezić I (2005) Spectral properties of dynamical systems, model reduction and decompositions. Nonlinear Dyn 41:309–325

Mezić I (2021) Koopman operator, geometry, and learning of dynamical systems. Not Am Math Soc 68:1087–1105

Navarra A, Tribbia J, Klus S (2021) Estimation of Koopman transfer operators for the equatorial pacific SST. J Atmos Sci 78:1227–1244

Ouala S, Fablet R, Herzet C, Chapron B, Pascual A, Collard F, Gaultier L (2018) Neural network based Kalman filters for the spatio-temporal interpolation of satellite-derived sea surface temperature. Remote Sens 10

Reich S, Cotter C (2015) Probabilistic forecasting and Bayesian data assimilation. Cambridge University Press

Roullet G, Gaillard T (2022) A fast monotone discretization of the rotating shallow water equations. J Adv Model Earth Syst 14

Rowley CW, Mezić I, Bagheri S, Schlatter P, Henningson DS (2009) Spectral analysis of nonlinear flows. J Fluid Mech 641:115–127

Sakov P, Bertino L (2010) Relation between two common localisation methods for the EnKF. Comput Geosci 15:225–237

Sakov P, Evensen G, Bertino L (2010) Asynchronous data assimilation with the EnKF. Tellus A 62:24–29

Schmid PJ (2010) Dynamic mode decomposition of numerical and experimental data. J Fluid Mech 656:5–28

Stein ML (1999) Interpolation of spatial data. Springer series in statistics, Springer, New York

Thiry L, Li L, Roullet G, Mémin E (2023) Mqgeometry-1.0: a multi-layer quasi-geostrophic solver on non-rectangular geometries. EGUsphere 2023:1–25

Thiry L, Li L, Mémin E, Roullet G (2024) A unified formulation of quasi-geostrophic and shallow water equations via projection. J Adv Model Earth Syst 16:e2024MS004510

Tu JH, Rowley CW, Luchtenburg DM, Brunton SL, Kutz JN (2014) On dynamic mode decomposition: theory and applications. J Comput Dyn 1:391–421

Wan L, Bertino L, Zhu J (2010) Assimilating altimetry data into a HYCOM model of the pacific: ensemble optimal interpolation versus ensemble Kalman filter. J Atmos Oceanic Tech 27:753–765

Whitaker JS, Hamill TM (2002) Ensemble data assimilation without perturbed observations. Mon Weather Rev 130:1913–1924

Williams MO, Rowley CW, Kevrekidis IG (2015) A Kernel-based method for data-driven Koopman spectral analysis. J Comput Dyn 2:247–265

Yang Y, Robinson C, Heitz D, Mémin E (2015) Enhanced ensemble-based 4DVar scheme for data assimilation. Comput Fluids 115:201–210

Zhen Y, Chapron B, Mémin E, Peng L (2022) Eigenvalues of autocovariance matrix: a practical method to identify the Koopman eigenfrequencies. Phys Rev E 105:034205

Chapter 6
A Stochastic Ekman–Stokes Model for Coupled Ocean–Wave–Atmosphere Dynamics

Long Li, Etienne Mémin, and Bertrand Chapron

Abstract Accurate representation of atmosphere–ocean boundary layers, including the interplay of turbulence, surface waves, and air–sea fluxes, remains a challenge in geophysical fluid dynamics, particularly for climate simulations. This study introduces a stochastic coupled Ekman–Stokes model (SCESM) developed within the physically consistent Location Uncertainty framework, explicitly incorporating random turbulent fluctuations and surface wave effects. The SCESM integrates established parameterizations for air–sea fluxes, turbulent viscosity, and Stokes drift, and its performance is rigorously assessed through ensemble simulations compared against observations from the LOTUS field experiment. A performance ranking analysis quantifies the impact of different model components, highlighting the critical role of explicit uncertainty representation in both oceanic and atmospheric dynamics for accurately capturing system variability. Among the tested configurations, the full model version—including both Stokes drift and wave-induced mixing—shows the best agreement with observations. Wave-induced mixing terms improve model performance, while wave-dependent surface roughness enhances air–sea fluxes but reduces the relative influence of wave-driven mixing. This fully coupled stochastic framework provides a foundation for advancing boundary layer parameterizations in large-scale climate models.

6.1 Introduction

Numerical modeling of geophysical fluid dynamics, particularly in climate systems, presents significant challenges due to the multi-scale nature of these systems and the complex nonlinear interactions that govern their behavior. Accurately simulating

L. Li (✉) · E. Mémin · B. Chapron
ODYSSEY Team, Centre Inria de l'Université de Rennes, Rennes, France
e-mail: long.li@inria.fr

L. Li · E. Mémin
IRMAR, UMR CNRS 6625, Rennes, France

B. Chapron
Laboratoire d'Océanographie Physique et Spatiale, Ifremer, Plouzané, France

© The Author(s) 2026

B. Chapron et al. (eds.), *Stochastic Transport in Upper Ocean Dynamics IV*, Mathematics of Planet Earth 15, https://doi.org/10.1007/978-3-032-12749-5_6

large-scale atmospheric and oceanic flows while maintaining computational efficiency requires the careful modeling or parameterization of subgrid-scale processes. Many of these processes are intermittent, nonlinear, and random, underscoring the need for innovative approaches to represent uncertainties. As highlighted by Hasselmann (1976), Majda et al. (1999), Palmer (2019), stochastic modeling has emerged as a powerful tool for improving the representation of unresolved dynamics, thereby enhancing the accuracy of weather and climate predictions.

In this study, we adopt a physically consistent stochastic framework known as modeling under location uncertainty (LU), originally introduced by Mémin (2014). This approach preserves fundamental physical properties, such as energy conservation (Bauer et al. 2020), while naturally incorporating modeling errors and approximations to improve the representation of unresolved scales and their effects. A hierarchy of stochastic geophysical models has been systematically developed and validated within this framework, effectively capturing mesoscale and submesoscale dynamics within large-scale atmospheric and oceanic flows. Examples include the shallow water model (Brecht et al. 2021), the quasi-geostrophic model (Li et al. 2023), and the primitive hydrostatic model (Tucciarone et al. 2025). By integrating stochasticity, LU-based models improve uncertainty representation and enable coarse-grid simulations to reproduce the long-term statistical properties of high-resolution reference flows.

The LU framework also advances stochastic formulations for wave–current and air–sea interactions by capturing the impact of unresolved small-scale fluctuations on large-scale dynamics. Bauer et al. (2020) demonstrates that wave–current interactions, including the Coriolis–Stokes and vortex forces in the Craik–Leibovich system, can be interpreted as statistical effects of unresolved flow inhomogeneity within a stochastic flow representation. Building on this, Li et al. (2025) propose a stochastic formulation of the Ekman–Stokes layer (also referred to as the Wavy Ekman Layer in previous studies, e.g., McWilliams et al. 2012), which explicitly incorporates spatial variations of the unresolved velocity field. This allows for a more realistic and dynamic response to Stokes vortex forces, which—when spatial variations are neglected—reduce to a quasi-hydrostatic background perturbation. However, much of the dynamical impact of these forces emerges precisely from their spatial structure, as highlighted in McWilliams and Fox-Kemper (2013) and Suzuki and Fox-Kemper (2016). Numerical results in Li et al. (2025) show that the proposed stochastic model produces greater variability, increased kinetic energy, and more extreme events compared to standard models. Sensitivity analyses in that study highlight the influence of transient winds and surface waves, which enhance the dispersion of realizations while maintaining a balanced representation of errors. However, the model treats wind as a prescribed Gaussian process, neglecting its interaction with currents and waves and thereby failing to capture wave-modulated feedback on wind stress.

Expanding upon this foundation, the present work develops a fully coupled stochastic model that integrates turbulence, surface wave effects, and air–sea fluxes, explicitly capturing the interactions between random wind, waves, and currents. This one-dimensional (1D) vertical modeling approach is essential for realistic coarse-scale climate simulations, as it provides a physically consistent framework for parameterizing unresolved boundary layer turbulence and generating vertical profiles at the model's resolution grid points. In this context, we adopt classical bulk flux parameterizations (Fairall et al. 1996, 2003) to represent turbulent air–sea exchanges. These schemes remain widely used in climate models due to their empirical foundation and computational simplicity. While originally developed for deterministic, time-averaged conditions, we apply them here within a stochastic framework as an initial step toward extending traditional parameterizations to account for intrinsic variability. This provides a practical baseline for future developments involving more physically consistent coupling strategies. To evaluate the performance of the stochastic configurations, we compare model outputs against current profile observations from the LOTUS (Long-Term Upper-Ocean Study) experiment (Price et al. 1987). This dataset provides high-quality Eulerian current measurements that are minimally affected by mooring motion, making it a robust benchmark for model validation. LOTUS data have been widely used in the literature for evaluating upper-ocean boundary layer processes (Large et al. 1994; Price and Sundermeyer 1999; Lewis and Belcher 2004; Polton et al. 2005).

The paper is structured as follows: Sect. 6.2 describes the stochastic coupled Ekman–Stokes model (SCESM) and its parameterizations. Section 6.3 presents numerical results and comparisons with observational data. Finally, Sect. 6.4 concludes the study and discusses potential avenues for future research.

6.2   Models

In this section, we introduce the proposed stochastic framework for the coupled ocean–atmosphere Ekman layers, along with well-established parameterizations for air–sea fluxes, turbulent viscosity, and Stokes drift.

6.2.1   Stochastic Formulation for Air–Sea Coupled Ekman Layers

As illustrated in Fig. 6.1, the vertical domain is decomposed as follows: the atmospheric boundary layer (ABL) is defined in $\Omega^a = [\delta^a, H^a]$ (with $\delta^a \sim 10\,\mathrm{m}$ and $H^a \sim 1000\,\mathrm{m}$), the ocean boundary layer (OBL) is defined in $\Omega^o = [H^o, \delta^o]$ (with $\delta^o \sim -1\,\mathrm{m}$ and $H^o \sim -100\,\mathrm{m}$), while the air–sea interface consists of the atmospheric surface boundary layer (ASBL) in $[0, \delta^a]$ and the ocean surface boundary layer (OSBL) in $[\delta^o, 0]$.

Following the derivation of the generalized stochastic Craik–Leibovich equations (Bauer et al. 2020) and the generalized stochastic Ekman–Stokes model (Li et al. 2025), we consider coupling two time-dependent Ekman layers for the atmosphere

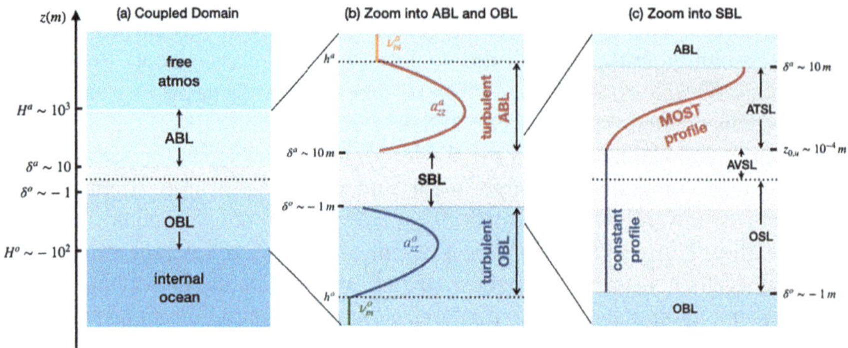

Fig. 6.1 Schematic of the coupled Ekman layers. **a** Vertical domain structure, including the atmospheric and oceanic boundary layers (ABL and OBL), and their near-surface sublayers. **b** Turbulent viscosity profiles in both ABL and OBL, defined from the near-surface depths δ^a and δ^o. **c** Classical one-sided bulk flux formulation (Pelletier et al. 2021) used for surface exchange computations within the surface boundary layer (SBL), which includes the ocean surface layer (OSL), atmospheric viscous sublayer (AVSL), and atmospheric turbulent sublayer (ATSL)

and ocean. Hereafter, this will be referred to as the stochastic coupled Ekman–Stokes model (SCESM), described by the following stochastic partial differential equations:

$$d\mathbf{u}^\alpha = \left(-if\left(\mathbf{u}^\alpha + \mathbf{u}_s^\alpha - \mathbf{u}_g^\alpha\right) + \partial_z\left(\nu^\alpha \partial_z(\mathbf{u}^\alpha + \mathbf{u}_s^\alpha)\right)\right) dt$$

$$- \left(\sigma_z^\alpha \partial_z(\mathbf{u}^\alpha + \mathbf{u}_s^\alpha) + if\boldsymbol{\sigma}_\mathbf{x}^\alpha\right) dB_t^\alpha, \quad z \in \mathring{\Omega}^\alpha, \ t > 0, \tag{6.2.1a}$$

$$\mathbf{u}^\alpha(z, 0) = \mathbf{u}_g^\alpha, \quad z \in \mathring{\Omega}^\alpha, \tag{6.2.1b}$$

$$\mathbf{u}^\alpha(H^\alpha, t) = \mathbf{u}_g^\alpha, \quad t > 0, \tag{6.2.1c}$$

$$\rho^\alpha \nu^\alpha \partial_z \mathbf{u}^\alpha(\delta^\alpha, t) = \boldsymbol{\tau}(t), \quad t > 0. \tag{6.2.1d}$$

Here, $\alpha \in \{a, o\}$ denotes atmospheric and oceanic components, respectively. The variable $\mathbf{u}^\alpha(z, t) = u^\alpha(z, t) + iv^\alpha(z, t)$ represents the horizontal velocity in complex notation, with i as the imaginary unit. The Coriolis frequency is denoted by f. The viscosity coefficient is given by $\nu^\alpha(z, t) = \nu_m^\alpha + a_{zz}^\alpha(z, t)$, incorporating both molecular and turbulent effects. The uniform density is denoted by ρ^α, and $\boldsymbol{\tau}(t) = \tau_x(t) + i\tau_y(t)$ is the time-dependent surface wind stress. The geostrophic velocity component $\mathbf{u}_g^\alpha = u_g^\alpha + iv_g^\alpha$ is assumed time- and depth-independent. This simplifying assumption is consistent with the reduced 1D formulation of the Ekman boundary layer. However, it prevents the model from capturing potential coupling between geostrophic flow and Stokes drift, a mechanism that has been shown to influence dynamics at submesoscales and, more weakly, at mesoscales (McWilliams and Fox-Kemper 2013; Suzuki and Fox-Kemper 2016). Such interactions, involving the interplay between geostrophic currents, internal gravity waves, and Stokes vortex forces, are important components of the full ocean boundary layer system. Their exclusion represents a limitation of the present formulation, which could be addressed in future extensions.

In Eq. (6.2.1a), the term $\boldsymbol{\sigma}^\alpha\,\mathrm{d}B_t^\alpha = (\sigma_{\mathbf{x}}^\alpha\,\mathrm{d}B_t^\alpha, \sigma_z^\alpha\,\mathrm{d}B_t^\alpha)$ represents a stochastic noise component originating from the Location Uncertainty (LU) framework (Mémin 2014; Resseguier et al. 2017; Bauer et al. 2020), which accounts for unresolved turbulent motions probabilistically. The noise is spatially correlated through the correlation operator $\boldsymbol{\sigma}^\alpha$, acting on a cylindrical Brownian motion B_t^α. In this framework, the Itô quadratic variation process (Bauer et al. 2020) associated with the vertical noise component is defined as

$$\frac{1}{2}\mathrm{d}\left\langle \int_0^\bullet \sigma_z^\alpha\,\mathrm{d}B_s^\alpha, \int_0^\bullet \sigma_z^\alpha\,\mathrm{d}B_s^\alpha \right\rangle_t = \frac{1}{2}\sigma_z^\alpha\sigma_z^\alpha\,\mathrm{d}t =: a_{zz}^\alpha\,\mathrm{d}t. \tag{6.2.2a}$$

Physically, a_{zz}^α plays the role of an eddy viscosity induced by unresolved turbulent fluctuations. The Itô–Stokes drift (introduced by Bauer et al. 2020), $\mathbf{u}_s = \partial_z a_{\mathbf{x}z}$ (within the Ekman layer scalings Li et al. 2025), captures the vertical inhomogeneity of co-quadratic variation between horizontal and vertical noise components:

$$\frac{1}{2}\mathrm{d}\left\langle \int_0^\bullet \sigma_{\mathbf{x}}^\alpha\,\mathrm{d}B_s^\alpha, \int_0^\bullet \sigma_z^\alpha\,\mathrm{d}B_s^\alpha \right\rangle_t = \frac{1}{2}\boldsymbol{\sigma}_{\mathbf{x}}^\alpha\sigma_z^\alpha\,\mathrm{d}t =: \boldsymbol{a}_{\mathbf{x}z}^\alpha\,\mathrm{d}t. \tag{6.2.2b}$$

In this formulation, the unresolved variability in the atmosphere and ocean is modeled through independent (cylindrical) Brownian motions B_t^a and B_t^o, representing intrinsic turbulence internal to each medium. This assumption enables a tractable stochastic representation of subgrid-scale processes. However, we acknowledge that in reality, coherent structures such as wave-phase-aligned turbulence or wave-induced surface-layer motions may introduce physical correlations across the interface (e.g., Qiao et al. 2016). Capturing such effects would require a more general framework with correlated noises, which represents an interesting direction for future developments.

The above provides a general description of our stochastic model. To specify further the noise, we may also formulate the following inverse problem: determining a specific noise such that the resulting statistical properties correspond to a prescribed turbulent viscosity $\widetilde{a}_{zz}^\alpha$ and a given Stokes drift $\widetilde{\mathbf{u}}_s^\alpha$. In particular, a simple projection-based construction was proposed by Li et al. (2025) in the ocean-only setting; this approach is adopted here in the coupled atmosphere–ocean context.

Given the turbulent viscosity $\widetilde{a}_{zz}^\alpha$, we adopt the following spectral decomposition for the vertical noise component:

$$\sigma_z^\alpha\,\mathrm{d}B_t^\alpha = \sqrt{2}\sum_{n>0}\left[(\widetilde{a}_{zz}^\alpha)^{1/2}, e_n^\alpha\right]e_n^\alpha\,\mathrm{d}\beta_{t,n}^\alpha, \tag{6.2.3a}$$

where $\{e_n^\alpha(z)\}_{n>0}$ denotes a set of localized basis functions with minimal overlapping support in the real-valued Hilbert space $L^2(\Omega^\alpha, \mathbb{R})$, equipped with the inner product $[f, g] = \int_{\Omega^\alpha} f(z)g(z)\,\mathrm{d}z$. The set $\{\beta_{t,n}^\alpha\}_{n>0}$ consists of independent one-dimensional real-valued Brownian motions. Furthermore, the two sets of Brownian motions, $\{\beta_{t,n}^o\}_{n>0}$ and $\{\beta_{t,n}^a\}_{n>0}$, are assumed to be independent.

Given the anti-derivative of the horizontal Stokes drift, defined as $\widetilde{\mathbf{U}}_s^\alpha(z) = \int_{H^o}^z \widetilde{\mathbf{u}}_s^\alpha(\zeta)\,\mathrm{d}\zeta$, where $\widetilde{\mathbf{u}}_s^\alpha(z) = \widetilde{u}_s^\alpha(z) + i\widetilde{v}_s^\alpha(z)$, the horizontal noise component can be constructed as

$$\boldsymbol{\sigma}_\mathbf{x}^\alpha\,\mathrm{d}B_t^\alpha = \sqrt{2}\sum_{n>0}\big[(\widetilde{a}_{zz}^\alpha)^{-1/2}\widetilde{\mathbf{U}}_s^\alpha, e_n^\alpha\big]e_n^\alpha\,\mathrm{d}\beta_{t,n}^\alpha, \tag{6.2.3b}$$

where the inner product is defined as $[\mathbf{U}, e_n] = [U, e_n] + i[V, e_n]$. The corresponding quadratic variation processes, defined in Eq. (6.2.2), are then given by

$$a_{zz}^\alpha = \sum_n \big[(\widetilde{a}_{zz}^\alpha)^{1/2}, e_n^\alpha\big]^2 (e_n^\alpha)^2, \tag{6.2.4a}$$

$$\boldsymbol{a}_{\mathbf{x}z}^\alpha = \sum_n \big[(\widetilde{a}_{zz}^\alpha)^{1/2}, e_n^\alpha\big]\big[(\widetilde{a}_{zz}^\alpha)^{-1/2}\widetilde{\mathbf{U}}_s^\alpha, e_n^\alpha\big] (e_n^\alpha)^2. \tag{6.2.4b}$$

By Parseval's theorem, these reconstructed functions are globally consistent with the given inputs $\widetilde{a}_{zz}^\alpha$ and $\widetilde{\mathbf{U}}_s^\alpha$, in the sense that

$$\int_{\Omega^\alpha} a_{zz}^\alpha(z)\,\mathrm{d}z = \int_{\Omega^\alpha} \widetilde{a}_{zz}^\alpha(z)\,\mathrm{d}z, \qquad \int_{\Omega^\alpha} \boldsymbol{a}_{\mathbf{x}z}^\alpha(z)\,\mathrm{d}z = \int_{\Omega^\alpha} \widetilde{\mathbf{U}}_s^\alpha(z)\,\mathrm{d}z. \tag{6.2.5}$$

Since the basis functions e_n are assumed to be localized (i.e., significant over small, distinct regions with negligible overlap), Parseval's identity applies approximately pointwise. This yields $a_{zz}^\alpha(z) \approx \widetilde{a}_{zz}^\alpha(z)$ and $\boldsymbol{a}_{\mathbf{x}z}^\alpha(z) \approx \widetilde{\mathbf{U}}_s^\alpha(z)$, so that the resulting Itô–Stokes drift approximates the given Stokes drift: $\mathbf{u}_s^\alpha = \partial_z \boldsymbol{a}_{\mathbf{x}z}^\alpha \approx \widetilde{\mathbf{u}}_s^\alpha$. This identification in the oceanic case has been illustrated in Li et al. (2025). For simplicity, we use the same notation for both the prescribed fields and their associated noise quadratic variations.

The Stokes drift is typically not included in the atmospheric momentum equation (Lewis and Belcher 2004), as it is generally much smaller than the geostrophic wind. Therefore, in the following, we neglect $\mathbf{u}_s^a$ by assuming negligible atmospheric horizontal noise ($\boldsymbol{\sigma}_\mathbf{x}^a\,\mathrm{d}B_t^a \approx 0$) and refer to the Stokes drift $\mathbf{u}_s$ exclusively in the ocean. As a result, the stochastic atmospheric Ekman model can read

$$\mathrm{d}\mathbf{u}^a = \Big(-if\big(\mathbf{u}^a - \mathbf{u}_g^a\big) + \partial_z\big(\nu^a \partial_z \mathbf{u}^a\big)\Big)\,\mathrm{d}t - \sigma_z^a \partial_z \mathbf{u}^a\,\mathrm{d}B_t^a. \tag{6.2.6}$$

Taking the expectation, the mean wind velocity $\mathbb{E}[\mathbf{u}^a]$ satisfies

$$\partial_t \mathbb{E}[\mathbf{u}^a] = -if\Big(\mathbb{E}[\mathbf{u}^a] - \mathbf{u}_g^a\Big) + \partial_z\Big(\mathbb{E}[\nu^a \partial_z \mathbf{u}^a]\Big), \tag{6.2.7}$$

where the last term is generally nonlinear, as the turbulent viscosity closure for a_{zz}^a (recalling that $\nu^a = \nu_m^a + a_{zz}^a$) typically depends on the pathwise solution $\mathbf{u}^a$, at least through air–sea coupling, as will be demonstrated later. The stochastic model (6.2.6) can therefore be interpreted as a generalization of the classical model, incorporating random fluctuation effects arising from unresolved smaller scales.

Similarly, taking the expectation of Eq. (6.2.1a) for the ocean component, the mean current velocity $\mathbb{E}[\mathbf{u}^o]$ satisfies

$$\partial_t \mathbb{E}[\mathbf{u}^o] = -if\left(\mathbb{E}[\mathbf{u}^o] + \mathbb{E}[\mathbf{u}_s] - \mathbf{u}_g^o\right) + \partial_z\left(\mathbb{E}\big[\nu^o \partial_z \mathbf{u}^o\big] + \mathbb{E}\big[\nu^o \partial_z \mathbf{u}_s\big]\right), \quad (6.2.8)$$

which remains a nonlinear equation—because of the dependency of μ^o on $\mathbf{u}^o$. In general, the Stokes drift may also exhibit stochastic variability, as will be shown later. Unlike the classical Ekman–Stokes model, which accounts only for the Coriolis–Stokes force $if\,\mathbb{E}[\mathbf{u}_s]$, the proposed mean model introduces an additional term, $\partial_z\big(\mathbb{E}[\nu^o \partial_z \mathbf{u}_s]\big)$, representing a *wave mixing* effect. This term initially arises from a change of variables in the derivation of the generalized stochastic Craik–Leibovich equation (Bauer et al. 2020).

To further examine its influence, one can vertically integrate the mean current velocity. The mean current transport, given by $\mathbb{E}[\mathbf{T}^o] = \int_{\Omega^o} \mathbb{E}[\mathbf{u}^o](z)\,\mathrm{d}z$, then satisfies

$$\partial_t \mathbb{E}[\mathbf{T}^o] = -if\left(\mathbb{E}[\mathbf{T}^o] + \mathbb{E}[\mathbf{T}_s] - \mathbf{T}_g^o\right) + \frac{1}{\rho^o}\left(\mathbb{E}[\tau] + \mathbb{E}[\tau_s]\right), \qquad (6.2.9)$$

where $\tau_s = \rho^o \nu^o \partial_z \mathbf{u}_s(\delta^o)$ represents an additional surface wave stress absent in the classical Ekman–Stokes model.

Returning to the general case, the SCESM (6.2.1), combined with the simplified atmospheric equation (6.2.6) and the specific noise representations (6.2.3), establishes a coupled framework for the atmospheric and oceanic boundary layers, incorporating both surface wave effects and random turbulent fluctuations. However, the SCESM does not account for stratification (buoyancy) effects, which will be investigated in future work. To complete the model, we present parameterizations for the air–sea momentum flux τ, the turbulent viscosity a_{zz}^α, and the Stokes drift $\mathbf{u}_s$ in the subsequent sections, all of which are based on well-established approaches.

6.2.2 Bulk Parameterization of Turbulent Air–Sea Fluxes

In this section, we briefly review the parameterization of turbulent fluxes for the ocean–atmosphere coupling. We begin by recalling the atmospheric surface boundary layer model, which follows the Monin–Obukhov similarity theory (MOST) (Monin and Obukhov 1954), a generalized law of the wall for stratified fluids (see Foken 2006 for a detailed introduction on the subject). In the classical one-sided bulk flux formulation (Pelletier et al. 2018), which remains standard in many climate models, the resolved wind $\mathbf{u}$, potential temperature θ, and humidity q at the atmospheric surface-layer depth δ^a can be expressed in terms of their Monin–Obukhov (M–O) scaling parameters (Fox-Kemper et al. 2022) (u_*, θ_*, q_*), as follows:

$$\left|[\mathbf{u}]_{\delta^o}^{\delta^a}\right| = \frac{u_*}{\kappa}\left(\ln\left(\frac{\delta^a}{z_{0,u}(u_*, \theta_*, q_*)}\right) - \psi_m\left(\frac{\delta^a}{L_O(u_*, \theta_*, q_*)}\right)\right), \qquad (6.2.10a)$$

$$[\theta]_{\delta^o}^{\delta^a} = \frac{\theta_*}{\kappa}\left(\ln\left(\frac{\delta^a}{z_{0,\theta}(u_*, \theta_*, q_*)}\right) - \psi_h\left(\frac{\delta^a}{L_O(u_*, \theta_*, q_*)}\right)\right), \qquad (6.2.10b)$$

$$[q]_{\delta^o}^{\delta^a} = \frac{q_*}{\kappa}\left(\ln\left(\frac{\delta^a}{z_{0,q}(u_*, \theta_*, q_*)}\right) - \psi_h\left(\frac{\delta^a}{L_O(u_*, \theta_*, q_*)}\right)\right), \qquad (6.2.10c)$$

where $[X]_{\delta^o}^{\delta^a} = X(\delta^a) - X(\delta^o)$ denotes the sea-surface-relative (under the ocean constant profiles assumption) value of the atmospheric quantity X, while $\kappa \approx 0.4$ is the von Kármán constant, $z_{0,u}$, $z_{0,\theta}$, and $z_{0,q}$ represent the surface roughness lengths for momentum, heat, and humidity, respectively. These roughness lengths depend on the scaling parameters (u_*, θ_*, q_*). Additionally, L_O is the Obukhov length, which also depends on the scaling parameters, and ψ_m and ψ_h are the stability functions for momentum and tracer. Explicit forms of these functions can be found in Beljaars and Holtslag (1991) and Grachev et al. (2000). For the neutral case (without stratification), the above relationships reduce to the classical law of the wall, which is simply a logarithmic profile.

Note that the classical one-sided bulk formulation (6.2.10) assumes surface values based on near-surface measurements or model outputs, effectively treating the ocean surface layer as vertically uniform. While this simplification was originally introduced for computational efficiency in coarse-resolution models, it can introduce biases in the near-surface structure—particularly when surface-intensified processes such as Stokes shear are active within the upper meter. A two-sided surface-layer parameterization was proposed by Pelletier et al. (2021) to address this issue. With modern ocean models now resolving the upper-ocean boundary layer at vertical scales of order 1 m or finer, and with increasing recognition of the role of near-surface dynamics, the adoption of more consistent formulations (e.g., Large et al. 2019; Fox-Kemper et al. 2022) provides a promising direction for future improvements.

The Obukhov length L_O depends on the friction velocity scale u_* and the atmospheric buoyancy flux scale B_f as follows:

$$L_O = \frac{u_*^3}{\kappa B_f}, \quad B_f = g u_*\left(\frac{\theta_*}{\theta_v(\delta_a)} + \frac{q_*}{q(\delta_a) + q_0}\right), \qquad (6.2.11)$$

where $g \approx 9.81$ m s^{-2} is the gravity acceleration, $\theta_v(z) = \theta(z)\left(1 + q(z)/q_0\right)$ is the virtual potential temperature, and $q_0 \approx 0.61$ kg kg^{-1} is the specific humidity of saturated air.

Equations (6.2.10) are nonlinear, depending on the unknown scaling parameters (u_*, θ_*, q_*). To solve these three variables, a fixed-point iterative algorithm can be used. This numerical counterpart of the nonlinear formulation is referred to as the "bulk formula". Various versions exist, differing mainly in the initialization of the algorithm, the parameterization of surface roughness lengths, and the choice of stability functions. In this study, we adopt the well-established COARE (Coupled Ocean–Atmosphere Response Experiment) algorithm (Fairall et al. 1996, 2003).

The wind stress τ, the sensible heat flux Q_{SH}, and the latent heat flux Q_{LH} across the air–sea interface are given by

$$|\tau| = \rho^a C_d U \left| [\mathbf{u}]_{\delta^o}^{\delta a} \right| = \rho^a u_*^2 \left| [\mathbf{u}]_{\delta^o}^{\delta a} \right| / U, \tag{6.2.12a}$$

$$Q_{SH} = \rho^a c_P C_h U [\theta]_{\delta^o}^{\delta a} = \rho^a c_P u_* \theta_*, \tag{6.2.12b}$$

$$Q_{LH} = \rho^a l_E C_e U [q]_{\delta^o}^{\delta a} = \rho^a l_E u_* q_*, \tag{6.2.12c}$$

where c_P is the specific heat of air and l_E is the latent heat of vaporization. The transfer coefficients (C_d, C_h, C_e) for momentum, heat, and humidity all depend on (u_*, θ_*, q_*), and are determined from (6.2.11) and (6.2.12):

$$C_d = \left(\frac{\kappa}{\ln(\delta_a/z_{0,u}) - \psi_m(\delta_a/L)} \right)^2, \tag{6.2.13a}$$

$$C_h = \frac{\kappa C_d^{1/2}(\delta_a)}{\ln(\delta_a/z_{0,\theta}) - \psi_h(\delta_a/L)}, \tag{6.2.13b}$$

$$C_e = \frac{\kappa C_d^{1/2}(\delta_a)}{\ln(\delta_a/z_{0,q}) - \psi_h(\delta_a/L)}. \tag{6.2.13c}$$

In Eq. (6.2.12), U denotes the scalar difference in velocity across the air–sea interface, incorporating a "gustiness" factor u_{gust}, defined as

$$U = \left(\left| [\mathbf{u}]_{\delta^o}^{\delta a} \right|^2 + u_{\text{gust}}^2 \right)^{1/2}, \tag{6.2.14a}$$

$$u_{\text{gust}} = \begin{cases} 1.2(B_f z_i)^{1/3}, & \text{if } B_f > 0 \\ 0.2, & \text{if } B_f \leq 0, \end{cases} \tag{6.2.14b}$$

where $z_i \approx 600$ m is the convective boundary layer depth. The gustiness parameter accounts for atmospheric stability and is included in the COARE algorithm to ensure non-zero momentum fluxes at low wind speeds, by parameterizing the convective effect on momentum transfer.

Finally, the surface roughness lengths are parameterized in COARE by

$$z_{0,u} = \alpha_{ch} \frac{u_*^2}{g} + 0.11 \frac{\nu_m^a}{u_*}, \tag{6.2.15a}$$

$$z_{0,\theta} = \min\left(1.15 \times 10^{-4}, 5.5 \times 10^{-5} R_r^{-0.6}\right), \quad R_r = z_{0,u} \frac{u_*}{\nu_m^a}, \tag{6.2.15b}$$

$$z_{0,\theta} = z_{0,q}. \tag{6.2.15c}$$

The momentum roughness in (6.2.15a) is separated into a rough-flow component using Charnock scaling (Charnock 1955) and a smooth-flow component with a fixed roughness Reynolds number. The Charnock coefficient α_{ch} can be parameterized in different ways to account for physical effects. One such approach, based on wind speed (Fairall et al. 2003), uses a piecewise increasing and affine function of U:

$$\alpha_{ch}(U) = \begin{cases} 0.011, & \text{if } U \le 10\,\text{m s}^{-1} \\ 0.011 + 0.007(U - 10)/8, & \text{if } U \in (10, 18)\,\text{m s}^{-1} \\ 0.018, & \text{if } U \ge 18\,\text{m s}^{-1}. \end{cases} \tag{6.2.16a}$$

Alternatively, the Charnock coefficient can be parameterized as a function of wave age (Maat et al. 1991):

$$\alpha_{ch}(u_*, C_p) = A \left(\frac{u_*}{C_p}\right)^B, \tag{6.2.16b}$$

where C_p is the phase speed of the waves at the spectral peak. The constants $A = 0.114$ and $B = 0.622$ are used in COARE 3.5 (Edson et al. 2013).

Another approach is to parameterize surface roughness based on wave slope, as in the sea state- and wave age-dependent formulation (Donelan 1990):

$$\alpha_{ch}(u_*, C_p, H_s) = A' H_s \left(\frac{u_*}{C_p}\right)^{B'} \frac{g}{u_*^2}, \tag{6.2.16c}$$

where H_s is the significant wave height. The constants $A' = 0.091$ and $B' = 2$ are used in COARE 3.5 (Edson et al. 2013).

In this work, we employ the COARE (fixed-point) algorithm using the relative wind $[\mathbf{u}]_{\delta^o}^{\delta^a}(t)$ at each time step, while assuming constant stratification: the potential temperature difference $[\theta]_{\delta^o}^{\delta^a}$ and specific humidity difference $[q]_{\delta^o}^{\delta^a}$ are kept fixed throughout the simulation. This allows us to first determine the scaling parameters (u_*, θ_*, q_*) along with the intermediate transfer coefficients (C_d, C_h, C_e). Consequently, the wind stress is given by $\tau(t) = \rho^a\left(u_*^2[\mathbf{u}]_{\delta^o}^{\delta^a}/U\right)(t)$, which provides the surface boundary condition (6.2.1d) at each time step. This formulation assumes a stratified atmospheric surface layer, albeit with steady or time-averaged tracers, while considering the neutral SCESM (6.2.1). The heat fluxes in (6.2.12) will be incorporated in future investigations of stratified Ekman boundary layers, particularly when incorporating the evolution equations for temperature (McWilliams et al. 2009).

We remark that the COARE bulk algorithm is derived from time- and space-averaged observations and effectively parameterizes complex near-surface processes, including wave, spray, and bubble dynamics. In our stochastic framework, it is applied pathwise to the resolved variables, allowing stochastic fluctuations to influence the fluxes through time integration. This approach implies an instantaneous adjustment of the air–sea exchange layers to stochastic inputs—an assumption that may overlook the finite response times and nonlocal effects associated with wave-mediated momentum transfer. While practical and consistent with current coarse-resolution coupled models, this assumption could be relaxed in future work by incorporating stochastic extensions of similarity theory or asynchronous coupling strategies (Valcke 2021).

6.2.3 Turbulent Viscosity Closure and Stokes Drift

In this work, we implement the diagnostic K-Profile Parameterization (KPP) scheme (Large et al. 1994) for turbulent viscosity closure. While the classical bulk flux formulation (Fig. 6.1c) applies MOST extrapolation only on the atmospheric side, the turbulent viscosity in each domain is computed from KPP with M–O surface scaling. This ensures consistency in turbulent closure across both fluids, as shown schematically in Fig. 6.1b.

The KPP scheme is formulated in a nondimensional vertical coordinate $\zeta^\alpha = |z|/h^\alpha$, where h^α represents the turbulent boundary layer depth. The general formulation is expressed as

$$a_{zz}^\alpha(\zeta^\alpha) = h^\alpha w(\zeta^\alpha) G(\zeta^\alpha), \quad w(\zeta^\alpha) = \frac{\kappa u_*}{\phi_m(\zeta h^\alpha/L_O)}, \tag{6.2.17}$$

where w is the turbulent vertical velocity scale and G is a fourth-order polynomial (Large et al. 1994), ensuring that the viscosity and its gradient match specific values at the top and bottom of the boundary layer. The universal function ϕ_m is set to unity for neutral conditions, simplifying the formulation to match the profiles of O'Brien (1970).

In the stratified case, the boundary layer depth h^α depends on both velocity and buoyancy, with the bulk Richardson number (Large et al. 1994) governing the relationship. In the particular case of neutral conditions, the Ekman layer depth is defined by

$$h^\alpha(t) = \frac{c^\alpha}{|f|} u_*(t), \tag{6.2.18}$$

where the atmospheric constant $c^a = 0.2$ (Arya 1981) and the oceanic constant $c^o = -0.7$ (Large et al. 1994) are used. These constants are consistent with MOST.

It is worth noting that a prognostic turbulence scheme, based on the generic length scale theory (Umlauf and Burchard 2003), and specifically second-moment closure (Harcourt 2013, 2015) for Langmuir turbulence, which solves a turbulent kinetic energy (TKE) equation, could be explored in future studies following a similar framework.

In general, surface gravity waves exhibit a broad spectrum, leading to a complex vertical profile for the Stokes drift (Huang 1971; Jenkins 1989). For simplicity, we consider a steady, monochromatic deep-water wave in this study. The surface elevation, accurate to the leading order in wave steepness, is expressed as $\eta = \eta_0 \cos(kx - \omega t)$, where η_0 is the wave amplitude, k is the horizontal wavenumber, and $\omega = (gk)^{1/2}$ is the angular frequency, consistent with the deep-water dispersion relation. The corresponding horizontal components of the Stokes drift are approximated by Phillips (1977) as

$$\mathbf{u}_s(z) = U_s e^{2kz} e^{i\theta_s}, \tag{6.2.19}$$

where $U_s = \omega k \eta_0^2$ is the magnitude of the Stokes drift and θ_s represents the wave propagation direction. Following Li et al. (2025), the wave direction θ_s is parameterized by a normal distribution: $\theta_s \sim \mathcal{N}(\Theta_s, \Sigma_s)$, where the mean direction Θ_s is aligned with the geostrophic wind, and the small standard deviation Σ_s represents the angular uncertainty, accounting for the misalignment between wind and wave direction.

Note that the anti-derivative $\mathbf{U}_s$, used in Eq. (6.2.3b) for the Stokes drift profile, is given by $\mathbf{u}_s/(2k)$. For the monochromatic wave described in Eq. (6.2.19), the phase speed of the waves at the spectral peak and the significant wave height, used in the wave-age-dependent formulation (6.2.16b) and (6.2.16c) for surface roughness parameterization, are defined as (McWilliams et al. 2014)

$$C_p = \omega/k = \sqrt{g/k}, \quad H_s = 2\sqrt{2}\eta_0. \tag{6.2.20}$$

6.3 Results

In this section, we numerically investigate the proposed stochastic coupled Ekman–Stokes model (SCESM) with the presented parameterization through ensemble simulations.

Table 6.1 Model parameters

Symbol	Value	Description
H^a	1000 m	Upper bound of atmospheric domain
H^o	-100 m	Lower bound of oceanic domain
δ^a	10 m	Lower bound of atmospheric domain
δ^o	-1 m	Upper bound of oceanic domain
f	8.36×10^{-5} s^{-1}	Coriolis parameter
ν_m^a	1.5×10^{-5} m^2 s^{-1}	Air kinematic viscosity
ν_m^o	10^{-6} m^2 s^{-1}	Water kinematic viscosity
ρ^o	10^3 kg m^{-3}	Water density
ρ^a	1 kg m^{-3}	Air density
$\mathbf{u}_g^a$	$(9 + 0i)$ m s^{-1}	Geostrophic wind velocity
$\mathbf{u}_g^o$	0 m s^{-1}	Geostrophic current velocity
$\theta^a(\delta^a)$	299.65 K (26.5 °C)	Atmospheric potential temperature at δ^a
$\theta^o(\delta^o)$	301.15 K (28 °C)	Oceanic potential temperature at δ^o
$q(\delta^a)$	0%	Atmospheric specific humidity at δ^a
η_0	0.8 m	Surface wave amplitude
λ	60 m	Wavelength of surface wave
Θ_s	0°	Wave mean propagation direction
Σ_s	5°	Wave spreading angle

6.3.1 Model Configurations

The configuration parameters are presented in Table 6.1, with most values selected to be consistent with the observational data discussed later. The SCESM is solved numerically using a pseudo-spectral Chebyshev method combined with an implicit time-stepping scheme (Li et al. 2025). The ocean and atmosphere domains are discretized with 300 and 1000 Chebyshev nodes, respectively, and 500 ensemble members are used. The coupled model is integrated over 20 days using a uniform time step of $\Delta t^a = \Delta t^o = 300$ s. In this configuration, air–sea coupling is performed at each time step using the instantaneous surface wind and current. If a larger ocean time step was employed relative to the atmosphere, asynchronous or synchronous coupling strategies could be considered (Valcke 2021), and further improvements may be achieved by incorporating the Schwarz iterative method (Marti et al. 2021).

To assess the contributions of different modeling terms in the SCESM, we compare it against several reduced versions:

- **RAM**: Stochastic atmosphere model coupled with a deterministic ocean model without Stokes drift:

$$\mathrm{d}\mathbf{u}^a = \left(-if\left(\mathbf{u}^a - \mathbf{u}_g^a\right) + \partial_z\left(\nu^a \partial_z \mathbf{u}^a\right) \right)\mathrm{d}t - \sigma_z^a \partial_z \mathbf{u}^a\, \mathrm{d}B_t^a, \qquad (6.3.1\text{a})$$

$$\partial_t \mathbf{u}^o = -if\left(\mathbf{u}^o - \mathbf{u}_g^o\right) + \partial_z\left(\nu^o \partial_z \mathbf{u}^o\right). \qquad (6.3.1\text{b})$$

- **ROM**: Deterministic atmosphere model coupled with a stochastic ocean model without Stokes drift:

$$\partial_t \mathbf{u}^a = -if\left(\mathbf{u}^a - \mathbf{u}_g^a\right) + \partial_z\left(\nu^a \partial_z \mathbf{u}^a\right), \qquad (6.3.2\text{a})$$

$$\mathrm{d}\mathbf{u}^o = \left(-if\left(\mathbf{u}^o - \mathbf{u}_g^o\right) + \partial_z\left(\nu^o \partial_z \mathbf{u}^o\right) \right)\mathrm{d}t - \sigma_z^o \partial_z \mathbf{u}^o\, \mathrm{d}B_t^o. \qquad (6.3.2\text{b})$$

- **RCM**: Stochastic atmosphere model (6.3.1a) coupled with a stochastic ocean model (6.3.2b) without Stokes drift.
- **RCM-RS**: RCM with Stokes drift under random angular directions, but without wave mixing terms. That is, Eq. (6.3.1a) with

$$\mathrm{d}\mathbf{u}^o = \left(-if\left(\mathbf{u}^o + \mathbf{u}_s - \mathbf{u}_g^o\right) + \partial_z\left(\nu^o \partial_z \mathbf{u}^o\right) \right)\mathrm{d}t - \left(\sigma_z^o \partial_z \mathbf{u}^o + if\sigma_{\mathbf{x}}^o \right)\mathrm{d}B_t^o.$$
$$(6.3.3)$$

- **RCM-RS-WM**: RCM-RS with wave mixing terms. That is, Eq. (6.3.1a) with

$$\mathrm{d}\mathbf{u}^o = \left(-if\left(\mathbf{u}^o + \mathbf{u}_s - \mathbf{u}_g^o\right) + \partial_z\left(\nu^o \partial_z (\mathbf{u}^o + \mathbf{u}_s)\right) \right)\mathrm{d}t$$
$$- \left(\sigma_z^o \partial_z\left(\mathbf{u}^o + \mathbf{u}_s\right) + if\sigma_{\mathbf{x}}^o \right)\mathrm{d}B_t^o. \qquad (6.3.4)$$

All models are initialized with the same initial condition (6.2.1b) and use the same Dirichlet boundary condition (6.2.1c). We remark that Equations (6.3.3) and (6.3.4) span two simplified configurations—excluding or including Stokes shear in the turbulent boundary condition—to explore the sensitivity of the coupled system to different assumptions. In reality, the interplay between turbulence and wave-induced effects is more intricate. Turbulence tends to reduce the Lagrangian shear (Pearson 2018), and adjustments such as the Coriolis–Stokes interaction give rise to the so-called "Lagrangian Thermal Wind" balance (McWilliams and Fox-Kemper 2013; Haney et al. 2015; Suzuki and Fox-Kemper 2016). These processes can produce an "anti-Stokes" Eulerian shear that opposes the Stokes drift, resulting in a small net Lagrangian shear. While the COARE algorithm assumes that Eulerian velocity enters the surface flux formulation, the Eulerian current is not independent of wave effects—such as vortex, Coriolis–Stokes, and advection feedbacks—raising fundamental questions about whether Eulerian or Lagrangian formulations provide a better basis for bulk parameterization. This is an open debate in the literature (see

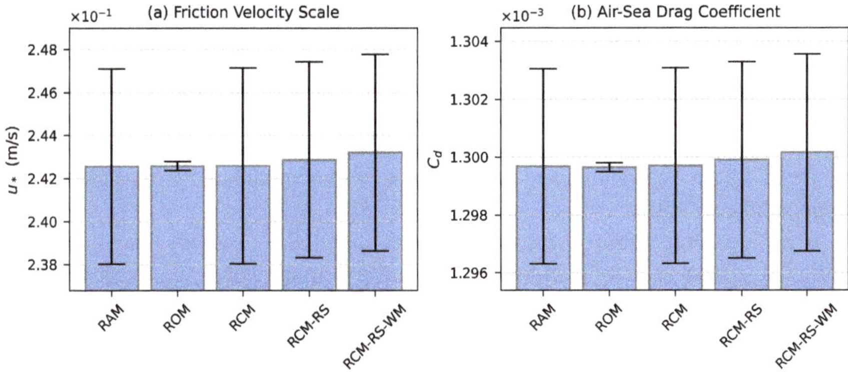

Fig. 6.2 Comparison of the ensemble mean (represented by bars) and spread (represented by error bars) for **a** friction velocity u_* and **b** air–sea momentum transfer coefficient C_d, across different random models (labeled on the x-axis). The ensemble mean and spread (mean(u) $\pm$ std(u)) are time-averaged over the last 10 days

Vanneste and Young 2022), and our two configurations are intended as limiting-case models within this ongoing discussion. A more comprehensive treatment involving wave-averaged closures and consistent shear-dependent fluxes is left for future work.

To distinguish the impact of different modeling terms from that of parameterization choices, we first conduct a comparative analysis using the same wind speed-dependent formulation (6.2.16a) for the surface roughness length. This ensures that the air–sea flux condition (6.2.1d) remains consistent across models.

For example, Fig. 6.2 demonstrates that all models yield similar ensemble mean values for the friction velocity u_* and the air–sea momentum transfer coefficient C_d, with RCM-RS and RCM-RS-WM exhibiting slight increases in these values. Additionally, the overall level of uncertainty is similar across models, except for ROM, which explicitly incorporates stochasticity only in the ocean. Unsurprisingly, this significantly reduces the uncertainty representation of air–sea fluxes.

In this context, we first compare the model performances against observations in Sect. 6.3.2, followed by a statistical analysis of model diagnostics in Sect. 6.3.3. Additional results incorporating wave-dependent formulations (6.2.16b) and (6.2.16c) for RCM-RS and RCM-RS-WM are presented in Sect. 6.3.4.

6.3.2   Comparison with LOTUS Observations

To assess the models' performance, we compare their outputs with observations from the Long-Term Upper-Ocean Study (LOTUS) experiment (Price et al. 1987; Price and Sundermeyer 1999), conducted in the western Sargasso Sea (34 °N, 70 °W) during the summer of 1982. This dataset provides 160 days of current profile measurements collected using vector-measuring instruments mounted on a stable platform to mitigate errors associated with mooring motion. The geostrophic velocity, assumed constant at a depth of 50 m, was removed to extract the wind-driven signal. Wind and current data were daily averaged to reduce high-frequency oscillations in the observed signals, rotated to align the wind direction with nominal north, and subsequently averaged over the 160-day period. Table 1 in Price et al. (1987) reports the mean values and confidence intervals for the downwind and crosswind current components at depths $z = (-5, -10, -15, -25)$ m.

To ensure comparability with observations, the model's eastward and northward (Eulerian) ocean velocity components (u^o and v^o) on the wind-relative coordinates are rotated into a wind-relative coordinate system. The downwind and crosswind velocity components, $u_\parallel$ and $u_\perp$, are defined as

$$\begin{pmatrix} u_\parallel \\ u_\perp \end{pmatrix} = \begin{pmatrix} \cos(\theta) & \sin(\theta) \\ \sin(\theta) & -\cos(\theta) \end{pmatrix} \begin{pmatrix} u^o \\ v^o \end{pmatrix}, \quad \theta = \arg(\boldsymbol{\tau}). \tag{6.3.5}$$

This rotation is applied to each random realization of the model outputs $\mathbf{u}^o$ and τ, without performing daily averaging of the data. We retain the full dataset to compute ensemble statistics before applying low-pass filtering or time-averaging for visualization.

Figure 6.3 qualitatively compares the ensemble mean and spread (defined as mean $\pm$ standard deviation) of the Ekman current profiles for different random models, alongside the mean and confidence intervals (CIs) of the observations. It shows that the RAM, which introduces explicit stochastic transport only in the atmosphere, produces a very low ensemble spread and fails to both encompass the mean of the observations and overlap with their CIs over depth. The ROM and RCM, which incorporate explicit stochastic transport in the ocean, increase the spread but still fail to align with the observations near the surface, particularly at $z = (-5, -10)$ m. The RCM-RS and RCM-RS-WM, which include the Coriolis–Stokes force, modify the shape of the current profiles and overlap most of the observation CIs across depth, particularly near the surface. Compared to the other random models, the RCM-RS-WM, which includes additional wave mixing effects, significantly increases the spread over depth and shifts the spread to the left for the downwind component and to the right for the crosswind component.

To further evaluate the models' performance against the LOTUS observations, we employ statistical metrics that account for observational uncertainties. Specifically, we generate observation samples following normal distributions, $u_\parallel^{\mathrm{obs}} \sim \mathcal{N}\left(\widehat{\mu}_\parallel, \widehat{\sigma}_\parallel^2\right)$ and $u_\perp^{\mathrm{obs}} \sim \mathcal{N}\left(\widehat{\mu}_\perp, \widehat{\sigma}_\perp^2\right)$, where the mean values $\widehat{\mu}$ and the standard errors $\widehat{s} = \alpha\widehat{\sigma}/\sqrt{n}$ are obtained from Table 1 of Price et al. (1987). The CI is given by $\mathrm{CI} = [\widehat{\mu} - \widehat{s}, \widehat{\mu} + \widehat{s}]$, where $n = 53$ represents the effective degrees of freedom used to estimate the standard error over the 160-day record. The confidence levels are set to 95% ($\alpha = 2$) for the downwind component and 90% ($\alpha = 1.7$) for the crosswind component. The corresponding sample standard deviations, $\widehat{\sigma}_\parallel$ and $\widehat{\sigma}_\perp$, are derived from these values.

We employ two statistical metrics for the model validation against LOTUS observations. The first is the Wasserstein distance (Villani 2009), which quantifies the

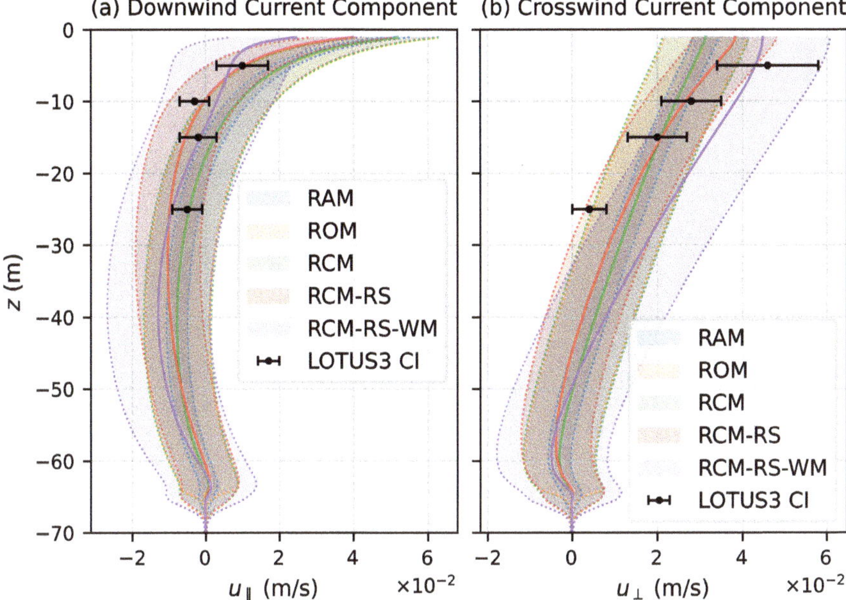

Fig. 6.3 Comparison of the ensemble mean (solid lines) and spread (shaded areas) for **a** downwind (Eulerian) Ekman current and **b** crosswind (Eulerian) Ekman current across different random models. These are compared to the confidence interval (CI, represented by black error bars) of the mean (centered point) LOTUS3 observations (Table 1 of Price et al. 1987) at near-surface depths. The ensemble mean and spread (mean(u) ± std(u)) are time-averaged over the last 10 days

similarity between two probability distributions. Specifically, we use the empirical cumulative distribution functions, $\widehat{F}_{X_e}$ and $\widehat{F}_{X_o}$, of the model ensemble X_e and the randomly generated observations X_o, respectively, to compute the p-th-order Wasserstein distance estimator:

$$\widehat{W}_p(X_e, X_o) = \left(\int_{\mathbb{R}} \left| \widehat{F}_{X_e}(x) - \widehat{F}_{X_o}(x) \right|^p \, dx \right)^{1/p}. \tag{6.3.6}$$

The second metric is the continuous ranked probability score (CRPS) (Weigel 2011), which evaluates ensemble forecast skill by measuring the integrated squared difference between the cumulative forecast and the observation. In this work, for each realization of the random observations, $x_o^{(i)}$, $i = 1, \ldots, n$, the CRPS is computed as

$$\mathrm{CRPS}\left(\widehat{F}_{X_e}, x_o^{(i)} \right) = \int_{\mathbb{R}} \left(\widehat{F}_{X_e}(x) - \mathrm{H}\left(x - x_o^{(i)} \right) \right)^2 \, dx, \tag{6.3.7a}$$

where H is the Heaviside step function. The final CRPS score is obtained by averaging over all sampled observations:

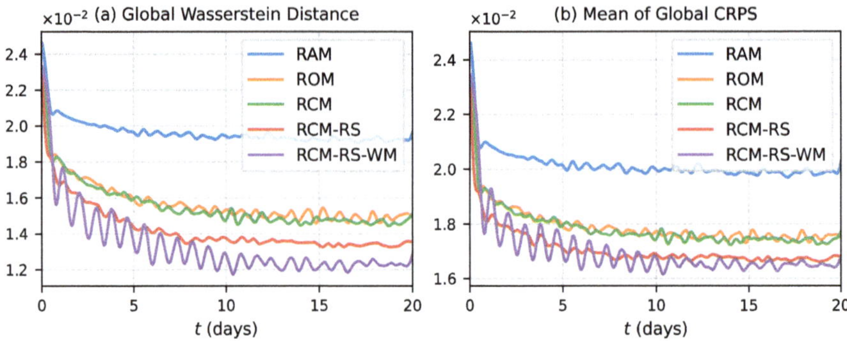

Fig. 6.4 Comparison of **a** the globally integrated Wasserstein distance and **b** the mean globally integrated continuous ranked probability score (CRPS) over time for different model ensembles against LOTUS3 observations. Observation samples are drawn from a normal distribution with empirical mean and standard deviation derived from Table 1 of Price et al. (1987). At each time, the CRPS is computed using an observation realization and averaged over samples. A half-day low-pass filter is applied to these time series

$$\overline{\mathrm{CRPS}} = \frac{1}{n} \sum_{i=1}^{n} \mathrm{CRPS}\left(\widehat{F}_{X_e}, x_o^{(i)}\right). \qquad (6.3.7b)$$

To compute these metrics, we independently draw observation samples (following the normal distributions described above) for each time step and depth. The scores are then averaged over depth, yielding time series of the scores for each model. A lower score indicates better model performance.

Figure 6.4 presents the temporal evolution of the globally integrated first-order ($p = 1$) Wasserstein distance and the mean globally integrated CRPS for each model ensemble against LOTUS observations. The results clearly demonstrate an increasing order of model performance:

$$\mathrm{RAM} < \mathrm{ROM} < \mathrm{RCM} < \mathrm{RCM\text{-}RS} < \mathrm{RCM\text{-}RS\text{-}WM}.$$

This ranking quantifies the contribution of each additional modeling term in the fully coupled stochastic system to the overall improvement of the model.

6.3.3 Statistical Diagnostics of Energy Budget and Wind Work

We begin by analyzing the ensemble decomposition of global energy for both atmospheric and oceanic components, specifically the mean kinetic energy (MKE) and eddy kinetic energy (EKE). The MKE is defined as $\mathrm{MKE}^{\alpha} = \rho^{\alpha}\|\overline{\mathbf{u}^{\alpha}}\|^2$, while the EKE is given by $\mathrm{EKE}^{\alpha} = \rho^{\alpha}\|\mathbf{u}^{\alpha} - \overline{\mathbf{u}^{\alpha}}\|^2$, where $\overline{u}$ denotes the ensemble mean of u.

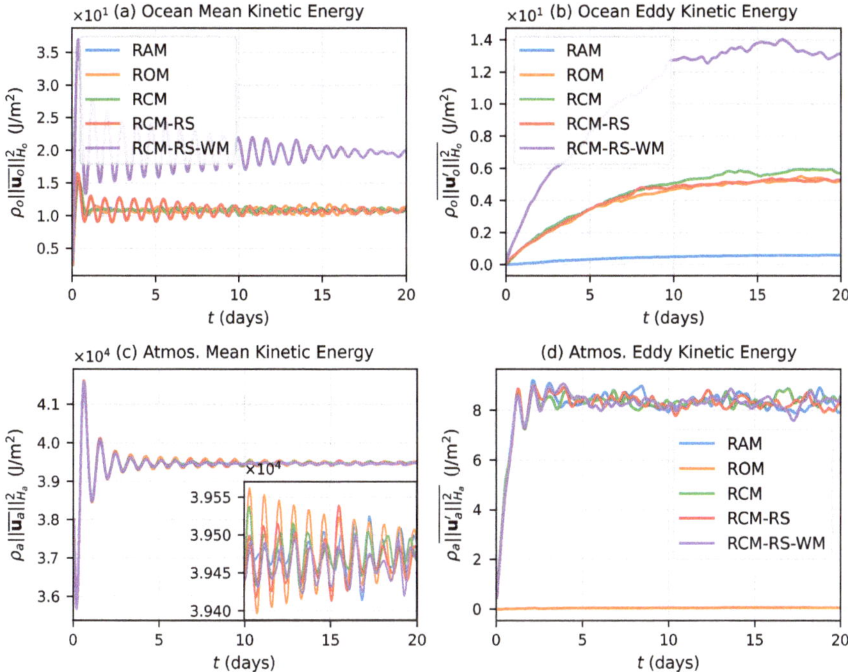

Fig. 6.5 Comparison of globally integrated mean kinetic energy (MKE, left) and eddy kinetic energy (EKE, right) for the ocean (top) and atmosphere (bottom) components across different random models (distinguished by color). MKE and EKE are defined in an ensemble sense as $\overline{\mathbf{u}} := \widehat{\mathbb{E}}[\mathbf{u}]$ and $\mathbf{u}' := \mathbf{u} - \widehat{\mathbb{E}}[\mathbf{u}]$. A half-day low-pass filter is applied to these time series. A zoomed-in view of atmospheric MKE over the last 10 days is included to better highlight the differences

Figure 6.5c shows that all models exhibit similar time-mean values of MKE^a, albeit with varying levels of oscillation around their respective means. In contrast, Fig. 6.5a demonstrates that the RCM-RS-WM significantly increases the MKE^o compared to all the other random models, due to the inclusion of additional wave mixing terms.

As illustrated in Fig. 6.5b, RAM, which introduces explicit stochastic transport only in the atmosphere, generates high EKE^a but very low EKE^o. Conversely, as shown in Fig. 6.5d, ROM, which introduces explicit stochastic transport only in the ocean, produces high EKE^o but very low EKE^a. The RCM model, which incorporates explicit stochastic transport in both the atmosphere and ocean, yields EKE^o levels comparable to ROM and EKE^a levels similar to RAM. These results clearly indicate that explicitly representing uncertainties in both components plays a crucial role in determining the variance levels of a coupled system. Furthermore, the two wave-dependent models exhibit EKE^a levels similar to those of RCM (Fig. 6.5d); however, the RCM-RS model slightly reduces EKE^o, whereas the RCM-RS-WM model significantly enhances it (Fig. 6.5c).

We next examine the wind contribution to the ensemble energy budget for both the atmosphere and ocean, characterized by the mean wind work $\mathrm{MWW}^\alpha = \overline{\boldsymbol{\tau}} \cdot \overline{\mathbf{u}^\alpha}(\delta^\alpha)$

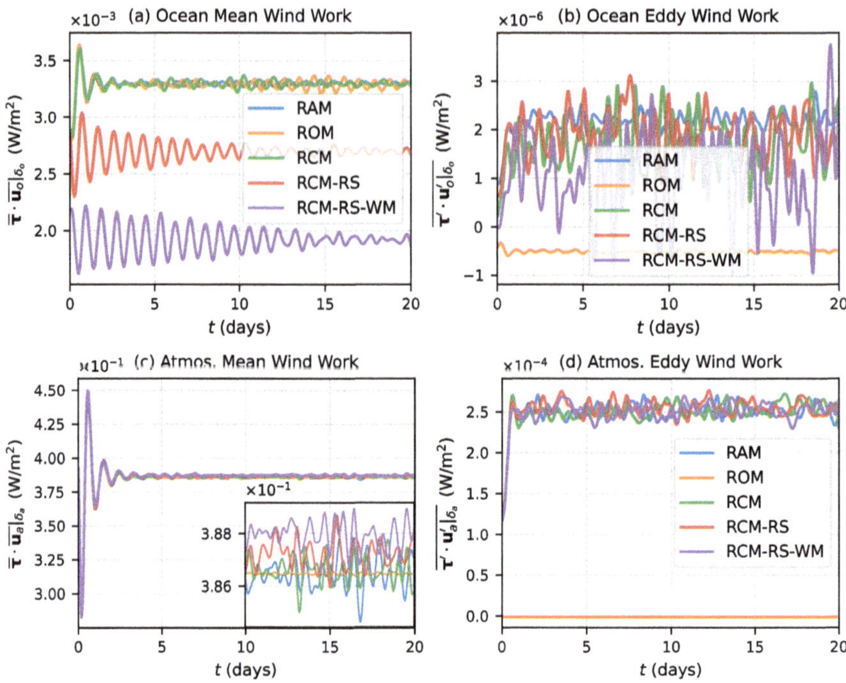

Fig. 6.6 Comparison of mean wind work (left) and eddy wind work (right) for the ocean (top) and atmosphere (bottom) components across different random models (distinguished by color)

and the eddy wind work $\mathrm{EWW}^{\alpha} = \overline{(\boldsymbol{\tau} - \overline{\boldsymbol{\tau}}) \cdot (\mathbf{u}^{\alpha} - \overline{\mathbf{u}^{\alpha}})}(\delta^{\alpha})$. Figure 6.6a indicates that surface waves significantly reduce the mean level of MWW^{o} over time while increasing its oscillation, as observed in the comparison of RCM-RS and RCM-RS-WM with RAM, ROM, and RCM. Furthermore, due to wave mixing effects, the reduction in MWW^{o} is more pronounced in RCM-RS-WM than in RCM-RS. Conversely, Fig. 6.6c illustrates that surface waves induce a slight increase in the mean level of MWW^{a}, an effect that becomes more evident when wave mixing terms are included.

As expected, ROM, which lacks explicit atmospheric uncertainty representation, exhibits an almost negligible EWW^{a} and a small negative EWW^{o}, as shown in Figure 6.6b and d. The other models exhibit similar levels of EWW^{a} (Fig. 6.6d). Additionally, the inclusion of wave mixing terms reduces the mean level of EWW^{o} over time while increasing its temporal variability (Fig. 6.6b).

6.3.4   Discussion of Wave-Dependent Surface Roughness Effects

We now examine the results obtained using the wave-dependent parameterization of surface roughness length for RCM-RS and RCM-RS-WM. Specifically, we focus on their impact on model performance relative to observations, employing the same statistical metrics as those presented in Sect. 6.3.2.

When adopting the wave age-dependent formulation (6.2.16b), Fig. 6.7 illustrates that both the mean friction velocity and air–sea transfer coefficients increase compared to those obtained using the wind speed-dependent formulation (6.2.16a). In this case, Fig. 6.8 demonstrates that the model performance ranking established in Sect. 6.3.2,

$$RAM < ROM < RCM < RCM\text{-}RS < RCM\text{-}RS\text{-}WM,$$

remains valid in terms of both the Wasserstein distance and CRPS. However, the relative improvement of RCM-RS-WM over RCM-RS is slightly less pronounced.

For the wave slope-dependent formulation (6.2.16c), Fig. 6.9 indicates a more significant increase in air–sea fluxes compared to the wave age-based case. Figure 6.10 demonstrates that the model performance ranking remains consistent in this scenario, at least in terms of the Wasserstein distance. However, the improvement of RCM-RS-WM relative to RCM-RS is noticeably less pronounced than in the previous cases, suggesting that stronger air–sea fluxes diminish the relative contribution of wave mixing terms.

Throughout this study, all model configurations rely on a common set of turbulence-related parameters—such as those in the KPP and COARE, formulations. This design choice facilitates fair comparisons, but may also bias performance assessments, particularly for models with added physical processes like wave-driven

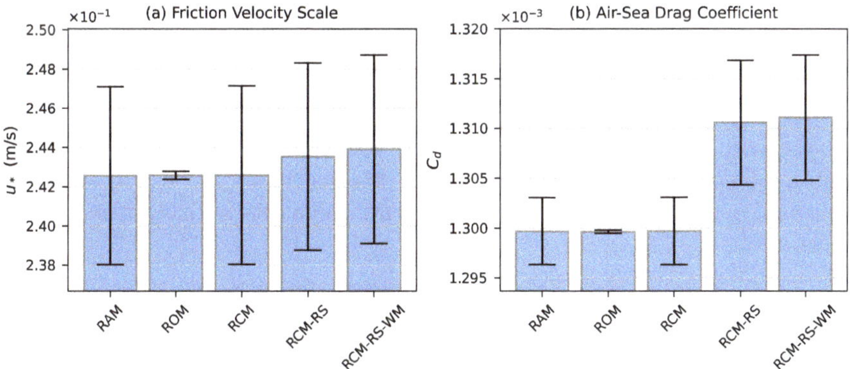

Fig. 6.7 Same as Fig. 6.2, but using the wave age-based roughness parameterization (6.2.16b) for RCM-RS and RCM-RS-WM

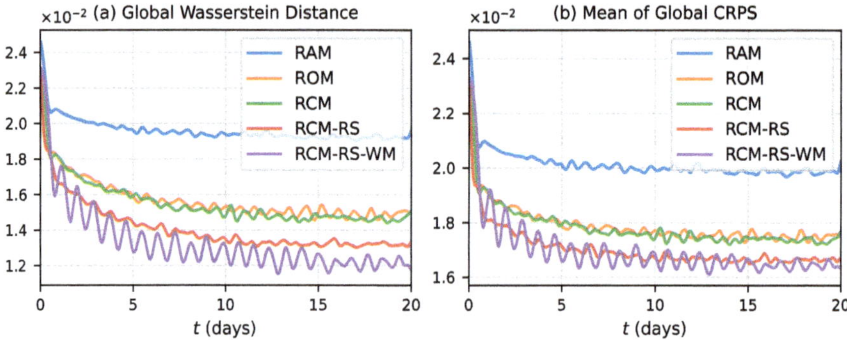

Fig. 6.8 Same as Fig. 6.4, but using the wave age-based roughness parameterization (6.2.16b) for RCM-RS and RCM-RS-WM

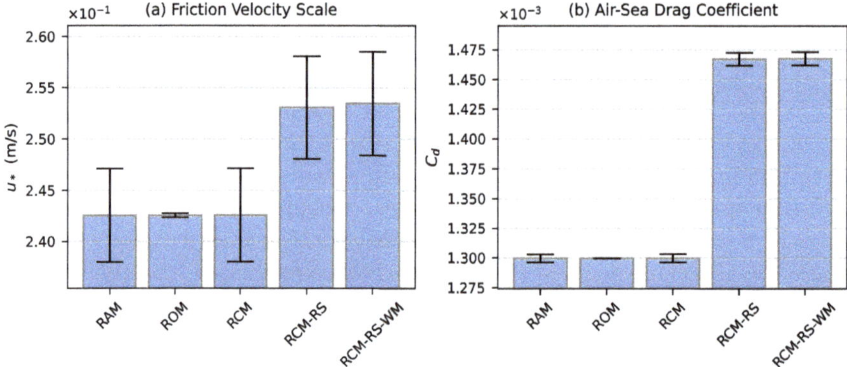

Fig. 6.9 Same as Fig. 6.2, but using the wave age-based roughness parameterization (6.2.16c) for RCM-RS and RCM-RS-WM

mixing (RCM-RS-WM). In such cases, model–observation mismatch may reflect parameter non-optimality rather than model structure alone. Moreover, the physical interpretation of these parameters may vary across models with distinct dynamics (e.g., inclusion of Stokes shear). A more consistent parameter tuning strategy, such as that proposed in McWilliams et al. (2012), would help adapt parameterizations to specific model assumptions. Future work may benefit from calibration against high-resolution large eddy simulation (LES) datasets, which provide full-depth current profiles, in contrast to the discrete-depth measurements available in the LOTUS experiment.

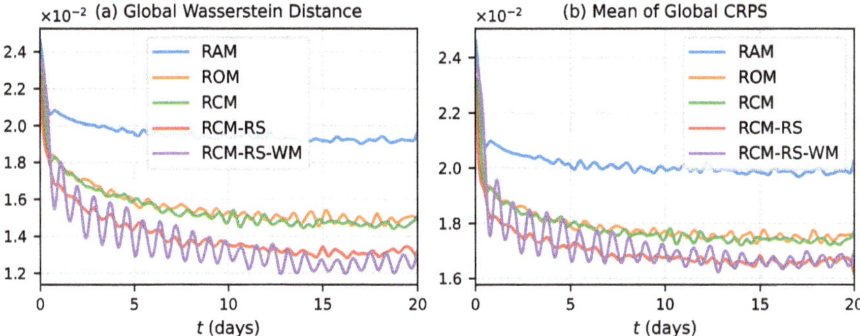

Fig. 6.10 Same as Fig. 6.4, but using the wave age-based roughness parameterization (6.2.16c) for RCM-RS and RCM-RS-WM

6.4   Conclusion

We have introduced a stochastic coupled Ekman–Stokes model (SCESM) to represent the atmosphere–ocean boundary layers while incorporating surface wave effects and turbulent fluctuations arising from unresolved scales. The SCESM, along with established parameterizations for air–sea fluxes, turbulent viscosity, and Stokes drift, was rigorously tested through ensemble simulations. Its performance was evaluated against LOTUS data using statistical metrics that account for observational uncertainties.

A performance ranking analysis quantified the impact of various modeling components within the fully coupled stochastic system. The results underscore the importance of explicitly representing uncertainties in both oceanic and atmospheric components to accurately capture the variance levels of the coupled system. Furthermore, while wave-dependent parameterizations of surface roughness enhance air–sea fluxes, they reduce the contribution of wave-induced mixing terms in the SCESM.

An important limitation of the present framework lies in the use of classical bulk flux parameterizations, which were originally designed for deterministic, time-averaged conditions. Future developments should consider more consistent formulations of surface coupling that explicitly incorporate uncertainty. Moreover, combining this framework with data assimilation and parameter estimation techniques would further enhance its predictive capability and realism.

Another limitation of the present validation is the lack of direct flux measurements in the LOTUS dataset. While comparisons of mean current profiles serve as a useful benchmark, they offer only indirect constraints on the underlying turbulent exchange mechanisms. More robust evaluation could be achieved using high-resolution coupled LES data (e.g., Sullivan et al. 2025) or modern observational platforms equipped with flux towers and wave-following buoys that measure momentum and heat fluxes.

Beyond these immediate improvements, several extensions are promising: incorporating buoyancy evolution via stochastic equations for stratified Ekman layers with time-varying stratification (McWilliams et al. 2009); introducing memory effects through second-order turbulence closures that solve a prognostic TKE equation including surface wave influences (Harcourt 2013); and implementing a spectral representation of surface waves with wave-induced momentum source terms (Perrie et al. 2003). Together, these developments could significantly advance our ability to simulate and predict coupled ocean–wave–atmosphere dynamics.

Data Availability Statement The PyTorch code **SCESM.py** used to reproduce the simulation data, along with the Python script **diag.py** which performs all diagnostics from the output data, and the Jupyter notebook **fig_stuod.ipynb**, which generates the figures presented in this study, are available at https://github.com/matlong/SCESM.

Acknowledgements The authors sincerely thank the two reviewers, Baylor Fox-Kemper and Charles Pelletier, for their insightful comments and suggestions, which have greatly helped improve the manuscript. We also acknowledge the support of the ERC EU project 856408-STUOD, which made this research possible.

References

Arya S (1981) Parameterizing the height of the stable atmospheric boundary layer. J Appl Meteor Climatol 20(10):1192–1202. https://doi.org/10.1175/1520-0450(1981)020<1192:PTHOTS>2.0.CO;2

Bauer W, Chandramouli P, Chapron B, Li L, Mémin E (2020) Deciphering the role of small-scale inhomogeneity on geophysical flow structuration: a stochastic approach. J Phys Oceanogr 50(4):983–1003. https://doi.org/10.1175/JPO-D-19-0164.1

Beljaars ACM, Holtslag AAM (1991) Flux parameterization over land surfaces for atmospheric models. J Appl Meteor Climatol 30:327–341. https://doi.org/10.1175/1520-0450(1991)030<0327:FPOLSF>2.0.CO;2

Brecht R, Li L, Bauer W, Mémin E (2021) Rotating shallow water flow under location uncertainty with a structure-preserving discretization. J Adv Model Earth Syst 13(12):e2021MS002492. https://doi.org/10.1029/2021MS002492

Charnock H (1955) Wind stress on a water surface. Quart J Roy Meteor Soc 81:639–640. https://doi.org/10.1002/qj.49708135027

Donelan MA (1990) Air–sea interaction, vol 9. Wiley, pp 239–292

Edson JB, Jampana V, Weller RA, Bigorre SP, Plueddemann AJ, Fairall CW, Miller SD, Mahrt L, Vickers D, Hersbach H (2013) On the exchange of momentum over the open ocean. J Phys Oceanogr 43(8):1589–1610. https://doi.org/10.1175/JPO-D-12-0173.1

Fairall CW, Bradley EF, Rogers DP, Edson JB, Young GS (1996) Bulk parameterization of air-sea fluxes for tropical ocean-global atmosphere coupled-ocean atmosphere response experiment. J Geophys Res 101(C2):3747–3764. https://doi.org/10.1029/95JC03205

Fairall CW, Bradley EF, Hare JE, Grachev AA, Edson JB (2003) Bulk parameterization of air-sea fluxes: Updates and verification for the COARE algorithm. J Climate 16:571–591. https://doi.org/10.1175/1520-0442(2003)016<0571:BPOASF>2.0.CO;2

Foken T (2006) 50 years of the Monin-Obukhov similarity theory. Bound-Lay Meteorol 119:431–447. https://doi.org/10.1007/s10546-006-9048-6

Fox-Kemper B, Johnson L, Qiao F (2022) Ocean near-surface layers. In: Ocean mixing, pp 65–94. Elsevier. https://doi.org/10.1016/B978-0-12-821512-8.00011-6

Grachev AA, Fairall CW, Bradley EF (2000) Convective profile constants revisited. Bound-Lay Meteorol 94:495–515. https://doi.org/10.1023/A:1002452529672

Haney S, Fox-Kemper B, Julien K, Webb A (2015) Symmetric and geostrophic instabilities in the wave-forced ocean mixed layer. J Phys Oceanogr 45(10):3033–3056. https://doi.org/10.1175/JPO-D-15-0044.1

Harcourt RR (2013) A second-moment closure model of Langmuir turbulence. J Phys Oceanogr 43(4):673–697. https://doi.org/10.1175/JPO-D-12-0105.1

Harcourt RR (2015) An improved second-moment closure model of Langmuir turbulence. J Phys Oceanogr 45(1):84–103. https://doi.org/10.1175/JPO-D-14-0046.1

Hasselmann K (1976) Stochastic climate models Part I. Theory Tellus 28(6):473–485. https://doi.org/10.1111/j.2153-3490.1976.tb00696.x

Huang NE (1971) Derivation of Stokes drift for a deep-water random gravity wave field. Deep-Sea Res 18(2):255–259. https://doi.org/10.1016/0011-7471(71)90115-X

Jenkins AD (1989) The use of a wave prediction model for driving a near- surface current model. Dtsch Hydrogr Z 42:134–149. https://doi.org/10.1007/BF02226291

Large WG, McWilliams JC, Doney SC (1994) Oceanic vertical mixing: A review and a model with a nonlocal boundary layer parameterization. Rev Geophys 32(4):63–403. https://doi.org/10.1029/94RG01872

Large WG, Patton EG, DuVivier AK, Sullivan PP, Romero L (2019) Similarity theory in the surface layer of large-eddy simulations of the wind-, wave-, and buoyancy-forced southern ocean. J Phys Oceanogr 49:2165–2187. https://doi.org/10.1175/JPO-D-18-0066.1

Lewis DM, Belcher SE (2004) Time-dependent, coupled, Ekman boundary layer solutions incorporating Stokes drift. Dyn Atmos Oceans 37(4):313–351. https://doi.org/10.1016/j.dynatmoce.2003.11.001

Li L, Deremble B, Lahaye N, Mémin E (2023) Stochastic data-driven parameterization of unresolved eddy effects in a baroclinic quasi-geostrophic model. J Adv Model Earth Syst 15(2):e2022MS003297. https://doi.org/10.1029/2022MS003297

Li L, Mémin E, Chapron B (2025) A generalized stochastic formulation of the Ekman-Stokes model with statistical analyses. J Phys Oceanogr

Maat N, Kraan C, Oost WA (1991) The roughness of wind waves. Bound-Lay Meteorol 54:89–103. https://doi.org/10.1007/BF00119414

Majda AJ, Timofeyev I, Vanden Eijnden E. Models for stochastic climate prediction. Proc Natl Acad Sci (USA) 96(26):14687–14691. https://doi.org/10.1073/pnas.96.26.14687

Marti O, Nguyen S, Braconnot P, Valcke S, Lemarié F, Blayo E (2021) A Schwarz iterative method to evaluate ocean- atmosphere coupling schemes: Implementation and diagnostics in IPSL-CM6-SW-VLR. Geosci Model Dev 14(5):2959–2975. https://doi.org/10.5194/gmd-14-2959-2021

McWilliams JC, Fox-Kemper B (2013) Oceanic wave-balanced surface fronts and filaments. J Fluid Mech 730:464–490. https://doi.org/10.1017/jfm.2013.348

McWilliams JC, Huckle E, Shchepetkin AF (2009) Buoyancy effects in a stratified Ekman layer. J Phys Oceanogr 39(10):2581–2599. https://doi.org/10.1175/2009JPO4130.1

McWilliams JC, Huckle E, Liang J-H, Sullivan PP (2012) The wavy Ekman layer: Langmuir circulations, breaking waves, and Reynolds stress. J Phys Oceanogr 42(11):1793–1816. https://doi.org/10.1175/JPO-D-12-07.1

McWilliams JC, Huckle E, Liang J-H, Sullivan PP (2014) Langmuir turbulence in swell. J Phys Oceanogr 44(3):870–890. https://doi.org/10.1175/JPO-D-13-0122.1

Mémin E (2014) Fluid flow dynamics under location uncertainty. Geophys Astrophys Fluid Dyn 108(2):119–146. https://doi.org/10.1080/03091929.2013.836190

Monin AS, Obukhov AM (1954) Basic laws of turbulent mixing in the surface layer of the atmosphere. Tr Geofiz Inst Akad Nauk SSSR 24:163–187

O'Brien JJ (1970) A note on the vertical structure of the eddy exchange coefficient in the planetary boundary layer. J Atmos Sci 27(8):1213–1215. https://doi.org/10.1175/1520-0469(1970)027<1213:ANOTVS>2.0.CO;2

Palmer TN (2019) Stochastic weather and climate models. Nat Rev Phys 1:463–471. https://doi.org/10.1038/s42254-019-0062-2

Pearson B (2018) Turbulence-induced anti-Stokes flow and the resulting limitations of large-eddy simulation. J Phys Oceanogr 48:117–122. https://doi.org/10.1175/JPO-D-17-0208.1

Pelletier C, Lemarié F, Blayo E (2018) Sensitivity analysis and metamodels for the bulk parameterization of turbulent air-sea fluxes. Quart J Roy Meteor Soc 144(736):658–669. https://doi.org/10.1002/qj.3233

Pelletier C, Lemarié F, Blayo E, Bouin MN, Redelsperger JL (2021) Two-sided turbulent surface-layer parameterizations for computing air-sea fluxes. Quart J Roy Meteor Soc 147(736):1726–1751. https://doi.org/10.1002/qj.3991

Perrie W, Tang CL, Hu Y, DeTracy BM (2003) The impact of waves on surface currents. J Phys Oceanogr 33:2126–2140. https://doi.org/10.1175/1520-0485(2003)033<2126:TIOWOS>2.0.CO;2

Phillips OM (1977) The dynamics of the upper ocean. Cambridge University Press

Polton JA, Lewis DM, Belcher SE (2005) The role of wave-induced Coriolis-Stokes forcing on the wind-driven mixed layer. J Phys Oceanogr 35(4):444–457. https://doi.org/10.1175/JPO2701.1

Price JF, Sundermeyer MA (1999) Stratified Ekman layers. J Geophys Res 104(C9):20467–20494. https://doi.org/10.1029/1999JC900164

Price JF, Weller RA, Schudlich RR (1987) Wind-driven ocean currents and Ekman transport. Science 238(4833):1534–1538. https://doi.org/10.1126/science.238.4833.1534

Qiao F, Yuan Y, Deng J, Dai D, Song Z (2016) Wave-turbulence interaction-induced vertical mixing and its effects in ocean and climate models. Philos Trans R Soc A 374(2065):20150201. https://doi.org/10.1098/rsta.2015.0201

Resseguier V, Mémin E, Chapron B (2017) Geophysical flows under location uncertainty, Part I, II & III. Geophys & Astro Fluid Dyn 111(3):149–227. https://doi.org/10.1080/03091929.2017.1310210

Sullivan PP, McWilliams JC, Patton EG (2025) An investigation of coupled atmospheric and oceanic boundary layers using large-eddy simulation. J Atmos Sci 82:829–846. https://doi.org/10.1175/JAS-D-24-0149.1

Suzuki N, Fox-Kemper B (2016) Understanding Stokes forces in the wave-averaged equations. J Geophys Res 121:3579–3596. https://doi.org/10.1002/2015JC011566

Tucciarone F, Li L, Mémin E, Chandramouli P (2025) Derivation and numerical assessment of a stochastic large-scale hydrostatic primitive model. J Adv Model Earth Syst 17:e2024MS004783. https://doi.org/10.1029/2024MS004783

Umlauf L, Burchard H (2003) A generic length-scale equation for geophysical turbulence models. J Mar Res 61:235–265

Valcke S (2021) Coupling algorithms and specific coupling features in CGCMs, Chap 9. World Scientific Publishing Co, pp 140–156. https://doi.org/10.1142/9789811232947_0009

Vanneste J, Young WR (2022) Stokes drift and its discontents. Philos Trans R Soc A 380(2225):20210032. https://doi.org/10.1098/rsta.2021.0032

Villani C (2009) The Wasserstein distances. Springer, Berlin, Heidelberg, pp 93–111. https://doi.org/10.1007/978-3-540-71050-9_6

Weigel AP (2011) Ensemble forecasts, Chap 8. Wiley, pp 141–166. https://doi.org/10.1002/9781119960003.ch8

Chapter 7
A Note on Stochastic Wave Action

Etienne Mémin, Louis Marié, and Bertrand Chapron

Abstract We explore a consistent derivation of the wave action conservation principle for stochastic flows. Beyond providing a proper stochastic wave action principle, this study highlights a stochastic form of the wavefront Hamilton–Jacobi equation. The stochastic framework follows the modeling under location uncertainty paradigm. Within this framework, the usual slow component of the underlying current and the fast wavy component are accordingly decomposed in terms of smooth-in-time resolved component and unresolved highly oscillating random field. The slow current is expressed as a two-dimensional evolution equation, potentially incorporating strong noise. The fast wavy components are associated with the random current but include their own noise contributions as well.

Keywords Wave current separation · Stochastic modeling · Fluid dynamics · Subgrid parametrization · Stochastic transport

7.1 Introduction

First introduced by Sturrock (1962), the conservation of the wave action, defined as the ratio of the wave energy to its intrinsic frequency (also referred to as Doppler shifted frequency), reads

$$\partial_t\Big(\frac{|\widetilde{\eta}|^2}{\overline{\lambda}}\Big) + \nabla_\mathbf{x} \cdot \Big(\frac{|\widetilde{\eta}|^2}{\overline{\lambda}}\big(\overline{\boldsymbol{u}} + \boldsymbol{C}_g\big)\Big) = 0, \tag{7.1}$$

E. Mémin (✉)
Odyssey, Centre Inria de l'Université de Rennes, IRMAR, Rennes, France
e-mail: etienne.memin@inria.fr

L. Marié · B. Chapron
LOPS Ifremer, Plouzané, France

B. Chapron
Odyssey, LOPS Ifremer, Centre Inria de l'Université de Rennes, Plouzané, France

© The Author(s) 2026

B. Chapron et al. (eds.), *Stochastic Transport in Upper Ocean Dynamics IV*, Mathematics of Planet Earth 15, https://doi.org/10.1007/978-3-032-12749-5_7

with $\overline{u}$ and C_g denoting the horizontal slow current and the wave group velocity, respectively. This equation, which has the form of a density evolution, was derived for slowly varying, infinitesimal waves by Whitham (1965), Hayes (1970), and Bretherton (1997), and later generalized through the generalized Lagrangian mean (GLM) framework (Andrews and Mcintyre 1978a, b). Notably, these derivations assume a slowly varying current, which is also taken to be irrotational, as in classical water wave theory.

In reality, however, these conditions are not fully met, and deviations from this conservation principle can be expected. Deviation from potential flow has been investigated in White (1999), as well as the effect of vertically sheared currents (Klein et al. 2022; Quinn et al. 2017; Voronovich 1976). In both cases, a correction term related to the vertical shear appears as a forcing term in the wave action conservation principle. The role of unresolved scales has been less studied. The influence of eddy viscosity has been numerically characterized in Klein et al. (2022), but no explicit contribution of small scales effect on the wave action conservation was derived. A wave action conservation involving a directional diffusion was obtained by Villas Bôas and Young (2020) following the diffusion approximation of McComas and Bretherton (1977) and Kafiabad et al. (2019) to take into account the scattering of internal gravity waves by mesoscale ocean turbulence. However, this dissipation term is not accompanied by a balanced fluctuation term. A numerical study on ray tracing for a 2D stochastic representation of the current was conducted in Resseguier et al. (2024), where the wave action conservation was conjectured to take the form of a stochastic transport equation, as derived in Bauer et al. (2020), Mémin (2014), and Resseguier et al. (2017a). In the present study, the form of this equation is rigorously derived and extended to vertically sheared current as well as to more general noises acting both in the current and wave components.

A stochastic formulation was also explored to incorporate topographic inhomogeneities of the sea bottom (Klyatskin and Koshel 2015), though in a different context—namely, the emergence of anomalous large waves. A geometric mechanics approach, extending the Generalized Lagrangian Mean (GLM) framework, was proposed in Holm et al. (2024) by incorporating noise in the current, with an arbitrary choice of setting decorrelated noises in both the transport equation and the phase evolution.

To our knowledge, no complete and precise formulation of the wave action principle has been provided in the presence of unresolved scales in both the current and the waves.

In this short note, we aim to explore a rigorous derivation of the wave action conservation principle in a setting where both the current and the waves exhibit a smooth-in-time component alongside a highly oscillatory component represented by a random field. To that end, we express the wave action derivation within a stochastic modeling framework derived from conservation principles (Mémin 2014). Following this modeling approach, we show that the conservation equation retains the same form as in the deterministic setting but includes additional rapidly oscillating forcing terms that capture nonlinear wave–wave and wave–current interactions. Formally, we obtain a random phase evolution, for a current $\overline{u}$ at the slow surface, $\overline{\eta}$, perturbed by a random velocity component—temporarily and informally denoted as $\boldsymbol{u}'$–which reads

$$\partial_t \mathcal{S} + (\overline{\boldsymbol{u}} + \boldsymbol{C}_g + \boldsymbol{u}') \cdot \nabla \mathcal{S} = \mathcal{J}\big(\partial_z(\overline{\boldsymbol{u}} + \boldsymbol{u}')\big) + \frac{1}{2}\sin(\mathcal{S})\mathcal{G}\big(\boldsymbol{u}' \cdot \nabla_{\mathbf{x}}\overline{\eta},\, \boldsymbol{u}' \cdot \nabla_{\mathbf{x}}\overline{\boldsymbol{u}},\, \boldsymbol{f}_o \times \boldsymbol{u}'\big). \quad (7.2)$$

This equation is coupled to a random wave action of the form

$$\partial_t\Big(\frac{|\widetilde{\eta}|^2}{\overline{\lambda}}\Big) + \nabla_{\mathbf{x}} \cdot \Big(\frac{|\widetilde{\eta}|^2}{\overline{\lambda}}(\overline{\boldsymbol{u}} + \boldsymbol{C}_g + \boldsymbol{u}')\Big) = \mathcal{F}\big(\nabla_{\mathbf{x}}\boldsymbol{u}',\, \nabla_{\mathbf{x}} \cdot \boldsymbol{u}'\big) - \cos(\mathcal{S})\mathcal{G}\big(\boldsymbol{u}' \cdot \nabla_{\mathbf{x}}\overline{\eta},\, \boldsymbol{u}' \cdot \nabla_{\mathbf{x}}\overline{\boldsymbol{u}},\, \boldsymbol{f}_o \times \boldsymbol{u}'\big).$$
$$(7.3)$$

Here, $\boldsymbol{f}_o$ is the Coriolis frequency directed along the vertical axis. The functions $\mathcal{J}$, $\mathcal{F}$, $\mathcal{G}$ as well as a precise form of the noise term associated to small-scale velocity component will be described in more detail later. The forcing $\mathcal{J}$ corresponds to contributions from rotational currents and vertically sheared flows. The term $\mathcal{F}$ is related to compressibility and deformation effects of the horizontal noise components. Additionally, $\mathcal{G}$ involves advection by the noise of the slow surface and current, together with a Coriolis correction of the noise. The modulation of the last term in the phase equation and in the wave action conservation arises because the noise term is kept general, with a nonlinear and intricate dependence on $\widetilde{\eta}$. In the case of a linear dependence, the phase modulation disappears.

The stochastic framework we rely on follows the location uncertainty (LU) approach (Mémin 2014), which is defined through a stochastic extension of the Reynolds transport theorem. This framework has the advantage of enabling formal derivations of stochastic flow dynamics by replicating, in this stochastic context, the key steps of deterministic derivations based on physical conservation laws. In particular, it preserves classical scaling and approximation strategies used to derive geophysical reduced-order models, while introducing an additional degree of freedom through the noise component.

This paradigm has been extensively applied to geophysical flows (Bauer et al. 2020; Brecht et al. 2021; Chandramouli et al. 2018, 2020; Chapron et al. 2018; Li et al. 2023; Tucciarone et al. 2023; Resseguier et al. 2017b,c) and wave models (Debussche et al. 2024b; Dinvay and Mémin 2022; Mémin et al. 2022), demonstrating its versatility—from idealized models to operational ocean simulation. These applications span from simplified theoretical models to realistic oceanic setups and large-eddy simulations.

The incorporation of noise terms, along with stochastic calculus rules, allows us to recover many terms that are typically introduced semi-empirically in large-scale flow models. Before deriving the stochastic wave action conservation law, we briefly review the LU framework.

7.2 Location Uncertainty

LU modeling relies on the decomposition of the flow into a smooth-in-time large-scale velocity and a highly oscillatory stochastic component:

$$\mathrm{d}X_t = v(X_t, t), \mathrm{d}t + \sigma_t(X_t, t), \mathrm{d}W_t. \tag{7.4}$$

Here, X_t is the Lagrangian displacement in the flow domain $\mathcal{D} \subset \mathbb{R}^d$ ($d = 2, 3$), with v the large-scale Eulerian velocity and $\sigma \mathrm{d}W_t$ the unresolved component. The unresolved component, also referred to as noise in the following, is assumed to be uncorrelated in time at the large-scale characteristic time. It is built from a functional (cylindrical) Wiener process via the integral operator σ_t, which imposes spatial correlations on the noise component. The correlation operator, σ_t, is defined from a positive definite, symmetric kernel $\widehat{\sigma}$, which we assume to be bounded in space and time:

$$(\sigma_t f)(x) := \int_{\mathcal{S}} \widehat{\sigma}(x, y, t) f(y) \, \mathrm{d}y, \; f \in H.$$

Due to these assumptions, the operator σ_t is Hilbert–Schmidt, symmetric, and positive definite. From the spectral theorem, it admits the following spectral decomposition:

$$\sigma_t W_t(x) = \sum_{i \in \mathbb{N}} \beta_t^i \varphi_i(x),$$

where $(\beta_i)_{i \in \mathbb{N}}$ is a sequence of independent standard Brownian motions, and $(\varphi_i)_{i \in \mathbb{N}}$ are the correlation operator eigenfunctions scaled by their eigenvalues, $\lambda_i^{1/2}$. Note that the correlation operator is not necessarily deterministic, and the noise can be extended to non-centered terms using Girsanov theorem, provided additional integrability conditions are satisfied (Li et al. 2023).

For simplicity, in the following we assume a divergence-free noise:

$$\nabla \cdot \hat{\sigma}(x, y, t) = 0, \; x, y \in \mathcal{S}, \; t \geq 0,$$

although a general compressible noise could be considered (Bauer et al. 2020).

Associated with σ_t, we define the (matrix) tensor a

$$a_{ij}(x, t) = \int_{\Omega} \check{\sigma}^{ik}(x, x', t) \check{\sigma}^{kj}(x', x, t) \mathrm{d}x' = \sum_{k=0}^{\infty} \varphi_k(x, t) \varphi_k^T(x, t). \tag{7.5}$$

The time integral of this quantity corresponds in the general case to the quadratic variation of the noise, which is a finite variation process when the correlation operator, σ_t, is random. If the operator is deterministic, it can be understood as the one point covariance tensor:

$$a_{ij}(x,t)\mathrm{d}t = \mathbb{E}\left((\boldsymbol{\sigma}_t \mathrm{d}\boldsymbol{W}_t)^i(x)(\boldsymbol{\sigma}_t\mathrm{d}\boldsymbol{W}_t)^j(x)\right), \tag{7.6}$$

and for that reason we refer to $\boldsymbol{a}$ as the variance tensor by abuse of language.

At this point, an important remark can be made. The unresolved noise term relies on the assumption of time decorrelation. This corresponds obviously to an idealized simplification of the interactions between large- and small-scale structures in real turbulent flows. However, as described below, this assumption, which enables differential formulations through stochastic calculus, rigorously introduces additional advection and diffusion terms that are correlated in time. These terms generalize the wave-induced Stokes drift and the Boussinesq eddy viscosity assumption. Notably, in the latter case, its introduction and derivation are rigorously justified. As for the former, it aligns with Generalized Lagrangian Mean (GLM) theory (Andrews and Mcintyre 1978b), another rigorous framework. This noise-induced drift arises more generally when the small-scale component is (statistically) inhomogeneous in space (Bauer et al. 2020). The relation of the Itô–Stokes drift and the Stokes drift for linear waves in the case of deep-water dispersion relationship is described in Appendix A.

Furthermore, the LU setting, by relying on a transport noise with an induced balanced diffusion term, incorporates directly the fluctuation dissipation relationship of statistical physics within this large-scale representation.

Stochastic Reynolds transport theorem Given the form of the stochastic flow (7.4), the derivation of stochastic dynamics within the LU setting relies on a stochastic representation of the Reynolds transport theorem (SRTT) (Bauer et al. 2020; Mémin 2014; Resseguier et al. 2017d). This theorem, which is the cornerstone of fluid dynamics models derivation, provides the rate of change of a random scalar q within a volume $V(t)$, transported by the stochastic flow (7.4). For incompressible unresolved flows, (i.e., $\nabla \cdot \boldsymbol{\sigma}_t = 0$), the SRTT reads

$$\mathrm{d}\left(\int_{V(t)} q(x,t)\,\mathrm{d}x\right) = \int_{V(t)}\left(\mathbb{D}_t q + q\nabla\cdot(\boldsymbol{v}-\boldsymbol{v}_a)\mathrm{d}t\right)\mathrm{d}x, \tag{7.7a}$$

$$\mathbb{D}_t q = \mathrm{d}_t q + (\boldsymbol{v}-\boldsymbol{v}_a)\cdot\nabla q\,\mathrm{d}t + \boldsymbol{\sigma}_t\mathrm{d}\boldsymbol{W}_t\cdot\nabla q - \frac{1}{2}\nabla\cdot(\boldsymbol{a}\nabla q)\,\mathrm{d}t, \tag{7.7b}$$

where $\mathrm{d}_t q(x,t) = q(x,t+\mathrm{d}t) - q(x,t)$ stands for the forward time-increment of q at a fixed point x. The operator $\mathbb{D}_t$ is introduced as the stochastic transport operator, and plays the role of the material derivative. Recall that $\boldsymbol{v}$ is the large-scale velocity used in (7.4) and $\boldsymbol{a}$ is defined in (7.5).

The advection process, when enriched by stochasticity, undergoes a correction that reshapes its dynamics. The effective transport component acquires a drift, $\boldsymbol{v}_a = -\frac{1}{2}\nabla\cdot\boldsymbol{a}$, termed *Itô–Stokes drift* (ISD) (Bauer et al. 2020). This drift encapsulates the subtle statistical imprint of small-scale inhomogeneities on the evolution of the transported scalar field. In the case of homogeneous noise, where the variance tensor $\boldsymbol{a}$ is spatially uniform, the drift vanishes entirely. As noted in Bauer et al. (2020), this drift elegantly generalizes the Stokes drift—characteristic of wave orbital motion—extending its scope to stochastic transport dynamics.

The third term in (7.7b) aligns with physical intuition, representing the advection of the scalar q by the unresolved, fluctuating velocity field. Through the quadratic variation of this term, energy is continuously backscattered into the tracer field, acting as a subtle yet persistent mechanism of energy transfer. However, this energy intake is counterbalanced by the loss incurred in the final diffusion term. The latter, reminiscent of a generalized Boussinesq eddy viscosity, casts the variance tensor, a, in the role of an effective viscosity—with units of m^2/s. This precise balance of energy gain and dissipation in the LU setting provides a direct illustration of the fluctuation–dissipation theorem, where the stochastic input is intrinsically offset by dissipative mechanisms. In the stochastic transport operator expression (7.7b), the stochastic integral is understood in the Itô sense. For the following analysis, it is helpful to transition to the Stratonovich interpretation of the integral.

LU Stratonovich expression The Stratonovich integral follows the same product rule as deterministic calculus, making it especially practical for many applications. However, unlike the Itô integral, it is not a martingale and is not of zero mean. Despite this distinction, the two stochastic integrals are equivalent in the sense that one can transition between them, provided the underlying stochastic processes are sufficiently regular for the Stratonovich integral to be well defined. Notably, the Stratonovich integral requires stronger regularity assumptions than the Itô integral. As detailed in Bauer et al. (2020), the stochastic flow (7.4) and the associated transport operator can be reformulated under the following equivalent expressions (subject to some regularity conditions on the transported quantity):

$$d\mathbf{x}_t = \boldsymbol{v}dt - \frac{1}{2}\nabla \cdot \boldsymbol{a}dt + \boldsymbol{\sigma}_t \circ d\boldsymbol{W}_t, \tag{7.8}$$

where the stochastic integral is expressed in the Stratonovich sense, denoted by the $\circ$ symbol. As outlined previously, in this formulation, the noise term is no longer of zero expectation. Let us note that accordingly, the Itô–Stokes drift can be interpreted as a mean drift emerging as a difference between the Itô integral and the Stratonovich integral. In appendix, we show its relation to the classical Stokes drift in the case of a monochromatic wave with deep-water dispersion relationship.

With this representation of the stochastic flow, the transport operator corresponds to the Lagrangian (material) derivative associated with the effective transport velocity (Bauer et al. 2020):

$$\mathbb{D}_t^\circ q = d_t \circ q + (\boldsymbol{v} - \boldsymbol{v_a}) \cdot \nabla q \, dt + \boldsymbol{\sigma}_t \circ d\boldsymbol{W}_t \cdot \nabla q, \tag{7.9}$$

where $d_t \circ q \triangleq \theta\big(x, t + (dt/2)\big) - \theta\big(x, t - (dt/2)\big)$ stands for the centered time-increment.

The mathematical study of stochastic models involving advection noise terms, often referred to as transport noise, has been the subject of intensive research works in recent years. The stochastic representation of flow dynamics and the rigorous analysis of these models are actively being investigated by multiple research groups.

For recent contributions, see, for example, Agresti et al. (2022), Brzeźniak and Slavík (2021), Carigi and Luongo (2023), Crisan et al. (2019), Debussche and Pappalettera (2023), Debussche et al. (2023), Flandoli and Luo (2021), Flandoli and Pappalettera (2021), Flandoli and Russo (2023),Flandoli et al. (2021), Galeati (2020), Galeati and Luo (2023), Goodair et al. (2022), Lang et al. (2023), and references therein. The LU stochastic Navier–Stokes equations have been shown to be well posed in 2D and admit a weak (martingale) solution in 3D (Debussche et al. 2023). This analysis has been recently extended to the primitive equations (Debussche et al. 2024a). In some sense, the analytical properties of the deterministic models continue to hold in this stochastic setting. Importantly, for these LU models, it has been proven that as stochastic noise diminishes, solutions converge to the deterministic case, unlike ad hoc multiplicative noise models that may exhibit significant deviations from the deterministic system for small noise levels as experimentally shown in Chapron et al. (2018).

7.3 Stochastic Representation of Incompressible Euler Equations with Free Surface

The transport operator and the stochastic representation of the Reynolds transport theorem can be used almost verbatim as in the classical deterministic setting to derive stochastic versions of any fluid dynamics model. The expression of the Euler equations in a rotating frame, when split in horizontal and vertical components $v = (u, w)^T$, reads as the result of such derivation as

$$\mathbb{D}^{\circ}_{t} u + f e_z \times (u \mathrm{d}t + \boldsymbol{\sigma} \circ \mathrm{d}W^{h}_{t}) = -\nabla_h p \mathrm{d}t - \nabla_h \mathrm{d}p^{\sigma}_{t}, \tag{7.10a}$$

$$\mathbb{D}^{\circ}_{t} w = -\partial_z(p \mathrm{d}t + \mathrm{d}p^{\sigma}_{t}) + g \mathrm{d}t, \tag{7.10b}$$

$$\mathrm{d}_t \circ \eta + \left((u_{|\eta} - u_{a|\eta})\mathrm{d}t + \boldsymbol{\sigma} \circ \mathrm{d}W^{h}_{t|\eta}\right) \cdot \nabla_h \eta = (w_{|\eta} - w_{a|\eta})\mathrm{d}t + \boldsymbol{\sigma} \circ \mathrm{d}W^{z}_{t|\eta}, \tag{7.10c}$$

$$\nabla \cdot v = 0, \quad \nabla \cdot v_a = 0, \quad \nabla \cdot \boldsymbol{\sigma} \circ \mathrm{d}W_t = 0, \tag{7.10d}$$

$$p_{|\eta} = 0 \quad p^{\sigma}_{|\eta} = 0. \tag{7.10e}$$

In the above equation, the superscript h refers to the horizontal vector components, while the superscript z indicates the vertical component. The variable η represents the free surface elevation relative to a resting reference and f is the Coriolis frequency. The pressure, divided by the (constant) volumetric mass, is split into two components: a continuous part p and a white noise part p^{σ}. As usual, g denotes the gravitational acceleration constant. The operator ∇_h stands for the horizontal components of the gradient. The noise is assumed to have a scaling such that the divergence of Itô–Stokes drift is negligible, implying noise with diffusivity smaller than $\mathcal{O}(1)$ (Li et al. 2023).

To separate the flow into fast and slow components, we associate the slow component with the underlying mean current and the fast variables with gravity waves. The slow component, denoted by $\overline{u}$, is assumed to depend slowly on time and space and takes the form: $\overline{u}(\epsilon x, \epsilon z, \epsilon t)$, where $\epsilon \ll 1$ reflects the slow variation scale. The slow horizontal spatial and temporal variables are represented by $\mathbf{x}$, τ, respectively, with $\mathbf{x} = \epsilon x$ and $\tau = \epsilon t$.

The wave components, denoted by $\widetilde{v}$, depend on the fast variables (x, z, t). The amplitudes of the slow and fast variables are represented by $\overline{A}$ and $\widetilde{A}$.

The differentiable and martingale components of the flow decompose as

$$u = \overline{A}\,\overline{u}(\epsilon x, \epsilon z, \epsilon t) + \widetilde{A}\widetilde{u}(x, z, t), \tag{7.11a}$$

$$w = \overline{A}\epsilon\,\overline{w}(\epsilon x, z, \epsilon t) + \widetilde{A}\widetilde{w}(x, z, t), \tag{7.11b}$$

$$\sigma \circ \mathrm{d}W_t^h = \overline{A}^{\frac{1}{2}}\,\overline{\sigma}_t \circ \mathrm{d}W_t^h(\epsilon x, \epsilon z) + \widetilde{A}^{\frac{1}{2}}\epsilon^{\frac{1}{2}}\widetilde{\sigma}_t \circ \mathrm{d}W_t^h(x, z), \tag{7.11c}$$

$$\sigma \circ \mathrm{d}W_t^z = \overline{A}^{\frac{1}{2}}\,\epsilon^{\frac{1}{2}}\overline{\sigma}_t \circ \mathrm{d}W_t^z(\epsilon x, z) + \widetilde{A}^{\frac{1}{2}}\epsilon^{\frac{1}{2}}\widetilde{\sigma}_t \circ \mathrm{d}W_t^z(x, z). \tag{7.11d}$$

The horizontal slow components are assumed to vary smoothly in space and depth, while the vertical slow components which are of order ϵ may exhibit rapid variation along the vertical direction. As shown below, this behavior arises directly from the incompressibility constraint. A similar decomposition into slow and fast variables is applied to all state variables, including the correlation tensor of the noise components. For the pressure variable, similar to the vertical velocity, the slow component may vary rapidly along z, taking the form $\overline{p}(\epsilon x, z, \epsilon t))$. The system (7.10) is thus separated into slow and fast components corresponding to the respective amplitudes $\overline{A}$ and $\widetilde{A}$.

At this point, it is important to highlight that the unresolved noise terms affecting both the slow current component and the waves rely on the assumption of time decorrelation. This assumption is, of course, an idealized simplification of the interactions between currents and waves. However, as previously described in Section 2, it enables differential formulations through stochastic calculus while rigorously introducing additional advection and diffusion terms that remain correlated in time. These terms include effects similar to wave-induced Stokes drift and eddy-induced diffusion. Additionally, the decorrelation assumption directly incorporates the fluctuation–dissipation relationship from statistical physics within this large-scale representation.

In the decomposition above, the noise term on the slow component can be associated with submesoscale effects or significant deviations from the potential current underlying wave motions. The noise on the fast wave component can be interpreted as arising from faster wave components or transient waves that violate the small-amplitude or small-slope assumptions.

7.3.1 *Dynamics of the Slow Variables*

For such a decomposition, in order to derive a dynamics for the slow current, we retain the terms in $\overline{A}$ and $\overline{A}^{1/2}$, and neglect terms in ϵ^2. It is assumed that ϵ scales as the Coriolis frequency $\epsilon \sim f$, and we observe that due to incompressibility, the slow vertical components $\overline{w}$ and $\overline{\sigma} \circ \mathrm{d}\boldsymbol{W}^z_{\tau|_\eta}$ are of order ϵ and $\epsilon^{3/2}$, respectively (accounting for the rescaling by $\tau = \epsilon t$ of the Brownian). This makes the term $\overline{w}\partial_z \overline{\boldsymbol{u}}$ in the horizontal momentum of order ϵ^2, while the corresponding noise $\overline{\sigma} \circ \mathrm{d}\boldsymbol{W}^z_{\tau|_\eta}\partial_z\overline{\boldsymbol{u}}$ is of order $\epsilon^{5/2}$. The slow Itô–Stokes drift is also of order ϵ^2 (and its vertical component of order ϵ^3). Let us note that as the ISD is divergence-free. The hydrostatic balance emerges immediately after neglecting the vertical velocity acceleration associated with a scaling of ϵ^3. This balance can be safely split into its differentiable finite variation and martingale components. This leads to the following relations:

$$\overline{p} = g(z - \bar\eta) \text{ and } \overline{p}^\sigma_t = 0 \tag{7.12}$$

with a zero surface condition for the martingale pressure. The pressure at the interface is assumed to be zero (in perfect equilibrium with atmospheric pressure). This constraint leads to

$$p_{|_\eta} = \overline{p}_{|_\eta} + \widetilde{p}_{|_\eta} = 0. \tag{7.13}$$

Linearizing $\overline{p}_{|_\eta}$ around $\bar\eta$ and neglecting higher-order terms leads to

$$p_{|_\eta} = \overline{p}_{|_{\bar\eta}} + \widetilde\eta\partial_z\overline{p}_{|_{\bar\eta}} + \widetilde{p}_{|_{\bar\eta}} = 0. \tag{7.14}$$

This yields for the slow pressure at the interface $\overline{p}_{|_{\bar\eta}} = 0$ and for the fast pressure component

$$\widetilde{p}_{|_{\bar\eta}} - \widetilde\eta g = 0. \tag{7.15}$$

For the martingale pressure, we obtain in a similar way $\overline{p}^\sigma_{|_{\bar\eta}} = 0$ and $\widetilde{p}^\sigma_{|_{\bar\eta}} = 0$, since $\overline{p}^\sigma = 0$. The surface velocity is linearized around the slow surface component, $\bar\eta$, such that $\overline{\boldsymbol{u}}_{|_\eta} = \overline{\boldsymbol{u}}_{|_{\bar\eta}} + \epsilon\widetilde\eta\partial_z\overline{\boldsymbol{u}}_{|_{\bar\eta}}$ with a similar terms for the vertical components and for the noise surface variables.

Gathering all these elements and neglecting the remaining terms of order ϵ, the dynamics of the slow variables is finally described by the following system:

$$\mathrm{d}_\tau \circ \overline{\boldsymbol{u}} + \left(\overline{\boldsymbol{u}}\mathrm{d}\tau + \overline{\sigma} \circ \mathrm{d}\boldsymbol{W}^h_\tau\right) \cdot \nabla_\mathbf{x}\overline{\boldsymbol{u}} + e_z \times (\overline{\boldsymbol{u}}\mathrm{d}\tau + \overline{\sigma} \circ \mathrm{d}\boldsymbol{W}^h_\tau) + g\nabla_\mathbf{x}\bar\eta\mathrm{d}\tau = 0 \tag{7.16a}$$

$$\mathrm{d}_\tau\bar\eta + (\overline{\boldsymbol{u}}_{|_{\bar\eta}}\mathrm{d}\tau + \overline{\sigma}\mathrm{d}\boldsymbol{W}^h_{\tau|_{\bar\eta}}) \cdot \nabla_\mathbf{x}\bar\eta = \overline{w}_{|_{\bar\eta}}\mathrm{d}\tau + \overline{\sigma}\mathrm{d}\boldsymbol{W}^z_{\tau|_{\bar\eta}}, \tag{7.16b}$$

$$\nabla_\mathbf{x} \cdot \overline{\boldsymbol{u}} - \partial_z\overline{w} = 0, \quad \nabla_\mathbf{x} \cdot \overline{\sigma} \circ \mathrm{d}\boldsymbol{W}^h_\tau + \partial_z\overline{\sigma} \circ \mathrm{d}\boldsymbol{W}^z_\tau = 0, \tag{7.16c}$$

where the noise correlation tensor, $\bar{\sigma}$, has been rescaled by $\sqrt{\epsilon/\overline{A}}$. This system remains close to what is obtained in the deterministic context. Nevertheless, this system introduces a new component of the advection of the slow current by the noise variable on the slow time scale. This term accounts for the unresolved effects of the fast components on the slow mean current. Its contribution acts on the slow time scale and increases the variance accordingly. Notably, there is no modified advection due to the noise component in the slow dynamics. With the scaling applied, the slow component of the noise can be understood as being quasi-homogeneous in space, leading to a negligible Itô–Stokes drift (ISD).

7.3.2 Dynamics of the Waves Variable

Considering now the fast variables, and keeping terms smaller than $\overline{A}$ and greater than $\widetilde{A}^2$, from (7.10) we have the horizontal momentum equations:

$$d_t\tilde{u} + ((\bar{v} - \epsilon\overline{v_a})dt + \bar{\sigma}_t \circ dW_t) \cdot \nabla\tilde{u} + \epsilon\big((\tilde{v} - \epsilon\tilde{v_a})dt + \tilde{\sigma}_t \circ dW_t\big) \cdot \nabla\bar{u}+$$
$$\epsilon e_z \times (\tilde{u}dt + \tilde{\sigma}_t \circ dW_t^h) = -\nabla_h\tilde{p}dt - \nabla_h d\tilde{p}_t^{\sigma}, \quad (7.17)$$

where the correlation operator $\tilde{\sigma}$ has been rescaled by $\sqrt{\epsilon}/\sqrt{\widetilde{A}}$.

Linearizing $\bar{u} - \epsilon\overline{u_a}$ at $\bar{\eta}$ gives $\bar{u} - \epsilon\overline{u_a} = (\bar{u}|_{\bar{\eta}} - \epsilon\overline{u_a}|_{\bar{\eta}})dt + \epsilon(z - \bar{\eta})\partial_z(\bar{u}|_{\bar{\eta}} - \epsilon\overline{u_a}|_{\bar{\eta}})$, proceeding in the same way for $\bar{\sigma} \circ dW_t^h$, and denoting $\mathbb{D}^{\circ}_{t|_{\bar{\eta}}} = d_t \circ + \big((\bar{u}|_{\bar{\eta}} - \epsilon\overline{u_a}|_{\bar{\eta}})dt + \bar{\sigma}|_{\bar{\eta}} \circ dW_t^h\big) \cdot \nabla_h$, the horizontal momentum can be written (dropping terms in ϵ^2) as

$$\mathbb{D}^{\circ}_{t|_{\bar{\eta}}}\tilde{u} + \epsilon\Big([(z - \bar{\eta})\partial_z((\bar{u}|_{\bar{\eta}} - \epsilon\overline{u_a}|_{\bar{\eta}})dt + \bar{\sigma}|_{\bar{\eta}} \circ dW_t^h)] \cdot \nabla_h\tilde{u} + ((\tilde{u} - \epsilon\tilde{u_a})dt + \tilde{\sigma} \circ dW_t^h) \cdot \nabla_x\bar{u}|_{\bar{\eta}}+$$
$$((\bar{w} - \epsilon\bar{w}_a)dt + \bar{\sigma}d \circ W_t^z)\partial_z\tilde{u} + ((\tilde{w} - \epsilon\tilde{w}_a)dt + \tilde{\sigma}d \circ W_t^z)\partial_z\bar{u}|_{\bar{\eta}} + e_z \times (\tilde{u}dt + \tilde{\sigma} \circ dW_t^h)\Big) = -\nabla_h\tilde{p}dt - \nabla_h d\tilde{p}_t^{\sigma}, \quad (7.18)$$

and the vertical momentum equations are given by

$$\mathbb{D}^{\circ}_{t|_{\bar{\eta}}}\tilde{w} + \epsilon\Big([(z - \bar{\eta})\partial_z((\bar{u}|_{\bar{\eta}} - \epsilon\overline{u_a}|_{\bar{\eta}})dt + \bar{\sigma}|_{\bar{\eta}} \circ dW_t^h] \cdot \nabla_h\tilde{w} + \epsilon((\tilde{u} - \epsilon\tilde{u_a})dt + \tilde{\sigma} \circ dW_t^h) \cdot \nabla_x\bar{w}+$$
$$((\bar{w} - \epsilon\bar{w}_a)dt + \bar{\sigma}d \circ W_t^z)\partial_z\tilde{w} + ((\tilde{w} - \epsilon\tilde{w}_a)dt + \tilde{\sigma}d \circ W_t^z)\partial_z\bar{w})\Big) = -\partial_z\tilde{p}dt - \partial_z d\tilde{p}_t^{\sigma}. \quad (7.19)$$

The fast component of the free surface is

$$\mathbb{D}^{\circ}_{t|_{\bar{\eta}}}\tilde{\eta} + \epsilon\Big([((\tilde{u}|_{\bar{\eta}} - \epsilon\overline{u_a}|_{\bar{\eta}}) + \epsilon\tilde{\eta}\partial_z(\bar{u}|_{\bar{\eta}} - \overline{u_a}|_{\bar{\eta}}))dt + (\bar{\sigma}|_{\bar{\eta}} + \epsilon\tilde{\eta}\partial_z\bar{\sigma}|_{\bar{\eta}}) \circ dW_t^h] \cdot \nabla_x\bar{\eta}\Big)$$
$$= (\tilde{w}|_{\bar{\eta}} - \epsilon\tilde{w}_a|_{\bar{\eta}} + \epsilon\tilde{\eta}\partial_z\bar{w}|_{\bar{\eta}})dt + (\tilde{\sigma}|_{\bar{\eta}} + \epsilon\tilde{\eta}\partial_z\tilde{\sigma}|_{\bar{\eta}}) \circ dW_t^z. \quad (7.20)$$

The incompressibility constraint reads

$$\nabla \cdot (\widetilde{\boldsymbol{u}} - \epsilon \widetilde{\boldsymbol{u}}_a) + \partial_z(\widetilde{w} - \epsilon \widetilde{w}_a) = 0, \quad \nabla \cdot \widetilde{\boldsymbol{\sigma}}\mathrm{d}\boldsymbol{W}_t^h + \partial_z\widetilde{\boldsymbol{\sigma}}\mathrm{d}W_t^z = 0, \tag{7.21}$$

and for the fast surface pressure component we have from (7.15):

$$\widetilde{p}_{|\overline{\eta}} - \widetilde{\eta}g = 0, \quad \widetilde{p}^{\sigma}_{|\overline{\eta}} = 0. \tag{7.22}$$

Denoting the total pressure as $\mathrm{d}\pi = p\mathrm{d}t + \mathrm{d}p_t^{\sigma}$, and defining the transport operator without ISD as $D^{\circ}_{t|\overline{\eta}} = \mathrm{d}_t \circ + \left(\overline{\boldsymbol{u}}_{|\overline{\eta}}\mathrm{d}t + \overline{\boldsymbol{\sigma}}_{|\overline{\eta}} \circ \mathrm{d}\boldsymbol{W}_t\right) \cdot \nabla$, we drop the terms in ϵ^2 and rearrange the system to isolate terms of order ϵ on the right-hand side. This leads to the following system governing the fast component:

$$D^{\circ}_{t|\overline{\eta}}\widetilde{\boldsymbol{u}} + \nabla_h\mathrm{d}\widetilde{\pi} = -\epsilon\Big([(z - \overline{\eta})\partial_z(\overline{\boldsymbol{u}}_{|\overline{\eta}}\mathrm{d}t + \overline{\boldsymbol{\sigma}}_{|\overline{\eta}} \circ \mathrm{d}\boldsymbol{W}_t^h) - \overline{\boldsymbol{u}_a}_{|\overline{\eta}}\mathrm{d}t] \cdot \nabla_h\widetilde{\boldsymbol{u}}$$
$$+ \left(\widetilde{\boldsymbol{u}}\mathrm{d}t + \widetilde{\boldsymbol{\sigma}} \circ \mathrm{d}\boldsymbol{W}_t^h\right) \cdot \nabla_{\mathbf{x}}\overline{\boldsymbol{u}}_{|\overline{\eta}} + \left(\overline{w}_{|\overline{\eta}}\mathrm{d}t + \overline{\boldsymbol{\sigma}}_{|\overline{\eta}} \circ \mathrm{d}W_t^z\right)\partial_z\widetilde{\boldsymbol{u}}$$
$$+ \left((\widetilde{w}\mathrm{d}t + \widetilde{\boldsymbol{\sigma}} \circ \mathrm{d}W_t^z)\partial_z\overline{\boldsymbol{u}}_{|\overline{\eta}} + e_z \times (\widetilde{\boldsymbol{u}}\mathrm{d}t + \widetilde{\boldsymbol{\sigma}} \circ \mathrm{d}\boldsymbol{W}_t^h)\right)\Big), \tag{7.23a}$$

$$D^{\circ}_{t|\overline{\eta}}\widetilde{w} + \partial_z\mathrm{d}\widetilde{\pi} = -\epsilon\Big([(z - \overline{\eta})\partial_z(\overline{\boldsymbol{u}}_{|\overline{\eta}}\mathrm{d}t + \overline{\boldsymbol{\sigma}}_{|\overline{\eta}} \circ \mathrm{d}\boldsymbol{W}_t^h) - \overline{\boldsymbol{u}_a}_{|\overline{\eta}}\mathrm{d}t] \cdot \nabla_h\widetilde{w}$$
$$+ \left(\overline{w}\mathrm{d}t + \overline{\boldsymbol{\sigma}}\mathrm{d} \circ W_t^z\right)\partial_z\widetilde{w} + \left(\widetilde{w}\mathrm{d}t + \widetilde{\boldsymbol{\sigma}}\mathrm{d} \circ W_t^z\right)\partial_z\overline{w})\Big), \tag{7.23b}$$

$$D^{\circ}_{t|\overline{\eta}}\widetilde{\eta} - \widetilde{w}_{|\overline{\eta}}\mathrm{d}t - \widetilde{\boldsymbol{\sigma}}_{|\overline{\eta}} \circ \mathrm{d}W_t^z = -\epsilon\Big([((\widetilde{\boldsymbol{u}}_{|\overline{\eta}} - \overline{\boldsymbol{u}_a}_{|\overline{\eta}})\mathrm{d}t + \overline{\boldsymbol{\sigma}}_{|\overline{\eta}} \circ \mathrm{d}\boldsymbol{W}_t^h)] \cdot \nabla_{\mathbf{x}}\overline{\eta} - \overline{\boldsymbol{u}_a}_{|\overline{\eta}}\mathrm{d}t \cdot \nabla_h\widetilde{\eta}$$
$$- \widetilde{\eta}\partial_z\overline{w}_{|\overline{\eta}}\mathrm{d}t + \widetilde{w}_a{}_{|\overline{\eta}}\mathrm{d}t - \widetilde{\eta}\partial_z\overline{\boldsymbol{\sigma}}_{|\overline{\eta}} \circ \mathrm{d}W_t^z\Big), \tag{7.23c}$$

$$\nabla \cdot \widetilde{\boldsymbol{u}} + \partial_z\widetilde{w} = 0, \tag{7.23d}$$

$$(\nabla \cdot \widetilde{\boldsymbol{u}}_a + \partial_z\widetilde{w}_a) = 0, \tag{7.23e}$$

$$\nabla \cdot \widetilde{\boldsymbol{\sigma}} \circ \mathrm{d}\boldsymbol{W}_t^h + \partial_z\widetilde{\boldsymbol{\sigma}} \circ \mathrm{d}W_t^z = 0, \; \widetilde{p}_{|\overline{\eta}} - \widetilde{\eta}g = 0, \quad \widetilde{p}^{\sigma}_{|\overline{\eta}} = 0. \tag{7.23f}$$

This system is a linear system with forcing on the right-hand side. The solution depends on the fast and slow scales variables: $x, t, \mathbf{x} = \epsilon x$, and $\tau = \epsilon t$. We assume that the solution of this linear system can be decomposed as

$$\widetilde{g}(x, \mathbf{x}, z, t, \tau) = \widetilde{g}_o(x, \mathbf{x}, z, t, \tau) + \epsilon\widetilde{g}'(x, \mathbf{x}, z, t, \tau), \tag{7.24}$$

where $\widetilde{g}_o(x, \mathbf{x}, z, t, \tau)$ represents a dominant solution of the homogeneous system (neglecting forcing terms), and $\epsilon\widetilde{g}'(x, \mathbf{x}, z, t, \tau)$ is a small correction that accounts for the spatial and temporal variations induced by the forcing terms.

7.3.2.1 Solution of the Fast System at the Lowest Order

The fast system, neglecting the forcing terms, is expressed at the lowest order as follows:

$$D^{\circ}_{t|\overline{\eta}}\widetilde{\boldsymbol{u}}_o + \nabla_h\mathrm{d}\widetilde{\pi}_o = 0 \tag{7.25a}$$

$$D^{\circ}_{t|\overline{\eta}}\widetilde{w}_o + \partial_z\mathrm{d}\widetilde{\pi}_o = 0 \tag{7.25b}$$

$$D^{\circ}_{t|\overline{\eta}}\,\widetilde{\eta}_o - \widetilde{w}_{o|\overline{\eta}}\mathrm{d}t - \widetilde{\sigma}_{o|\overline{\eta}} \circ \mathrm{d}W^z_t = 0, \tag{7.25c}$$

$$\nabla \cdot \widetilde{\boldsymbol{u}}_o + \partial_z \widetilde{w}_o = 0, \quad \nabla \cdot \widetilde{\boldsymbol{\sigma}}_o \circ \mathrm{d}W^h_t + \partial_z \widetilde{\sigma} \circ \mathrm{d}W^z_t = 0, \tag{7.25d}$$

$$\widetilde{p}_{o|\overline{\eta}} - \widetilde{\eta}_o g = 0, \quad \widetilde{p}^{\sigma}_{o|\overline{\eta}} = 0. \tag{7.25e}$$

We seek a solution in the form of traveling waves with a slowly varying envelope:

$$\mathcal{E}(\epsilon \boldsymbol{x}, \epsilon z, \epsilon t)e^{i\mathcal{S}},$$

where $\mathcal{S}(\boldsymbol{x}, t)$ represents the phase of the waves. The wavelength and frequency of the waves are then defined by

$$\mathbf{k} = \nabla \mathcal{S}, \text{ and } \mathrm{d}\omega = -\mathrm{d}_t \mathcal{S}. \tag{7.26}$$

If $\mathcal{S}$ is smooth enough these definitions imply that

$$\mathrm{d}_t \mathbf{k} = \nabla \mathrm{d}\omega, \text{ and } \partial_i k^j = \partial_j k^i. \tag{7.27}$$

The frequency and wavelength are random variables, driven by semi-martingales. The dispersion relationships are derived by substituting the traveling wave solution into the governing system and replacing $\mathrm{d}_t \mathcal{S}$ and $\nabla \mathcal{S}$ by $\mathrm{d}\omega$ and $\mathbf{k}$, respectively. This yields

$$\left(-i\mathrm{d}\omega + i\mathbf{k} \cdot (\overline{\boldsymbol{u}}_{|\overline{\eta}}\mathrm{d}t + \overline{\boldsymbol{\sigma}}_{|\overline{\eta}} \circ \mathrm{d}W_t)\right)\widetilde{\boldsymbol{u}}_o + i\mathbf{k}\mathrm{d}\widetilde{\pi}_o = 0, \tag{7.28a}$$

$$\left(-i\mathrm{d}\omega + i\mathbf{k} \cdot (\overline{\boldsymbol{u}}_{|\overline{\eta}}\mathrm{d}t + \overline{\boldsymbol{\sigma}}_{|\overline{\eta}} \circ \mathrm{d}W_t)\right)\widetilde{w}_o + \partial_z \mathrm{d}\widetilde{\pi}_o = 0, \tag{7.28b}$$

$$\left(-i\mathrm{d}\omega + i\mathbf{k} \cdot (\overline{\boldsymbol{u}}_{|\overline{\eta}}\mathrm{d}t + \overline{\boldsymbol{\sigma}}_{|\overline{\eta}} \circ \mathrm{d}W_t)\right)\widetilde{\eta}_o - \widetilde{w}_{o|\overline{\eta}}\mathrm{d}t - \widetilde{\sigma}_{o|\overline{\eta}} \circ \mathrm{d}W^z_t = 0, \tag{7.28c}$$

$$i\mathbf{k} \cdot \widetilde{\boldsymbol{u}}_o + \partial_z \widetilde{w}_o = 0, \quad i\mathbf{k} \cdot \widetilde{\boldsymbol{\sigma}}_o \circ \mathrm{d}W^h_t + \partial_z \widetilde{\boldsymbol{\sigma}}_o \circ \mathrm{d}W^z_t = 0, \tag{7.28d}$$

$$\widetilde{p}_{o|\overline{\eta}} - \widetilde{\eta}_o g = 0, \quad \widetilde{p}^{\sigma}_{o|\overline{\eta}} = 0. \tag{7.28e}$$

Denote by $\mathrm{d}\lambda = \mathrm{d}\omega - \mathbf{k} \cdot (\overline{\boldsymbol{u}}_{|\overline{\eta}}\mathrm{d}t + \overline{\boldsymbol{\sigma}}_{|\overline{\eta}} \circ \mathrm{d}W_t)$ the shifted frequency rate in a reference frame that is randomly advected by $\overline{\boldsymbol{u}}_{|\overline{\eta}}\mathrm{d}t + \overline{\boldsymbol{\sigma}}_{|\overline{\eta}} \circ \mathrm{d}W_t$. This shifted frequency corresponds to the intrinsic frequency at which a buoy, advected by the slow component of the current, perceives the underlying waves. Using this, we have

$$-i\mathrm{d}\lambda\,\widetilde{\boldsymbol{u}}_o + i\mathbf{k}\mathrm{d}\widetilde{\pi}_o = 0, \tag{7.29a}$$

$$-i\mathrm{d}\lambda\,\widetilde{w}_o + \partial_z \mathrm{d}\widetilde{\pi}_o = 0, \tag{7.29b}$$

$$-i\mathrm{d}\lambda\,\widetilde{\eta}_o - \widetilde{w}_{o|\overline{\eta}}\mathrm{d}t - \widetilde{\sigma}_{o|\overline{\eta}} \circ \mathrm{d}W^z_t = 0, \tag{7.29c}$$

$$i\mathbf{k} \cdot \widetilde{\boldsymbol{u}}_o + \partial_z \widetilde{w}_o = 0, \tag{7.29d}$$

$$i\mathbf{k} \cdot \widetilde{\boldsymbol{\sigma}}_o \circ \mathrm{d}W^h_t + \partial_z \widetilde{\boldsymbol{\sigma}}_o \circ \mathrm{d}W^z_t = 0, \tag{7.29e}$$

$$\widetilde{p}_{o|\overline{\eta}} - \widetilde{\eta}_o g = 0, \tag{7.29f}$$

$$\widetilde{p}^{\sigma}_{o\,|\overline{\eta}} = 0. \tag{7.29g}$$

By combining the vertical derivative of the second equation and the horizontal divergence of the first equation, and incorporating the incompressibility constraint, we arrive at a homogeneous screened Poisson equation for the pressure. The resulting system can be expressed as follows:

$$\left(\partial_{zz} - |\mathbf{k}|^2\right)\mathrm{d}\tilde{\pi}_o = 0, \tag{7.30a}$$

$$- i\mathrm{d}\lambda\,\tilde{w}_o + \partial_z\mathrm{d}\tilde{\pi}_o = 0, \tag{7.30b}$$

$$\mathrm{d}\lambda\,\tilde{\eta}_o - i\tilde{w}_{o|\overline{\eta}}\mathrm{d}t - i\tilde{\boldsymbol{\sigma}}_{o|\overline{\eta}} \circ \mathrm{d}W_t^z = 0, \tag{7.30c}$$

$$i\mathbf{k}\cdot\tilde{\boldsymbol{u}}_o + \partial_z\tilde{w}_o = 0, \quad i\mathbf{k}\cdot\tilde{\boldsymbol{\sigma}}_o \circ \mathrm{d}W_t^h + \partial_z\tilde{\boldsymbol{\sigma}}_o \circ \mathrm{d}W_t^z = 0, \tag{7.30d}$$

$$\tilde{p}_{o|\overline{\eta}} - \tilde{\eta}_o g = 0, \quad \tilde{p}_{o|\overline{\eta}}^{\sigma} = 0. \tag{7.30e}$$

The screened Poisson equation admits a solution of the form of $\tilde{\pi} \sim \exp(|\mathbf{k}|z)$. Separating the finite variation part and martingale part of $\mathrm{d}\lambda = \bar{\lambda}\mathrm{d}t + \lambda^{\sigma}\mathrm{d}W_t$, where the finite variation part $\bar{\lambda}$ incorporates the Itô correction term when transitioning from a Stratonovich integral to an Ito integral, and noting that $\partial_z\tilde{\pi} \sim |\mathbf{k}|\tilde{\pi}$, we use Equations ((7.30b))–((7.30e)), separated into finite variation and martingale components. At the lowest order, this leads to finite variation and martingale dispersion relationships for the fast components:

$$\bar{\lambda}^2 = g|\mathbf{k}| \text{ and } \lambda^{\sigma} = 0. \tag{7.31}$$

The martingale dispersion relation is derived from (7.30b), applied at the surface: $|\mathbf{k}|\tilde{p}_{o|\overline{\eta}}^{\sigma}\mathrm{d}W_t = i\lambda^{\sigma}\mathrm{d}W_t\tilde{w}_{o|\overline{\eta}}$, which cancels due to (7.30e), leading to $\lambda^{\sigma} = 0$. Furthermore, from (7.30c), we obtain

$$\tilde{\boldsymbol{\sigma}}_{o|\overline{\eta}} \circ \mathrm{d}W_t^z = 0 \implies \tilde{\sigma}_{o\,|\overline{\eta}}^{z\cdot} = \tilde{\sigma}_{o\,|\overline{\eta}}^{\cdot z} = 0, \tag{7.32}$$

which yields also that the Itô Stokes drift vertical component cancels:

$$\tilde{w}_{\tilde{a}_{o|\overline{\eta}}} = 0. \tag{7.33}$$

For waves propagating in the direction $\mathbf{k}$, the evolution of the phase function is obtained from the definition of the shifted frequency rate, $d\lambda$, and the phase derivative definition (7.26). The phase evolution is finally described by the first-order Hamilton–Jacobi equation:

$$\mathrm{d}_t\mathcal{S} + (\overline{u}_{|\overline{\eta}}\mathrm{d}t + \overline{\boldsymbol{\sigma}}_{|\overline{\eta}} \circ \mathrm{d}W_t^h)\cdot\nabla\mathcal{S} + \sqrt{g|\nabla\mathcal{S}|}\mathrm{d}t = 0. \tag{7.34}$$

This rigorously derived stochastic formulation constitutes a natural extension of the deterministic Eulerian dynamics for the phase evolution driven by a random current. This phase evolution equation constitutes a justification of the equation intuited in Resseguier et al. (2024). It is also worth outlining that similar to the induced diffusion considered in Villas Bôas and Young (2020) for the wave action, this stochastic setting introduces a directional diffusion associated to the horizontal small-scale component $\overline{\boldsymbol{\sigma}}_{|\widetilde{\eta}} \circ \mathrm{d} \boldsymbol{W}_t^h$ when expressed in Itô representation (see (7.7b)).

Gathering the above results, the expression of the solutions of the slow components can be inferred. Recalling that $\widetilde{\eta} = \mathcal{E} e^{i\mathcal{S}}$, the other slow components of the wave variables are given by

$$
\begin{aligned}
&\partial_z \widetilde{p}_o^{\sigma} = 0 \implies \widetilde{p}_o^{\sigma} = 0, \\
&\widetilde{p}_o = g\mathcal{E} e^{|\mathbf{k}|(z-\widetilde{\eta})} e^{i\mathcal{S}}, \\
&\widetilde{w}_o = -i\frac{|\mathbf{k}|}{\widetilde{\lambda}} \widetilde{p}_o, \\
&\widetilde{\boldsymbol{u}}_o = \frac{\mathbf{k}}{\widetilde{\lambda}} \widetilde{p}_o = \overline{\lambda}\frac{\mathbf{k}}{k}\mathcal{E} e^{|\mathbf{k}|(z-\widetilde{\eta})} e^{i\mathcal{S}}.
\end{aligned}
\tag{7.35}
$$

The incompressibility condition for the noise along with the null vertical component of the noise wave at the surface (7.32) gives

$$
\widetilde{\boldsymbol{\sigma}}_o \mathrm{d} W_t^z = -i\mathbf{k} \cdot \int_z^{\overline{\eta}} \widetilde{\boldsymbol{\sigma}}_o \mathrm{d} W_t^h \mathrm{d} z'.
\tag{7.36}
$$

Additionally, due to the incompressibility of the noise, we also have

$$
\boldsymbol{\nabla}_h \cdot \widetilde{\boldsymbol{\sigma}}_o \mathrm{d} W_t^h = 0.
\tag{7.37}
$$

It can be remarked that the wavelength depends on time and space through the Hamilton–Jacobi equation driving $\mathcal{S}$. The variation of the wavelength vector can be expressed in terms of a forced transport equation.

7.3.2.2 Variation of the Local Wave Length

Going back to the phase definition relations (7.26) we have

$$
\mathrm{d}_t \mathbf{k} + \boldsymbol{\nabla} \mathrm{d}\omega = 0.
\tag{7.38}
$$

The local wavelength is related to the slow component of the flow through

$$
\mathrm{d}\omega = \overline{\lambda}(|\mathbf{k}|)\mathrm{d} t + \mathbf{k} \cdot (\overline{\boldsymbol{u}}_{|\widetilde{\eta}}\mathrm{d} t + \overline{\boldsymbol{\sigma}}_{|\widetilde{\eta}} \circ \mathrm{d} \boldsymbol{W}_t^h),
\tag{7.39}
$$

where the dependence of the intrinsic frequency on the wavelength is specifically indicated. As a consequence we have

$$
\mathrm{d}_t \mathbf{k} + \boldsymbol{\nabla}\big(\overline{\lambda}(|\mathbf{k}|)\mathrm{d} t + \mathbf{k} \cdot (\overline{\boldsymbol{u}}_{|\widetilde{\eta}}\mathrm{d} t + \overline{\boldsymbol{\sigma}}_{|\widetilde{\eta}} \circ \mathrm{d} \boldsymbol{W}_t^h)\big) = 0.
\tag{7.40}
$$

The random wavelength adapts itself to minimize the random scalar potential: $H = \overline{\lambda}(|\mathbf{k}|)\mathrm{d} t + \mathbf{k} \cdot (\overline{\boldsymbol{u}}_{|\widetilde{\eta}}\mathrm{d} t + \overline{\boldsymbol{\sigma}}_{|\widetilde{\eta}} \circ \mathrm{d} \boldsymbol{W}_t)$. Recalling that the phase function, $\mathcal{S}$, within the WKBJ approximation yields $\partial_{x_i} k_j = \partial_{x_j} k_i$ and denoting the group velocity as $\boldsymbol{C}_g = \nabla_{\mathbf{k}}\overline{\lambda} = (\partial_{k_i}\overline{\lambda})_{i=x,y}$, and assuming the wavelength depends on the slow spatial variable, this equation can be written as

$$\mathrm{d}_t\mathbf{k} + \boldsymbol{C}_g \cdot \nabla_{\mathbf{x}}\mathbf{k}\mathrm{d}t + \nabla_{\mathbf{x}}\big(\mathbf{k}\cdot(\overline{\boldsymbol{u}}_{|\overline{\eta}}\mathrm{d}t + \overline{\boldsymbol{\sigma}}_{|\overline{\eta}}\circ\mathrm{d}W_t^h)\big) = 0, \qquad (7.41)$$

which corresponds to the stochastic transport (without the ISD term) by the group velocity, forced by a gradient term depending on the angle between the current velocity and the wavelength vector. Assuming that $\mathbf{k}$ evolves slowly in time under the condition of an appropriate smooth initial condition, we obtain

$$\mathrm{d}_\tau\mathbf{k} + \boldsymbol{C}_g \cdot \nabla_{\mathbf{x}}\mathbf{k}\mathrm{d}\tau + \nabla_{\mathbf{x}}\big(\mathbf{k}\cdot(\overline{\boldsymbol{u}}_{|\overline{\eta}}\mathrm{d}\tau + \sqrt{\tau}\overline{\boldsymbol{\sigma}}_{|\overline{\eta}}\circ\mathrm{d}W_\tau^h)\big) = 0. \qquad (7.42)$$

The gradient forcing term depends on the angle between the wavelength vector and the random current. The wavelength is nonlinearly transported when the wavelength vector is orthogonal to the random current. Note again that formulated in the Itô setting this equation would involve an additional noise-induced diffusion. When expressing the expectation of the wavelength this diffusion cannot be discarded.

7.3.2.3   Solution of the Fast Components at Next Higher Order

Let us now consider the higher-order wave fluctuation terms. In system (7.23), we retain only terms of order at least ϵ in the wave components, as the lower-order terms satisfy the homogeneous equation. For both components, we have $\mathrm{d}_t f = \mathrm{d}_t f(t) + \epsilon\mathrm{d}_t f(\tau)$, $\nabla f = \nabla f + \epsilon\nabla_{\mathbf{x}}f$, and $D_{t_{|\overline{\eta}}}^\circ f := D_{t_{|\overline{\eta}}}^\circ f + \epsilon D_{\tau_{|\overline{\eta}}}^\circ$, where the transport at the slow time scale reads $\epsilon D_{\tau_{|\overline{\eta}}}^\circ = \mathrm{d}_\tau\circ+\big(\overline{\boldsymbol{u}}_{|\overline{\eta}}\mathrm{d}\tau + \sqrt{\tau}\overline{\boldsymbol{\sigma}}_{|\overline{\eta}}\circ\mathrm{d}W_\tau\big)\cdot\nabla$. We also note that $\widetilde{\eta} = e^{i\mathcal{S}} + \epsilon\widetilde{\eta}'$. Substituting these relations gives the following system:

$$\epsilon D_{t_{|\overline{\eta}}}^\circ\widetilde{\boldsymbol{u}}' + \epsilon\nabla_h\mathrm{d}\widetilde{\pi}' = -D_{\tau_{|\overline{\eta}}}^\circ\widetilde{\boldsymbol{u}}_o - \nabla_{\mathbf{x}}\mathrm{d}\widetilde{\pi}_{o_\tau}$$

$$-\Big([(z-\overline{\eta})\partial_z\big(\overline{\boldsymbol{u}}_{|\overline{\eta}}\mathrm{d}\tau + \overline{\boldsymbol{\sigma}}_{|\overline{\eta}}\circ\mathrm{d}W_\tau^h\big) - \overline{\boldsymbol{u_a}}_{|\overline{\eta}}\mathrm{d}\tau]\cdot\nabla_h\widetilde{\boldsymbol{u}}_o$$

$$+\big(\widetilde{\boldsymbol{u}}_o\mathrm{d}\tau + \widetilde{\boldsymbol{\sigma}}_o\circ\mathrm{d}W_\tau^h\big)\cdot\nabla_{\mathbf{x}}\overline{\boldsymbol{u}} + \big(\overline{w}\mathrm{d}\tau + \overline{\boldsymbol{\sigma}}\circ\mathrm{d}W_\tau^z\big)\partial_z\widetilde{\boldsymbol{u}}_o$$

$$+\big(\widetilde{w}_o\mathrm{d}\tau + \widetilde{\boldsymbol{\sigma}}_o\circ\mathrm{d}W_\tau^z\big)\partial_z\overline{\boldsymbol{u}}_o + e_z\times\big(\widetilde{\boldsymbol{u}}_o\mathrm{d}\tau + \widetilde{\boldsymbol{\sigma}}_o\circ\mathrm{d}W_\tau^h\big)\Big), \quad (7.43\mathrm{a})$$

$$\epsilon D_{t_{|\overline{\eta}}}^\circ\widetilde{w}' + \epsilon\partial_z\mathrm{d}\widetilde{\pi}' = -D_{\tau_{|\overline{\eta}}}^\circ\widetilde{w}_o - \Big([(z-\overline{\eta})\partial_z\big(\overline{\boldsymbol{u}}_{|\overline{\eta}}\mathrm{d}\tau + \overline{\boldsymbol{\sigma}}_{|\overline{\eta}}\circ\mathrm{d}W_\tau^h\big) - \overline{\boldsymbol{u_a}}_{|\overline{\eta}}\mathrm{d}\tau]\cdot\nabla_h\widetilde{w}_o$$

$$+\big(\overline{w}\mathrm{d}\tau + \overline{\boldsymbol{\sigma}}\mathrm{d}\circ W_\tau^z\big)\partial_z\widetilde{w}_o + \big(\widetilde{w}_o\mathrm{d}\tau + \widetilde{\boldsymbol{\sigma}}_o\mathrm{d}\circ W_\tau^z\big)\partial_z\overline{w}\big)\Big), \quad (7.43\mathrm{b})$$

$$\epsilon D_{t_{|\overline{\eta}}}^\circ\widetilde{\eta}' - \widetilde{w}'_{|\overline{\eta}}\mathrm{d}\tau = -D_{\tau_{|\overline{\eta}}}^\circ\widetilde{\eta}_o - \Big([((\widetilde{\boldsymbol{u}}_{|\overline{\eta}} - \overline{\boldsymbol{u_a}}_{|\overline{\eta}})\mathrm{d}\tau + \overline{\boldsymbol{\sigma}}_{|\overline{\eta}}\circ\mathrm{d}W_\tau^h)]\cdot\nabla_{\mathbf{x}}\overline{\eta}$$

$$-\overline{\boldsymbol{u_a}}_{|\overline{\eta}}\mathrm{d}\tau\cdot\nabla_h\widetilde{\eta}_o - \widetilde{\eta}\partial_z\overline{w}_{|\overline{\eta}}\mathrm{d}\tau\Big), \qquad (7.43\mathrm{c})$$

$$\epsilon\nabla\cdot\widetilde{\boldsymbol{u}}' + \epsilon\partial_z\widetilde{w}' = -\epsilon\nabla_{\mathbf{x}}\cdot\widetilde{\boldsymbol{u}}_o, \qquad (7.43\mathrm{d})$$

$$\epsilon\nabla\cdot\widetilde{\boldsymbol{\sigma}}'\mathrm{d}W_t^h + \epsilon\partial_z\widetilde{\boldsymbol{\sigma}}'\mathrm{d}W_t^z = -\epsilon\nabla_{\mathbf{x}}\cdot\widetilde{\boldsymbol{\sigma}}_o\mathrm{d}W_t^h, \qquad (7.43\mathrm{e})$$

$$\epsilon\widetilde{p}'_{|\overline{\eta}} - \epsilon\widetilde{\eta}'g = 0, \quad \epsilon\widetilde{p^{\boldsymbol{\sigma}}}'_{|\overline{\eta}} = 0. \qquad (7.43\mathrm{f})$$

We note that the fast wave noise term, $\boldsymbol{\sigma}' \mathrm{d}\boldsymbol{W}_t$, is of the order $\epsilon^{3/2}$. The noise correlation operators $\widetilde{\boldsymbol{\sigma}}, \overline{\boldsymbol{\sigma}}$ and their associated variance tensor are rescaled by factors of $\sqrt{\epsilon}$ and ϵ, respectively.

The system obtained above is a linear system with a right-hand-side forcing term that depends on the slow variations of the current and low-order wave components. It is noteworthy that the left-hand side of this system has exactly the same form as the homogeneous system governing the wavy components at lower order. To ensure the existence of a solution, the system must satisfy a solvability condition, commonly referred to as the Fredholm alternative. To express this solvability condition, we first rewrite the system in terms of traveling waves, as was done previously. The system can be formulated as

$$-i\lambda\widetilde{\boldsymbol{u}}'\mathrm{d}\tau + i\mathbf{k}\mathrm{d}\widetilde{\pi}'_\tau = \mathcal{R}_u \tag{7.44a}$$

$$-i\overline{\lambda}\widetilde{w}'\mathrm{d}\tau + \partial_z\mathrm{d}\widetilde{\pi}'_\tau = \mathcal{R}_w \tag{7.44b}$$

$$-i\overline{\lambda}\widetilde{\eta}'\mathrm{d}\tau - \widetilde{w}'_{|\overline{\eta}}\mathrm{d}\tau = \mathcal{R}_\eta \tag{7.44c}$$

$$i\mathbf{k}\cdot\widetilde{\boldsymbol{u}}' + \partial_z\widetilde{w}' = \mathcal{R}_d \tag{7.44d}$$

$$i\mathbf{k}\cdot\boldsymbol{\sigma}'\mathrm{d}\boldsymbol{W}^h_\tau + \partial_z\widetilde{\boldsymbol{\sigma}}'\mathrm{d}\boldsymbol{W}^z_\tau = \mathcal{R}_{d^\sigma} \tag{7.44e}$$

$$\widetilde{p}_{|\overline{\eta}} - \widetilde{\eta}g = 0, \quad \widetilde{p}^\sigma_{|\overline{\eta}} = 0. \tag{7.44f}$$

Here, the notation $\mathrm{d}\widetilde{\pi}'_\tau$ of the pressure process reflects a rescaling: the finite variation part is rescaled by ϵ, while the martingale part is rescaled $\sqrt{\epsilon}$ at time τ. The right-hand side terms above are given in (7.43).

As in the homogeneous system, the pressure π' alone determines all other variables. Following the same steps as in the homogeneous case, we can express the pressure through an inhomogeneous pressure equation:

$$\left(\partial_{zz} - |\mathbf{k}|^2\right)\mathrm{d}\widetilde{\pi}'_\tau = \mathcal{R}_\pi \text{ with } \mathcal{R}_\pi = \partial_z\mathcal{R}_w + i\mathbf{k}\cdot\mathcal{R}_u + i\overline{\lambda}\mathcal{R}_d\mathrm{d}\tau. \tag{7.45}$$

The upper boundary condition is given by

$$\partial_z\mathrm{d}\widetilde{\pi}'_{\tau|\overline{\eta}} - |\mathbf{k}|\widetilde{p}'_{\tau|\overline{\eta}}\mathrm{d}\tau = \mathcal{D}_\pi \text{ with } \mathcal{D}_\pi = \mathcal{R}_{w|\overline{\eta}} - i\overline{\lambda}\mathcal{R}_\eta, \tag{7.46}$$

using the dispersion relation (7.31). This boundary condition can be expanded in terms of $\widetilde{p}'_{|\overline{\eta}}$ and $\widetilde{p}'^\sigma_{|\overline{\eta}}$ as

$$\partial_z\widetilde{p}'_{|\overline{\eta}}\mathrm{d}\tau + \sqrt{\epsilon}\partial_z\widetilde{p}'^\sigma_{|\overline{\eta}} \circ \mathrm{d}\boldsymbol{W}_\tau - |\mathbf{k}|\widetilde{p}'_{|\overline{\eta}}\mathrm{d}\tau = \mathcal{R}_{w|\overline{\eta}} - i\overline{\lambda}\mathcal{R}_\eta. \tag{7.47}$$

7.3.3 Solvability Condition

According to the Fredholm alternative, a nontrivial solution of the homogeneous system can exist only if the solution of (7.44) lies in the complementary space of the solution space of the homogeneous equation—specifically, in the kernel space of the adjoint of the screened Poisson operator. To establish this, we have for all f, g

$$\int_{-\infty}^{\overline{\eta}} f\big(\partial_{zz} - |k|^2\big) g \, \mathrm{d}z = \big[f\partial_z g - g\partial_z f\big]_{-\infty}^{\overline{\eta}} + \int_{-\infty}^{\overline{\eta}} g\big(\partial_{zz} - |k|^2\big) f \, \mathrm{d}z. \qquad (7.48)$$

For functions of the form $f = e^{|\mathbf{k}|(z-\overline{\eta})}$ and functions g decreasing sufficiently fast at ∞, we derive the condition:

$$\int_{-\infty}^{\overline{\eta}} e^{|\mathbf{k}|(z-\overline{\eta})}\big(\partial_{zz} - |k|^2\big) g \, \mathrm{d}z = \big(\partial_z g - \|\mathbf{k}\| g\big)_{|z=\overline{\eta}},$$

from which we obtain, using (7.45) and (7.46), the condition for π':

$$\int_{-\infty}^{\overline{\eta}} e^{|\mathbf{k}|(z-\overline{\eta})} \mathscr{R}_\pi \, \mathrm{d}z = \mathscr{D}_\pi, \qquad (7.49)$$

where the expression of $\mathscr{R}_\pi$ and $\mathscr{D}_\pi$ are provided in (7.45) and (7.46), respectively.

7.3.3.1 Simplifications

Computing (7.48) and simplifying the expressions on both sides, we obtain an explicit equation governing the slow dynamics of the wave field:

$$\tilde{\eta}_o D^{\circ}_{\tau_{|\overline{\eta}}} \overline{\lambda} + 2\overline{\lambda} D^{\circ}_{\tau_{|\overline{\eta}}} \tilde{\eta}_o - i2\overline{\lambda}\tilde{\eta}_o\big(\overline{\boldsymbol{u_a}}_{|\overline{\eta}} \mathrm{d}\tau \cdot \mathbf{k}\big) + \overline{\lambda}\Big(\big((\overline{\lambda}\tilde{\eta}_o \frac{\mathbf{k}}{|\mathbf{k}|} - \overline{\boldsymbol{u_a}}_{|\overline{\eta}}\big)\mathrm{d}\tau + \overline{\boldsymbol{\sigma}}_{|\overline{\eta}} \circ \mathrm{d}\boldsymbol{W}^h_\tau\big) \cdot \nabla_{\mathbf{x}}\overline{\eta}\Big)$$

$$= \frac{g}{2|\mathbf{k}|}\frac{\tilde{\eta}_o}{\overline{\lambda}} \mathbf{k} \cdot \Big(i|\mathbf{k}|\partial_z\big(2\overline{\boldsymbol{u}}_{|\overline{\eta}}\mathrm{d}\tau + \overline{\boldsymbol{\sigma}}_{|\overline{\eta}} \circ \mathrm{d}\boldsymbol{W}^h_\tau\big) + \mathbf{k}\big(2\partial_z\overline{w}_{|\overline{\eta}}\mathrm{d}\tau + \partial_z\overline{\boldsymbol{\sigma}}_{|\overline{\eta}} \circ \mathrm{d}\boldsymbol{W}^z_\tau\big)$$

$$+ \Big(D^{\circ}_{\tau_{|\overline{\eta}}}\mathbf{k} - \mathbf{k} \cdot \nabla_{\mathbf{x}}\overline{\boldsymbol{u}}_{|\overline{\eta}}\mathrm{d}\tau - \mathbf{k} \times \boldsymbol{e}_z \mathrm{d}\tau\big) + 2\overline{\lambda}|\mathbf{k}|\nabla_{\mathbf{x}}\overline{\eta}\mathrm{d}\tau - \mathbf{k}\frac{\overline{\lambda}^2}{|\mathbf{k}|^2}\nabla_{\mathbf{x}} \cdot (\frac{\mathbf{k}}{\overline{\lambda}})\mathrm{d}\tau + \frac{\overline{\lambda}}{|\mathbf{k}|} \cdot \nabla_{\mathbf{x}}|\mathbf{k}|\mathrm{d}\tau\Big)$$

$$- \frac{1}{2|\mathbf{k}|}\mathbf{k} \cdot \big(\tilde{\boldsymbol{\sigma}}_{\mathbf{x}|\overline{\eta}} \circ \mathrm{d}\boldsymbol{W}^h_\tau \cdot \nabla_{\mathbf{x}}\overline{\boldsymbol{u}}_{|\overline{\eta}} - \boldsymbol{e}_z \times \tilde{\boldsymbol{\sigma}}_{\mathbf{x}|\overline{\eta}} \circ \mathrm{d}\boldsymbol{W}^h_\tau\big) - \frac{g}{|\mathbf{k}|}\mathbf{k} \cdot \nabla_{\mathbf{x}}\tilde{\eta}_o\mathrm{d}\tau. \qquad (7.50)$$

From this equation, we infer two evolution equations for the frequency and amplitude of the wave field.

7.3.4 Frequency Correction

It can be observed that imaginary terms remain in the right-hand side of the wave field evolution (7.50). These terms drive a phase change due to the vertical shear/vorticity of the slow current component and to the angle between the wavelength vector and the ISD. Multiplying this equation by the conjugate of $\widetilde{\eta}_o$ and subtracting by the conjugate of the above equation multiplied by $\widetilde{\eta}_o$ we obtain, after rearrangements,

$$
\mathrm{d}_\tau^\circ \mathcal{S} + \left((\overline{\boldsymbol{u}}_{|\overline{\eta}} - \overline{\boldsymbol{u}_a}_{|\overline{\eta}})\mathrm{d}\tau + \boldsymbol{C}_g\mathrm{d}\tau + \overline{\boldsymbol{\sigma}}_{|\overline{\eta}} \circ \mathrm{d}\boldsymbol{W}_\tau\right) \cdot \nabla\mathcal{S} = \frac{1}{4}\widehat{\boldsymbol{e}_\mathbf{k}} \cdot \partial_z\left(2\overline{\boldsymbol{u}}_{|\overline{\eta}}\mathrm{d}\tau + \overline{\boldsymbol{\sigma}}_{|\overline{\eta}} \circ \mathrm{d}\boldsymbol{W}_\tau^h\right) +
$$
$$
+ \frac{1}{2}\sin(\mathcal{S})\left(\left(-\overline{\boldsymbol{u}_a}_{|\overline{\eta}}\mathrm{d}\tau + \overline{\boldsymbol{\sigma}}_{|\overline{\eta}} \circ \mathrm{d}\boldsymbol{W}_\tau^h\right) \cdot \nabla_\mathbf{x}\overline{\eta} + \frac{1}{2\overline{\lambda}}\widehat{\boldsymbol{e}_\mathbf{k}} \cdot \left(\widetilde{\boldsymbol{\sigma}}_{\mathbf{x}|\overline{\eta}} \circ \mathrm{d}\boldsymbol{W}_\tau^h \cdot \nabla_\mathbf{x}\overline{\boldsymbol{u}}_{|\overline{\eta}} - e_z \times \widetilde{\boldsymbol{\sigma}}_{\mathbf{x}|\overline{\eta}} \circ \mathrm{d}\boldsymbol{W}_\tau^h\right)\right) \tag{7.51}
$$

with the wavelength unitary vector $\widehat{\boldsymbol{e}_\mathbf{k}} = \mathbf{k}/|\mathbf{k}|$, and where we recall that $\widetilde{\eta}_o = \mathcal{E}e^{i\mathcal{S}}$ with $\nabla\mathcal{S} = \boldsymbol{k}$, while $\boldsymbol{C}_g = \frac{\overline{\lambda}}{2|\mathbf{k}|^2}\mathbf{k}$ denotes the group velocity.

We observe that the phase field is transported by the slow component of the random flow corrected by the ISD. This transport expressed in Stratonovich involves an implicit diffusion term that can be recovered transitioning to an Itô representation. The phase transport is forced first with a shear term $\widehat{\boldsymbol{e}_\mathbf{k}} \cdot \partial_z\left(2\overline{\boldsymbol{u}}_{|\overline{\eta}}\mathrm{d}\tau + \overline{\boldsymbol{\sigma}}_{|\overline{\eta}} \circ \mathrm{d}\boldsymbol{W}_\tau^h\right)$. This term corresponds to a forcing by a sheared random flow (Quinn et al. 2017) or to current with non-zero vorticity (both at large and small scales) (White 1999). The last forcing takes the form of a nonlinear oscillatory noise term, which amplifies an anisotropy between trough and crest:

$$
\mathrm{d}\mathcal{F}^\sigma(\mathcal{S}) = \frac{1}{2}\sin(\mathcal{S})\left(\left(-\overline{\boldsymbol{u}_a}_{|\overline{\eta}}\mathrm{d}\tau + \overline{\boldsymbol{\sigma}}_{|\overline{\eta}} \circ \mathrm{d}\boldsymbol{W}_\tau^h\right) \cdot \nabla_\mathbf{x}\overline{\eta} + \frac{1}{2\overline{\lambda}}\widehat{\boldsymbol{e}_\mathbf{k}} \cdot \left(\widetilde{\boldsymbol{\sigma}}_{\mathbf{x}|\overline{\eta}} \circ \mathrm{d}\boldsymbol{W}_\tau^h \cdot \nabla_\mathbf{x}\overline{\boldsymbol{u}}_{|\overline{\eta}} - e_z \times \widetilde{\boldsymbol{\sigma}}_{\mathbf{x}|\overline{\eta}} \circ \mathrm{d}\boldsymbol{W}_\tau^h\right)\right). \tag{7.52}
$$

This term arises as no specific assumption has been done on the noise form. In particular, if the horizontal noise was assumed to be proportional to $\widetilde{\eta}_o$ this modulation term would disappear from the phase evolution.

When the spatial gradient of the phase field is negligible, the source terms can be directly interpreted as a correction to the intrinsic frequency, which depends on the noise component and on the vertical shear of the flow:

$$
\mathrm{d}\omega = \overline{\lambda}\mathrm{d}t - \frac{\epsilon}{4}\widehat{\boldsymbol{e}_\mathbf{k}} \cdot \partial_z\left(2\overline{\boldsymbol{u}}_{|\overline{\eta}}\mathrm{d}\tau + \overline{\boldsymbol{\sigma}}_{|\overline{\eta}} \circ \mathrm{d}\boldsymbol{W}_\tau^h\right) -
$$
$$
\frac{\epsilon}{2}\sin(\mathcal{S})\left(\left(-\epsilon\overline{\boldsymbol{u}_a}_{|\overline{\eta}}\mathrm{d}\tau + \overline{\boldsymbol{\sigma}}_{|\overline{\eta}} \circ \mathrm{d}\boldsymbol{W}_\tau^h\right) \cdot \nabla_\mathbf{x}\overline{\eta} +
$$
$$
\frac{\epsilon^{\frac{3}{2}}}{2\overline{\lambda}}\widehat{\boldsymbol{e}_\mathbf{k}} \cdot \left(\widetilde{\boldsymbol{\sigma}}_{\mathbf{x}|\overline{\eta}} \circ \mathrm{d}\boldsymbol{W}_\tau^h \cdot \nabla_\mathbf{x}\overline{\boldsymbol{u}}_{|\overline{\eta}} - e_z \times \widetilde{\boldsymbol{\sigma}}_{\mathbf{x}|\overline{\eta}} \circ \mathrm{d}\boldsymbol{W}_\tau^h\right)\right). \tag{7.53}
$$

We note that the first forcing term is in full agreement with White (1999). It corresponds to the correction due the current vorticity and similarly, in our case, to the noise current vorticity. This forcing term is of order ϵ^2 for irrotational current and noise current.

For slow and smooth currents (i.e., small value of ϵ), the modification of the frequency is insensitive to the velocity shear and to the random perturbation. The last group of terms acts as an amplitude modulation of an oscillatory forcing in the frequency shift. This modulation is primarily weighted by the advection of the surface elevation by the noise. To a lesser extent the advection of the slow flow by the noise and the Coriolis correction associated with the random wave perturbation component contribute as well. This wave forcing effect can be understood as resulting from a nonlinear wave–wave interaction between the noise and the wave. However, this forcing term requires to solve the phase evolution equation. With a small-slope assumption of the slow component of the sea surface this latter equation can be further simplified as

$$
\omega = \overline{\lambda} + \mathbf{k} \cdot \left(\overline{u} - \frac{\epsilon}{2|\mathbf{k}|} \partial_z \left(2\overline{u}_{|\overline{\eta}} \mathrm{d}\tau + \overline{\sigma}_{|\overline{\eta}} \circ \mathrm{d}W^h_\tau \right) \right) -
$$
$$
\frac{\epsilon}{2} \sin(\mathcal{S}) \left(\frac{\epsilon^{\frac{3}{2}}}{2\overline{\lambda}} \widehat{e}_{\mathbf{k}} \cdot \left(\widetilde{\sigma}_{\mathbf{x}|\overline{\eta}} \circ \mathrm{d}W^h_\tau \cdot \nabla_\mathbf{x} \overline{u}_{|\overline{\eta}} - e_z \times \widetilde{\sigma}_{\mathbf{x}|\overline{\eta}} \circ \mathrm{d}W^h_\tau \right) \right). \quad (7.54)
$$

For a very smooth/slow component (i.e., $\epsilon \ll 1$—and small Rossby number), neglecting second-order terms, we then have

$$
\omega = \overline{\lambda} + \mathbf{k} \cdot \left(\overline{u} - \frac{\epsilon}{2|\mathbf{k}|} \partial_z \left(2\overline{u}_{|\overline{\eta}} \mathrm{d}\tau + \overline{\sigma}_{|\overline{\eta}} \circ \mathrm{d}W^h_\tau \right) \right), \quad (7.55)
$$

which corresponds to the classical frequency correction equation (White 1999) with an additional random term related to the vertical shear of the slow current uncertainty. This intuitive extension is valid within a small-slope assumption and smooth phase flow (i.e., with negligible gradient). For a rougher slow component the wave forcing term in (7.53) remains.

7.3.5 Amplitude Equation

The evolution equation on the wave amplitude is obtained multiplying (7.50) by $\widetilde{\eta}_o$ and adding the complex conjugate. After some rearrangements and simplifications associated, in particular, to a small-slope assumption the stochastic wave action equation is obtained. It reads

$$
\mathrm{d}_\tau \left(\frac{|\widetilde{\eta}_o|^2}{\overline{\lambda}} \right) + \nabla_\mathbf{x} \cdot \left(\frac{|\widetilde{\eta}_o|^2}{\overline{\lambda}} \left((\overline{u}_{|\overline{\eta}} + C_g)\mathrm{d}\tau + \frac{1}{2}\overline{\sigma}_{|\overline{\eta}} \circ \mathrm{d}W^h_\tau \right) \right) =
$$
$$
\frac{|\widetilde{\eta}_o|^2}{2\overline{\lambda}} \left(\widehat{e}_\mathbf{k} \cdot \nabla_\mathbf{x}\overline{\sigma}_{|\overline{\eta}} \circ \mathrm{d}W_\tau \cdot \widehat{e}_\mathbf{k} - \nabla_\mathbf{x} \cdot \overline{\sigma}_{|\overline{\eta}} \circ \mathrm{d}W^h_\tau \right)
$$
$$
- \cos \mathcal{S} \left(\overline{\lambda}(-\overline{u a}_{|\overline{\eta}} \mathrm{d}\tau + \overline{\sigma}_{|\overline{\eta}} \circ \mathrm{d}W^h_\tau) \cdot \nabla_\mathbf{x}\overline{\eta} + \frac{1}{2\overline{\lambda}^2} e_\mathbf{k} \cdot \left(\widetilde{\sigma}_{\mathbf{x}|\overline{\eta}} \circ \mathrm{d}W^h_\tau \cdot \nabla_\mathbf{x} \overline{u}_{|\overline{\eta}} - e_z \times \widetilde{\sigma}_{\mathbf{x}|\overline{\eta}} \circ \mathrm{d}W^h_\tau \right) \right). \quad (7.56)
$$

In the same way as in Villas Bôas and Young (2020) the wave action involves an implicit noise-induced diffusion associated to the Stratonovich expression of the noise. This diffusion term associated to the Itô representation reads

$$\nabla_{\mathbf{x}} \cdot \left(\overline{\boldsymbol{a}}_{|\overline{\eta}} \nabla_{\mathbf{x}} \frac{|\widetilde{\eta}_o|^2}{\overline{\lambda}} \right)$$

where the (slow) horizontal variance tensor $\overline{\boldsymbol{a}}_{|\overline{\eta}}$ imposes inhomogeneous diffusion directions.

Notably, in the absence of noise, we recover the usual wave action relation:

$$d_\tau \left(\frac{|\widetilde{\eta}_o|^2}{\overline{\lambda}} \right) + \nabla_{\mathbf{x}} \cdot \left(\frac{|\widetilde{\eta}_o|^2}{\overline{\lambda}} \left(\overline{\boldsymbol{u}}_{|\overline{\eta}} + \boldsymbol{C}_g \right) d\tau \right) = 0. \tag{7.57}$$

However, in the presence of strong noise, both the forcing term on the right-hand side of (7.56) and the stochastic advection on the left-hand side must be accounted for. Similar to the frequency correction, the forcing terms introduce an oscillatory component whose amplitude depends on noise contributions associated with the advection of the slow component of the free surface and the horizontal velocity, combined with a Coriolis correction for the noise. For noise proportional to $\widetilde{\eta}_o$ the modulation component would boil down to 1, leaving a non-oscillatory forcing related to the slow surface advection by the noise and the slow current advection.

The first forcing term captures a deformation effect induced by the small-scale deformation tensor, as well as a compression effect linked to the divergence of the small-scale horizontal components.

7.4　Conclusion

In this work, we examined the wave action principle in the presence of currents exhibiting rough temporal variations. These small-scale fluctuations are represented as random fields that decorrelate in time at the characteristic timescale of the slow, large-scale components. This framework effectively models unresolved eddies and waves, without assuming the current to be potential.

Our derivation shows that for regular deterministic currents, wave action is approximately conserved. This deterministic setting corresponds to Taylor's frozen turbulence, where the transport field closely resembles a stationary mean flow. However, when unresolved random components become significant, additional forcing terms emerge, leading to a breakdown of wave action conservation.

These forcing terms encapsulate the influence of non-uniform currents on wave dynamics, as well as nonlinear wave–wave interactions. Furthermore, we identify specific terms responsible for introducing anisotropy between troughs and crests in surface waves.

Acknowledgements The authors acknowledge the support of the ERC EU project 856408-STUOD. They warmly thank Ruiao Hu, Darryl Holm, Oliver Street, and Valentin Resseguier for fruitful discussions and valuable advice. We are grateful to the anonymous reviewers for their suggestions and remarks.

Appendix A: Relation Between Stokes Drift and Itô–Stokes Drift for Linear Waves

In this appendix, we show how the Itô–Stokes drift is related to the Stokes drift associated to a monochromatic wave of wavelength $\mathbf{k}$, frequency ω, and amplitude η^2. The incompressible noise is defined from a random, φ_σ, scalar potential solution of a Poisson equation, $\Delta\varphi_\sigma$, with null boundary condition. This random potential is defined as

$$\varphi_\sigma = \frac{\omega}{|\mathbf{k}|}\eta e^{|\mathbf{k}|z}\sin(\mathcal{S}(\mathbf{x}_t, t)), \tag{7.58}$$

where the phase $\mathcal{S}(\mathbf{x}_t, s)$ is a random process, driven by the martingale $\mathrm{d}\mathcal{S}$. The Lagrangian coordinates are given by the integration of the potential gradient. We have

$$\mathbf{x}_t = \mathbf{x}_0 + \int_0^t \frac{\eta}{|\mathbf{k}|}e^{|\mathbf{k}|z}\cos(\mathcal{S}(\mathbf{x}_s, s))\frac{\partial \mathcal{S}}{\partial \mathbf{x}}\mathrm{d}\mathcal{S} = \mathbf{x}_0 + \frac{1}{|\mathbf{k}|}\eta e^{|\mathbf{k}|z}\mathbf{k}\sin(\mathcal{S}(\mathbf{x}_t, t)), \tag{7.59}$$

and

$$z_t = z + \int_0^t \frac{\eta}{|\mathbf{k}|}e^{|\mathbf{k}|z}\sin(\mathcal{S}(\mathbf{x}_s, s))\mathrm{d}\mathcal{S} = z_0 - \eta e^{|\mathbf{k}|z}\cos(\mathcal{S}(\mathbf{x}_t, t)), \tag{7.60}$$

where we recall that $\omega = \mathrm{d}_t\mathcal{S}$ and $\mathbf{k} = \nabla\mathcal{S}$. The corresponding divergence-free noise is defined as the centered integral of the potential gradient. The horizontal noise components are

$$\boldsymbol{\sigma} \circ B_t^h = \frac{1}{|\mathbf{k}|}\mathbf{k}\int_{t-\epsilon}^{t+\epsilon}\eta e^{|\mathbf{k}|z}\cos(\mathcal{S}(\mathbf{x}_s, s))\mathrm{d}\mathcal{S}, \tag{7.61}$$

while the vertical component reads

$$\boldsymbol{\sigma} \circ B_t^z = \omega\int_{t-\epsilon}^{t+\epsilon}\eta e^{|\mathbf{k}|z}\sin(\mathcal{S}(\mathbf{x}_s, s))\mathrm{d}\mathcal{S}. \tag{7.62}$$

The Itô–Stokes corresponds to the mean drift between the Itô and Stratonovich representation. To infer the Itô–Stokes drift we need thus to transition the above expression to (decentered) Itô integrals. We have

$$\mathbb{E}\boldsymbol{\sigma} \circ B_t^h = \frac{1}{|\mathbf{k}|}\eta\mathbf{k}\mathbb{E}\left(\int_t^{t+\epsilon} e^{|\mathbf{k}|z}\cos(\mathcal{S}(\mathbf{x}_s,t))\mathrm{d}\mathcal{S} + \int_{t-\epsilon}^t e^{|\mathbf{k}|z}\cos(\mathcal{S}(\mathbf{x}_s,t))\mathrm{d}\mathcal{S}\right).$$
$$(7.63)$$

The first integral cancels as it is the expectation of an Itô integral. The expression can be approximated for a fixed point at time t, as

$$\mathbb{E}\boldsymbol{\sigma} \circ B_t^h \approx \frac{1}{|\mathbf{k}|}\eta\mathbf{k}\mathbb{E}\left(\int_{t-\epsilon}^t e^{|\mathbf{k}|z}\cos(\mathcal{S}(\mathbf{x}_t,t))\mathrm{d}\mathcal{S}+\right.$$
$$\int_{t-\epsilon}^t \mathbf{k}^T e^{|\mathbf{k}|z}\sin(\mathcal{S}(\mathbf{x}_t,t))(\mathbf{x}_t - \mathbf{x}_{t-\epsilon})\mathrm{d}\mathcal{S}-$$
$$\left.\int_{t-\epsilon}^t |\mathbf{k}|e^{|\mathbf{k}|z}\cos(\mathcal{S}(\mathbf{x}_t,t))\mathrm{d}\mathcal{S}(z_t - z_{t-\epsilon})\right). \quad (7.64)$$

Using the Itô isometry with the quadratic variation $\omega \mathrm{d}s = \mathrm{d}\langle \mathcal{S}, \mathcal{S}\rangle$ and noticing that the first integral is again an Itô integral, we have

$$\mathbb{E}\boldsymbol{\sigma} \circ B_t^h \approx \omega\mathbf{k}\eta^2 e^{2|\mathbf{k}|z}\mathbb{E}\left(\sin^2(\mathcal{S}(\mathbf{x}_t,t)) + \cos^2(\mathcal{S}(\mathbf{x}_t,t))\right), \tag{7.65}$$

which scales as the classical Stokes drift expression associated to monochromatic surface wave with deep-water condition. For the vertical component, we obtain in the same way

$$\mathbb{E}\boldsymbol{\sigma} \circ B_t^z \approx 0. \tag{7.66}$$

Let us note that time correlation is essential in this recovering of the Stokes drift. As a matter of fact for the same noise expressed in Itô form, the horizontal Itô–Stokes is null. Only the vertical component remains due to inhomogeneity in the vertical direction.

References

Agresti A, Hieber M, Hussein A, Saal M (2022) The stochastic primitive equations with transport noise and turbulent pressure. Stochastics and partial differential equations: analysis and computations

Andrews DG, Mcintyre ME (1978a) On wave-action and its relatives. J Fluid Mech 4:647–664

Andrews DG, Mcintyre ME (1978b) An exact theory of nonlinear waves on a lagrangian-mean flow. J Fluid Mech 89(4):609–646

Bauer W, Chandramouli P, Chapron B, Li L, Mémin E (2020) Deciphering the role of small-scale inhomogeneity on geophysical flow structuration: a stochastic approach. J Phys Oceanogr 50(4):983–1003

Bauer W, Chandramouli P, Li L, Mémin E (2020) Stochastic representation of mesoscale eddy effects in coarse-resolution barotropic models. Ocean Model 151:101646

Brecht R, Li L, Bauer W, Mémin E (2021) Rotating shallow water flow under location uncertainty with a structure-preserving discretization. J Adv Model Earth Syst 13(12):e2021MS002492

Bretherton FP, Garrett C, Lighthill M (1997) Wavetrains in inhomogeneous moving media. Proc R Soc Lond Ser A Math Phys Sci 302(1471):529–554

Brzeźniak Z, Slavík J (2021) Well-posedness of the 3D stochastic primitive equations with multiplicative and transport noise. J Diff Equ 296:617–676

Carigi G, Luongo E (2023) Dissipation properties of transport noise in the two-layer quasi-geostrophic model. J Math Fluid Mech 25(2):28

Chandramouli P, Heitz D, Laizet S, Mémin E (2018) Coarse large-eddy simulations in a transitional wake flow with flow models under location uncertainty. Comput & Fluids 168:170–189

Chandramouli P, Mémin E, Heitz D (2020) 4D large scale variational data assimilation of a turbulent flow with a dynamics error model. J Comput Phys 412:109446

Chapron B, Dérian P, Mémin E, Resseguier V (2018) Large-scale flows under location uncertainty: a consistent stochastic framework. Q J R Meteorol Soc 144(710):251–260

Crisan D, Flandoli F, Holm D (2019) Solution properties of a 3D stochastic Euler fluid equation. J Nonlinear Sci 29(3):813–870

Debussche A, Pappalettera U (2023) Second order perturbation theory of two-scale systems in fluid dynamics. arXiv:2206.07775

Debussche A, Hug B, Mémin E (2023) A consistent stochastic large-scale representation of the navier-stokes equations. J Math Fluid Mech 25(1):19

Debussche A, Mémin E, Moneyron A (2024a) Some properties of a non-hydrostatic stochastic oceanic primitive equations model. In: Chapron B, Crisan D, Holm D, Mémin E, Coughlan J-L (eds) Stochastic transport in upper ocean dynamics III. Springer Nature, pp 161–182. https://inria.hal.science/hal-04632530

Debussche A, Mémin E, Moneyron A (2024b) Derivation of stochastic models for coastal waves. Stoch Transp Upper Ocean Dyn III:183–222

Dinvay E, Mémin E (2022) Hamiltonian formulation of the stochastic surface wave problem. Proc R Soc A 1–26

Dynkin EB (1962) Energy and momentum in the theory of waves in plasmas. In: Plasma hydromagnetics. Sixth lockheed symposium on magnetohydrodynamics. Stanford University Press, pp 47–57

Flandoli F, Luo D (2021) High mode transport noise improves vorticity blow-up control in 3D Navier-Stokes equations. Probab Theory Relat Fields 180(1):309–363

Flandoli F, Pappalettera U (2021) 2D Euler equations with stratonovich transport noise as a large-scale stochastic model reduction. J Nonlinear Sci 31(1):24

Flandoli F, Russo F (2023) Reduced dissipation effect in stochastic transport by Gaussian noise with regularity greater than 1/2. arXiv:2305.19293

Flandoli F, Galeati L, Luo D (2021) Delayed blow-up by transport noise. Comm Part Diff Equ 46(9):1757–1788

Galeati L (2020) On the convergence of stochastic transport equations to a deterministic parabolic one. Stoch Part Diff Equ Anal Comput 8(4):833–868

Galeati L, Luo D (2023) Weak well-posedness by transport noise for a class of 2D fluid dynamics equations. arXiv:2305.08761

Goodair D, Crisan D, Lang O (2022) Existence and uniqueness of maximal solutions to spdes with applications to viscous fluid equations. arXiv:2209.09137

Hayes WD (1970) Conservation of action and modal wave action. Proc R Soc Lond A 320(320):187–208

Holm D, Hu R, Street O (2024) On the interactions between mean flows and inertial gravity waves in the wkb approximation. In: Chapron B, Dan Crisan D, Holm E. Mémin, Radomska A (eds) Stochastic transport in upper ocean dynamics II. Springer Nature Switzerland, Cham, pp 111–141

Kafiabad H, Savva M, Vanneste J (2019) Diffusion of inertia-gravity waves by geostrophic turbulence. J Fluid Mech 869:R7. https://doi.org/10.1017/jfm.2019.300

Klein O, Heifetz E, Toledo Y (2022) Surface gravity waves in the presence of vertically shearing current and eddy viscosity. Phys Fluids 34(10):106604

Klyatskin VI, Koshel KV (2015) Anomalous sea surface structures as an object of statistical topography. Phys Rev E 91:063003. https://doi.org/10.1103/PhysRevE.91.063003. https://link.aps.org/doi/10.1103/PhysRevE.91.063003

Lang O, Crisan D, Mémin E (2023) Analytical properties for a stochastic rotating shallow water model under location uncertainty. J Math Fluid Mech 25(2):29

Li L, Deremble B, Lahaye N, Mémin E (2023) Stochastic data-driven parameterization of unresolved eddy effects in a baroclinic quasi-geostrophic model. J Adv Model Earth Syst 15(2):e2022MS003297

McComas CH, Bretherton FP (1977) Resonant interaction of oceanic internal waves. J Geophys Res (1896–1977) 82(9):1397–1412. https://doi.org/10.1029/JC082i009p01397. https://agupubs.onlinelibrary.wiley.com/doi/abs/10.1029/JC082i009p01397

Mémin E (2014) Fluid flow dynamics under location uncertainty. Geophys & Astro Fluid Dyn 108(2):119–146

Mémin E, Li L, Lahaye N, Tissot G, Chapron B (2022) Linear wave solutions of a stochastic shallow water model. Stoch Transp Upper Ocean Dyn II:223–245

Quinn BE, Toledo Y, Shrira VI (2017) Explicit wave action conservation for water waves on vertically sheared flows. Ocean Modell 112:33–47. ISSN 1463-5003. https://doi.org/10.1016/j.ocemod.2017.03.003. https://www.sciencedirect.com/science/article/pii/S1463500317300288

Resseguier V, Mémin E, Chapron B (2017a) Geophysical flows under location uncertainty, Part I Random transport and general models. Geophys & Astro Fluid Dyn 111(3):149–176

Resseguier V, Mémin E, Chapron B (2017b) Geophysical flows under location uncertainty, Part II Quasi-geostrophy and efficient ensemble spreading. Geophys & Astro Fluid Dyn 111(3):177–208

Resseguier V, Mémin E, Chapron B (2017c) Geophysical flows under location uncertainty, Part III SQG and frontal dynamics under strong turbulence conditions. Geophys & Astro Fluid Dyn 111(3):209–227

Resseguier V, Mémin E, Chapron B (2017d) Geophysical flows under location uncertainty, Part I Random transport and general models. Geophys & Astro Fluid Dyn 111(3):149–176

Resseguier V, Hascoët E, Chapron B (2024) Wave propagation in random two-dimensional turbulence: a multiscale approach. J Fluid Mech 997:A51. https://doi.org/10.1017/jfm.2024.769

Tucciarone FL, Mémin E, Li L (2023) Primitive equations under location uncertainty: analytical description and model development. In: Stochastic transport in upper ocean dynamics II. Springer Nature

Villas Bôas A, Young W (2020) Directional diffusion of surface gravity wave action by ocean macroturbulence. J Fluid Mech 890:R3. https://doi.org/10.1017/jfm.2020.116

Voronovich AG (1976) Propagation of surface internal waves in the approximation of geometric optics. Izv Akad Nauk Fiz Atmos Okeana 12(6):850–857

White BS (1999) Wave action on currents with vorticity. J Fluid Mech 386:329–344

Whitham GB (1965) A general approach to linear and non-linear dispersive waves using a lagrangian. J Fluid Mech 22(2):273–283

Chapter 8
A Stochastic Description of Eddy–Mean Flow Interactions

Mattéo Nex, Quentin Jamet, Etienne Mémin, and Florian Sévellec

Abstract Interactions between the mean flow and eddies play a central role in shaping oceanic circulation by redistributing heat, momentum, and energy. Accurately representing these processes in General Circulation Models (GCMs) remains a challenge due to limitations in subgrid-scale parameterizations. This study highlights the constraints imposed by small ensemble sizes on capturing the Reynolds stress tensor and diagnosing its associated work—key elements for assessing mean-to-eddy energy conversion rates. To address these challenges, we employ the Location Uncertainty (LU) framework, which introduces stochastic variability into the governing equations to represent unresolved turbulent effects. Using a 48-member ensemble simulation of the North Atlantic under realistic forcing, we compare deterministic and stochastic estimates of Reynolds stress work. The LU framework improves statistical convergence and stabilizes higher-order moments, leading to a more accurate representation of energy transfer statistics. Key contributions include a novel formulation of energy transfer equations and enhanced statistical diagnostics of Reynolds stress work within the LU framework. This study provides valuable dynamical diagnostics for identifying energetic patterns and regions of stability, offering new insights into eddy–mean flow interactions in the Gulf Stream.

8.1 Introduction

General Circulation Models (GCMs) are essential tools for studying oceanic and climate systems, enabling the simulation and analysis of complex geophysical interactions. The MIT General Circulation Model (MITgcm) (Marshall et al. 1997), used

M. Nex (✉) · E. Mémin
Odyssey, Centre Inria de l'Université de Rennes, IRMAR, Rennes, France
e-mail: matteo.nex@inria.fr

Q. Jamet
SHOM, Brest, France

F. Sévellec
Odyssey, CNRS, LOPS, Plouzané, France

B. Chapron et al. (eds.), *Stochastic Transport in Upper Ocean Dynamics IV*, Mathematics of Planet Earth 15, https://doi.org/10.1007/978-3-032-12749-5_8

in this study, is specifically designed to capture large-scale oceanic and atmospheric dynamics. GCMs play a crucial role in predicting climate variability, analyzing ocean currents, and assessing the impact of physical processes on the Earth system.

A key advancement in modern GCM applications is the use of ensemble simulations to account for the intrinsic variability and chaotic nature of geophysical flows (Palmer 2012; Leutbecher and Palmer 2008). Due to the sensitivity of such systems to initial conditions, even small perturbations can lead to markedly different outcomes. Ensemble simulations address this by running multiple instances of the model with slightly varied initial states, enabling a probabilistic characterization of system behavior. This approach facilitates the exploration of a range of possible outcomes and allows for the estimation of statistical moments without relying on ergodic assumptions. Ensembles are particularly valuable for quantifying uncertainties, assessing the likelihood of extreme events, and evaluating the robustness of climate projections. However, to ensure reliable interpretation, the empirical moments computed from ensemble simulations must be statistically converged. Without sufficient convergence, these estimates become unreliable and can lead to misleading conclusions. Achieving statistical convergence typically requires large ensemble sizes, which, in high-dimensional systems such as GCMs, can result in prohibitive computational costs.

Another major challenge in geophysical fluid simulations is the accurate representation of unresolved subgrid-scale (SGS) processes, due to the vast range of spatial and temporal scales involved. Resolving all relevant scales is computationally infeasible, necessitating the use of SGS parameterizations to model the influence of smaller, unresolved motions on the resolved flow. Central to these parameterizations is the Reynolds stress tensor, $\tau = \overline{u' \otimes u'}$, which capture the covariance of velocity fluctuations and their associated momentum transfer. Here, the $\otimes$ notation denotes the outer product of two vectors and the overbar represents a chosen averaging operator. This tensor plays a key role in governing how unresolved turbulence influences larger-scale dynamics, particularly in mediating energy transfers across scales. However, in practical applications, the Reynolds stress tensor is not directly available and must be approximated.

A common approach to approximating the Reynolds stress tensor involves mixing-length arguments, which relate it to an eddy viscosity coefficient intended to mimic the dissipative effects of turbulence. In its simplest form, eddy viscosity is prescribed constant in both time and space, typically tuned to ensure numerical stability (e.g., Adcroft 1995). This approach underpins widely used subgrid-scale parameterizations, such as those developed in Redi (1982), Gent and McWilliams (1990). More recent developments have focused on spatially and temporally varying eddy viscosities, as well as on incorporating energy backscatter mechanisms to better capture the bidirectional nature of energy transfer and improve physical realism (Bachman 2019; Eden and Greatbatch 2008; Jansen et al. 2019; Marshall et al. 2012). Evaluating the performance of SGS closures typically relies on reference simulations, as in Eden and Greatbatch (2008), where a non-eddy-resolving simulation employing an active SGS closure is compared against a high-resolution twin simulation (with the SGS closure disabled), filtered onto the coarse grid. While effective, this strategy requires to predefine the scale of the "eddies" that have to be parameterized. From a physical perspective, it is often challenging to disentangle the effects of genuinely unresolved turbulent scales from those of marginally resolved coherent structures. The latter may violate the assumptions of homogeneity and isotropy that underlie many SGS models, potentially introducing non-local effects that standard parameterizations fail to capture (Grooms et al. 2013; Kang and Curchitser 2015).

To address the limitations of spatial filtering diagnostics, recent work has introduced ensemble-based approaches that decompose the flow into an ensemble mean and fluctuations about that mean (Jamet et al. 2022; Li et al. 2023; Uchida et al. 2021, 2022). In this framework, the ensemble mean represents the component of the flow that is coherent across realizations and responds to common external forcing, while the deviations from this mean capture the stochastic, realization-specific components associated with nonlinear turbulent dynamics—commonly referred to as "eddies." From a parameterization perspective, both the ensemble mean and the eddy components possess distinct spatial and temporal scales that can be identified a posteriori (e.g., Jamet et al. 2022) to inform the construction of subgrid-scale models. Although the ensemble mean often exhibits larger-scale structures than the eddies, leading to a common—albeit somewhat imprecise—interpretation of the former as the "large-scale flow" and the latter as the "small-scale" or turbulent component. This ensemble-based decomposition enables the estimation of turbulent statistics without assuming spatial homogeneity, isotropy, or temporal stationarity. The strength of this approach lies in its generality and flexibility. However, as discussed above, generating large ensembles is computationally expensive. Most practical ensembles contain fewer than 100 members, which limits statistical convergence—particularly given the slow convergence rate of Monte Carlo methods, which scales as $1/\sqrt{n}$. For instance, achieving a precision of 1% would require approximately 10,000 samples.

In this study, we propose a novel approach to define robust statistical diagnostics from small ensembles in high-dimensional ocean simulations. Our method leverages the Location Uncertainty (LU) framework (Mémin 2014), which derives stochastic fluid equations via a probabilistic reformulation of the Reynolds transport theorem and fundamental conservation laws (Bauer et al. 2020a; Chapron et al. 2018; Resseguier et al. 2017a). By interpreting ensemble fluctuations as realizations of stochastic fields, the LU framework enables the generation of an arbitrary number of virtual samples for improved statistical estimation. We demonstrate the effectiveness of this framework using a 48-member eddy-resolving ensemble (at 1/12° resolution) of the Gulf Stream.

The paper is structured as follows: Sect. 8.2 presents the theoretical background for representing kinetic energy exchanges between the mean flow and eddies within an ensemble-averaged framework, followed by an overview of the LU formulation. Section 8.3 provides numerical results illustrating the improved stability of energy diagnostics achieved with the LU approach, alongside comparisons with traditional empirical estimates. Finally, Sect. 8.4 offers concluding remarks and outlines directions for future research.

8.2 Background and Methods

8.2.1 Model

This study is based on eddy-resolving ensemble simulations of the North Atlantic Basin, obtained by numerically integrating the hydrostatic primitive equations (HPE). Details of the numerical code and model configuration are provided in Sect. 8.3. The horizontal momentum equations of the HPE system are given by

$$\begin{cases} \partial_t \boldsymbol{u}_h = -\boldsymbol{\nabla} \cdot (\boldsymbol{u} \otimes \boldsymbol{u}_h) - f\boldsymbol{k} \times \boldsymbol{u}_h - \frac{1}{\rho_0}\boldsymbol{\nabla}_h P + \mathcal{F} + \mathcal{D}, \\ \boldsymbol{\nabla} \cdot \boldsymbol{u} = 0, \end{cases} \tag{8.1}$$

where P is the pressure, f is the Coriolis parameter, and $\rho_0 = 999.8$ kg m^{-3} is the reference seawater density under the Boussinesq approximation. The velocity field is denoted by $\boldsymbol{u} = (\boldsymbol{u}_h, w) = (u, v, w)$ the full 3D velocity field, and $\boldsymbol{u}_h$ its horizontal component. The terms $\mathcal{F}$ and $\mathcal{D}$ represent external forcing and dissipation, respectively. The vertical velocity component w is computed diagnostically from the continuity equation to ensure incompressibility. The full HPE system includes additional thermodynamic equations for temperature T and salinity S, as well as an equation of state, which are omitted here for clarity.

8.2.2 Eddy–Mean Flow Kinetic Energy Transfers

The analysis of interactions between small-scale fluctuations and the mean flow within the kinetic energy budget is central to the parameterization of subgrid-scale (SGS) eddies and the representation of velocity fluctuations. Numerical models are typically characterized by an effective resolution, enabling a decomposition of the flow into large-scale (resolved) and small-scale (unresolved) components—often defined via a wavenumber cutoff.

Oceanic flow energy can be decomposed into kinetic energy and potential energy. In stratified flows, potential energy is associated with buoyancy forces. Although both total and available potential energies are important concepts, we restrict our attention here to the horizontal kinetic energy:

$$K = \frac{\rho_0}{2}\boldsymbol{u}_h \cdot \boldsymbol{u}_h = \frac{\rho_0}{2}(u^2 + v^2), \tag{8.2}$$

Applying Reynolds decomposition $\boldsymbol{u}_h = \overline{\boldsymbol{u}_h} + \boldsymbol{u}_h'$—where $\overline{\cdot}$ denotes an averaging operator, left unspecified for now, and $\cdot'$ the deviation from the mean—yields

$$K = \underbrace{\frac{\rho_0}{2}\overline{\boldsymbol{u}_h} \cdot \overline{\boldsymbol{u}h}}_{\text{Mean kinetic energy } K_M} + \underbrace{\frac{\rho_0}{2}\overline{\boldsymbol{u}'_h \cdot \boldsymbol{u}h'}}_{\text{Eddy kinetic energy } K_E} + \underbrace{\rho_0\overline{\boldsymbol{u}_h} \cdot \boldsymbol{u}h'}_{\text{Cross term } K_C}. \tag{8.3}$$

In this study, we are primarily interested in the evolution of the mean kinetic energy (MKE), K_M. A budget for K_M can be derived by taking the dot product of the averaged momentum equation (8.1) with the mean velocity $\rho_0\,\overline{\boldsymbol{u}_h}$:

$$\rho_0\,\overline{\boldsymbol{u}_h} \cdot \overline{\partial_t \boldsymbol{u}_h} = \rho_0\,\overline{\boldsymbol{u}_h} \cdot \overline{\left(-\boldsymbol{\nabla} \cdot (\boldsymbol{u} \otimes \boldsymbol{u}_h) - f\boldsymbol{k} \times \boldsymbol{u}_h - \frac{1}{\rho_0}\boldsymbol{\nabla}_h P\right)},$$
$$\partial_t K_M = -\boldsymbol{\nabla} \cdot (\overline{\boldsymbol{u}}K_M) - \rho_0\,\overline{\boldsymbol{u}_h} \cdot \boldsymbol{\nabla} \cdot (\overline{\boldsymbol{u}' \otimes \boldsymbol{u}'_h}) - \boldsymbol{\nabla} \cdot (\overline{\boldsymbol{u}P}) - \overline{wb}. \tag{8.4}$$

The pressure term transforms as follows:

$$-\overline{\boldsymbol{u}_h} \cdot \boldsymbol{\nabla}_h \overline{P} = -\boldsymbol{\nabla} \cdot (\overline{\boldsymbol{u}P}) - \overline{wb}, \tag{8.5}$$

using incompressibility and the hydrostatic relation $\overline{b} = -\partial_z \overline{P}$. This yields the following form of the MKE budget:

$$\partial_t K_M = -\boldsymbol{\nabla} \cdot (\overline{\boldsymbol{u}}K_M) - \boldsymbol{\nabla} \cdot (\overline{\boldsymbol{u}P}) - \overline{wb} \underbrace{-\rho_0\,\overline{\boldsymbol{u}_h} \cdot \boldsymbol{\nabla} \cdot (\overline{\boldsymbol{u}' \otimes \boldsymbol{u}'_h})}_{\text{MEC}}. \tag{8.6}$$

The first two terms on the right-hand side of Eq. (8.6) represent the divergence of kinetic energy and pressure fluxes, respectively. The third term, involving $\overline{wb}$, captures the conversion between kinetic and potential energies, often referred to as buoyancy work. The final term, labeled MEC (mean-to-eddy conversion), quantifies the interaction between the mean flow and the Reynolds stress tensor. It represents the rate at which kinetic energy is transferred from the mean flow to the eddies. This term is also commonly referred to as the barotropic energy conversion (BTR) in the literature (see, e.g., Matsuta and Masumoto 2021).

The MEC term plays a central role in the development and evaluation of subgrid-scale (SGS) parameterizations. When using a spatial filtering approach (e.g., Aluie et al. 2018; Grooms et al. 2021), the averaging operator separates the flow into resolved (mean) and unresolved (eddy) components. In this context, K_M corresponds to the kinetic energy of the resolved flow, while $\overline{K_E}$ represents the kinetic energy associated with unresolved fluctuations. The MEC term quantifies the energy exchange between these scales, and its accurate estimation is therefore essential for properly capturing the dynamical influence of unresolved turbulence on the resolved flow.

We propose to diagnose the MEC term using an ensemble-based approach, in which the averaging operator is defined as an ensemble mean, and perturbations are interpreted as deviations of each ensemble member from this mean. This removes the need to specify an explicit spatial or temporal cutoff scale, enabling a more flexible and flow-driven characterization of the dynamics. As a result, the ensemble-based framework provides a robust means of diagnosing the spatiotemporal structure of the MEC term across a range of dynamical regimes.

A major limitation of the ensemble-based approach is the typically small number of available realizations. While ensemble analysis holds great promise for diagnosing scale interactions and informing subgrid-scale modeling, assembling sufficiently large ensembles remains computationally expensive. With a limited ensemble size, the empirical mean may converge poorly, leading to biased or noisy estimates of the Reynolds stress tensor. As we will show, the available dataset with $N = 48$ members does not achieve full convergence. As a result, variability across realizations is underrepresented, limiting the robustness of diagnostics based on the ensemble-averaged MEC.

8.2.3 *Location Uncertainty (LU)*

In the following sections, we introduce an alternative approach to better capture eddy–mean flow interactions, based on the stochastic framework of Location Uncertainty (LU) (Mémin 2014). This framework models velocity fluctuations as random variables, assuming they decorrelate over the characteristic timescale of the large-scale flow. This mathematical formulation allows LU to capture flow variability and the complex dynamics of eddy–mean flow interactions from a stochastic standpoint. As we will show, it enables the computation of stable statistical diagnostics using the same ensemble of simulations.

8.2.3.1 Particle Displacement

Location Uncertainty (LU) is a framework that reformulates the equations of fluid mechanics by incorporating stochastic variables and processes directly into their structure (see, for example, Bauer et al. 2020b, a; Brecht et al. 2021; Li et al. 2023; Resseguier et al. 2017a, b, c; Tucciarone et al. 2023). The core concept of LU is to augment the Lagrangian displacement of a fluid particle with a stochastic component that represents positional "uncertainty" or "fluctuation." Specifically, the displacement $\mathrm{d}X_t$ is decomposed into a smooth-in-time velocity part, $\boldsymbol{u}$, and a stochastic component that is correlated in space, but decorrelated in time:

$$X_t = X_{t_0} + \int_{t_0}^{t} \boldsymbol{u}(X_s, s)\mathrm{d}s + \int_{t_0}^{t} \boldsymbol{\sigma}_s \mathrm{d}\boldsymbol{B}_s, \tag{8.7}$$

where the stochastic integral in the last term is understood in the Itô sense. In differential form, this reads

$$\mathrm{d}X_t = \boldsymbol{u}_t \mathrm{d}t + \boldsymbol{\sigma}_t \mathrm{d}\boldsymbol{B}_t. \tag{8.8}$$

Here, the first term represents the smooth deterministic displacement, while the second term, $\sigma_t \mathrm{d}\boldsymbol{B}_t$, corresponds to the stochastic displacement accounting for unresolved turbulent effects. This stochastic component will henceforth be referred to as the "noise."

8.2.3.2 Noise Term $\sigma \mathrm{d}\boldsymbol{B}_t$

The term $\sigma \mathrm{d}\boldsymbol{B}_t$ represents small-scale uncertainty, encapsulating the spatially correlated variability of the flow. It is a vector with one component per spatial dimension, defined as

$$(\sigma_t \mathrm{d}\boldsymbol{B}_t)^{(i)}(\boldsymbol{x}) = \int_\Omega \breve{\sigma}_{ik} \mathrm{d}\boldsymbol{B}_t^k(\boldsymbol{x}, \boldsymbol{x}', t)\mathrm{d}\boldsymbol{x}', \tag{8.9}$$

where the Einstein summation convention over repeated indices is adopted. Here, σ acts as an integral operator on a well-defined functional Brownian motion $\mathrm{d}\boldsymbol{B}_t$, with the matrix-valued kernel $\breve{\sigma}(\boldsymbol{x}, \boldsymbol{x}', t)$ introducing spatial correlations that maintain coherence between points $\boldsymbol{x}$ and $\boldsymbol{x}'$. This formulation allows for an effective statistical representation of unresolved small-scale flows.

The kernel $\breve{\sigma}$ is assumed symmetric, positive-definite, and bounded in space and time. For simplicity, we consider a deterministic kernel here, though random kernels can also be incorporated (Li et al. 2023). Under these assumptions, the correlation operator is a positive-definite Hilbert–Schmidt operator. It admits a spectral decomposition into a set of eigenfunctions φ_n, $n \in \mathbb{N}$, with corresponding eigenvalues λ_n, forming an orthonormal basis. Consequently, the noise can be expressed as

$$\sigma_t \, \mathrm{d}\boldsymbol{B}_t(\boldsymbol{x}) = \sum_{n \in \mathbb{N}} \sqrt{\lambda_n} \varphi_n(\boldsymbol{x}, t) \, \mathrm{d}\beta_n, \tag{8.10}$$

where $(\beta_n)_{n \in}$ are independent standard Brownian motions.

8.2.3.3 Covariance Tensor

By construction, the noise term $\sigma_t \mathrm{d}\boldsymbol{B}_t$ is a centered Gaussian process whose covariance depends on spatial position and the squared kernel. Using Itô isometry, the covariance between two points $\boldsymbol{x}$ and $\boldsymbol{y}$ is given by

$$\mathrm{Cov}\big((\sigma \mathrm{d}B_t)^{(i)}(\boldsymbol{x}, t), (\sigma \mathrm{d}B_t)^{(j)}(\boldsymbol{y}, t')\big) = q_{ij}(\boldsymbol{x}, \boldsymbol{y}, t)\,\delta(t - t')\mathrm{d}t, \tag{8.11}$$

where q_{ij} denotes the (i, j)-th component of the two-point covariance tensor $\boldsymbol{Q}(\boldsymbol{x}, \boldsymbol{y}, t)$, and δ is the Dirac delta function. The one-point covariance tensor is defined as

$$\boldsymbol{a}(\boldsymbol{x}, t)\mathrm{d}t := \boldsymbol{Q}(\boldsymbol{x}, \boldsymbol{x}, t), \tag{8.12}$$

which represents a generalized eddy viscosity tensor associated with the unresolved small-scale flow component. It has units of m^2/s and, up to a decorrelation time τ, characterizes the variance of the noise. By slight abuse of terminology, a is often called the variance tensor of the noise. Its trace integrated over the domain Ω provides a bounded measure of the turbulent kinetic energy:

$$\frac{1}{2} \int_\Omega \mathrm{Tr}a(x, t)\, dx < +\infty. \tag{8.13}$$

This expression can also be formulated in terms of the eigenpairs of the correlation tensor:

$$\int_\Omega \mathrm{Tr}a(x, t) = \sum_{i\in\mathbb{N}} \check{\lambda}_i \mathrm{Tr}\left(\varphi_i(x)\varphi_i(x)^\mathsf{T}\right) dx, \tag{8.14}$$

where superscript $\cdot^T$ denotes vector transposition.

8.2.3.4 Transport Operator

The derivation of general stochastic flow dynamics within the LU framework is based on a stochastic extension of the Reynolds transport theorem (Mémin 2014). This theorem allows differentiation of a quantity within a volume transported by the stochastic flow and introduces a stochastic transport operator, which can be interpreted as a stochastic material derivative acting on a transported scalar field θ. This operator is defined as

$$\mathbb{D}_t\theta = \underbrace{d_t\theta}_{\theta(x,t+dt)-\theta(x,t)} + \underbrace{(u^*dt + \sigma_t d B_t) \cdot \nabla\theta}_{\text{Stochastic advection}} - \underbrace{\frac{1}{2}\nabla \cdot (a\nabla\theta)dt}_{\text{Stochastic diffusion}}. \tag{8.15}$$

This material derivative comprises three terms. The first term represents the time increment of θ at a fixed location. The second term is the stochastic advection, involving both the noise term and a modified large-scale drift velocity u^*. This drift includes a component u_S that can be interpreted as a generalization of the Stokes drift associated with orbital wave motion (Bauer et al. 2020a). For incompressible noise (i.e., when $\nabla \cdot \sigma = 0$), it takes the form:

$$u^* = u - u_S = \frac{1}{2}\nabla \cdot a. \tag{8.16}$$

The Stokes drift u_S is defined by the divergence of the variance tensor a. This tensor also appears in the last term of Eq. (8.15), which represents stochastic diffusion. It defines the principal axes of noise-induced diffusion (Resseguier et al. 2017d).

8.2.4 *LU HPE Horizontal Momentum Equations*

Using the stochastic transport operator, any fluid dynamics model can be derived by following the same procedure and approximations as in the deterministic setting. The horizontal momentum equations of the hydrostatic LU model are given by

$$\mathbb{D}_t \boldsymbol{u}_h = -f\boldsymbol{k} \times (\boldsymbol{u}_h \mathrm{d}t + \boldsymbol{\sigma}_t \mathrm{d}\boldsymbol{B}_t) - \frac{1}{\rho_0}\boldsymbol{\nabla}\mathrm{d}p, \tag{8.17}$$

with pressure expressed as $\mathrm{d}p = P\mathrm{d}t + \mathrm{d}P^\sigma$, where the last martingale pressure term satisfies $\mathrm{d}P^\sigma = 0$. For interested readers, the full set of stochastic primitive equations is provided in Tucciarone et al. (2023).

8.2.4.1 Stochastic Kinetic Energy Equation

To derive the kinetic energy equation within the LU framework, we follow a procedure analogous to the deterministic case. Specifically, we multiply the mean of the stochastic momentum equation (8.17) by the mean flow velocity, i.e., $\rho_0\,\overline{\boldsymbol{u}_h}\cdot\overline{\mathbb{D}_t\boldsymbol{u}_h}$. After several derivation steps—detailed in Appendix 8.4—we obtain the following equation for the mean kinetic energy K_M:

$$\partial_t K_M = -\boldsymbol{\nabla}\cdot(\overline{\boldsymbol{u}}\,K_M) - \boldsymbol{\nabla}\cdot(\overline{\boldsymbol{u}}\,\overline{P}) - \overline{w}\,\overline{b} + \boldsymbol{\nabla}\cdot(\boldsymbol{u}_s\,K_M) + \rho_0\,\overline{\boldsymbol{u}_h}\cdot\left(\frac{1}{2}\boldsymbol{\nabla}\cdot(\boldsymbol{a}\,\boldsymbol{\nabla}\,\overline{\boldsymbol{u}_h})\right). \tag{8.18}$$

The deterministic equation (8.6) and its stochastic counterpart (8.18) differ notably in their last two terms. To ensure that both formulations represent the same physical process—albeit with different statistical distributions—we define the stochastic representation of the mean–eddy correction (MEC) by equating these terms, through a specifically calibrated decorrelation time:

$$\underbrace{-\rho_0\,\overline{\boldsymbol{u}_h}\cdot\boldsymbol{\nabla}\cdot(\overline{\boldsymbol{u}'\otimes\boldsymbol{u}'_h})}_{MEC} = \underbrace{\boldsymbol{\nabla}\cdot(\boldsymbol{u}_s\,K_M)}_{\text{Advection of } K_M \text{ by } \boldsymbol{u}_s} + \rho_0\,\overline{\boldsymbol{u}_h}\cdot\underbrace{\left(\frac{1}{2}\boldsymbol{\nabla}\cdot(\boldsymbol{a}\,\boldsymbol{\nabla}\,\overline{\boldsymbol{u}_h})\right)}_{\text{Stochastic diffusion}}. \tag{8.19}$$

The right-hand side of this equation includes two notable contributions: the advection of mean kinetic energy by the Itô–Stokes drift, and a diffusion term arising from stochastic transport. In practice, this diffusive contribution can be efficiently approximated using a Milstein-type numerical scheme, as detailed in Appendix 8.4. This approach interprets the diffusion as a double advection of the mean velocity $\boldsymbol{u}_h$ by the noise $\boldsymbol{\sigma}_t\mathrm{d}\boldsymbol{B}_t$, a term we denote as $DADV$. When averaged over a sufficiently large ensemble, this double advection term converges to the diffusive expression:

$$DADV = \mathbb{E}\left[\rho_0\,\overline{\boldsymbol{u}_h}\cdot\left((\boldsymbol{\sigma}_t\,\mathrm{d}\boldsymbol{B}_t\cdot\boldsymbol{\nabla})\,(\boldsymbol{\sigma}_t\,\mathrm{d}\boldsymbol{B}_t\cdot\boldsymbol{\nabla})\,\overline{\boldsymbol{u}_h}\right)\right]. \tag{8.20}$$

To evaluate both terms on the right-hand side of (8.19), we need a method to sample the noise term and to compute the variance tensor $\boldsymbol{a}$, involved both in the diffusion and in the Itô-Stokes drift. We also need a method to specify the decorrelation time to ensure the coherence of the stochastic and deterministic expressions of the mean Kinetic energy (8.6) and (8.19). The following section describes the approaches used for this purpose.

8.2.4.2 Noise Ansatz

As noted earlier, the noise can be effectively constructed using a spectral representation of the correlation tensor. To this end, we define a set of time-varying empirical eigenfunctions $\{\varphi_k(\boldsymbol{x}, t)\}_{k=1}^{m}$ and corresponding eigenvalues $\lambda_k(t)$ following the Proper Orthogonal Decomposition (POD) approach of Uchida et al. (2022). At each time t, POD is applied to the ensemble fluctuation matrix

$$\boldsymbol{U} = [\boldsymbol{u}_1', \boldsymbol{u}_2', \ldots, \boldsymbol{u}_m'] \in \mathbb{R}^{n \times m},$$

where each column $\boldsymbol{u}_i' \in \mathbb{R}^n$ represents the horizontal velocity fluctuation of the i-th ensemble member relative to the ensemble mean. This yields a spectral representation of the noise, along the ensemble dimension m, at fixed time t:

$$\sigma_t \, \mathrm{d}B_t^x(\boldsymbol{x}) = \sqrt{\tau} \sum_{k=1}^{m} \sqrt{\lambda_k(t)} \, \varphi_k^x(\boldsymbol{x}, t) \, \mathrm{d}\beta_t^k, \quad \sigma_t \, \mathrm{d}B_t^y(\boldsymbol{x}) = \sqrt{\tau} \sum_{k=1}^{m} \sqrt{\lambda_k(t)} \, \varphi_k^y(\boldsymbol{x}, t) \, \mathrm{d}\beta_t^k.$$

$$(8.21)$$

Here, τ denotes the decorrelation time, which ensures that the correlation operator has the correct physical units of $m/\sqrt{s}$—since the noise represents a displacement and the eigenfunctions are velocity fields. An accurate estimation of τ—discussed in Appendix 8.4—is essential for the noise model to faithfully capture the statistical properties of the underlying velocity fluctuations as well their kinetic energy.

From a practical standpoint, we employ the snapshot POD method (Sirovich 1987), as detailed in Appendix 8.4. To preserve the full variability captured by the ensemble, we retain all $m - 1$ nontrivial eigenfunctions. This spectral representation of the noise provides a flexible basis that enables the fast generation of an arbitrary number of ensemble fluctuation samples.

The top panel of Fig. 8.1 shows a realization of the two horizontal noise components, normalized by $\sqrt{\tau}$ to express them in velocity-like units. The bottom panel displays the corresponding velocity fluctuations. Due to the construction method, the noise fields share structural features with the fluctuations—for instance, horizontal banding in the x-component and vertical banding in the y-component. However, the noise fields exhibit finer-scale variability, leading to less regular and more intricate patterns compared to the smoother, large-scale structure observed in the velocity fluctuations.

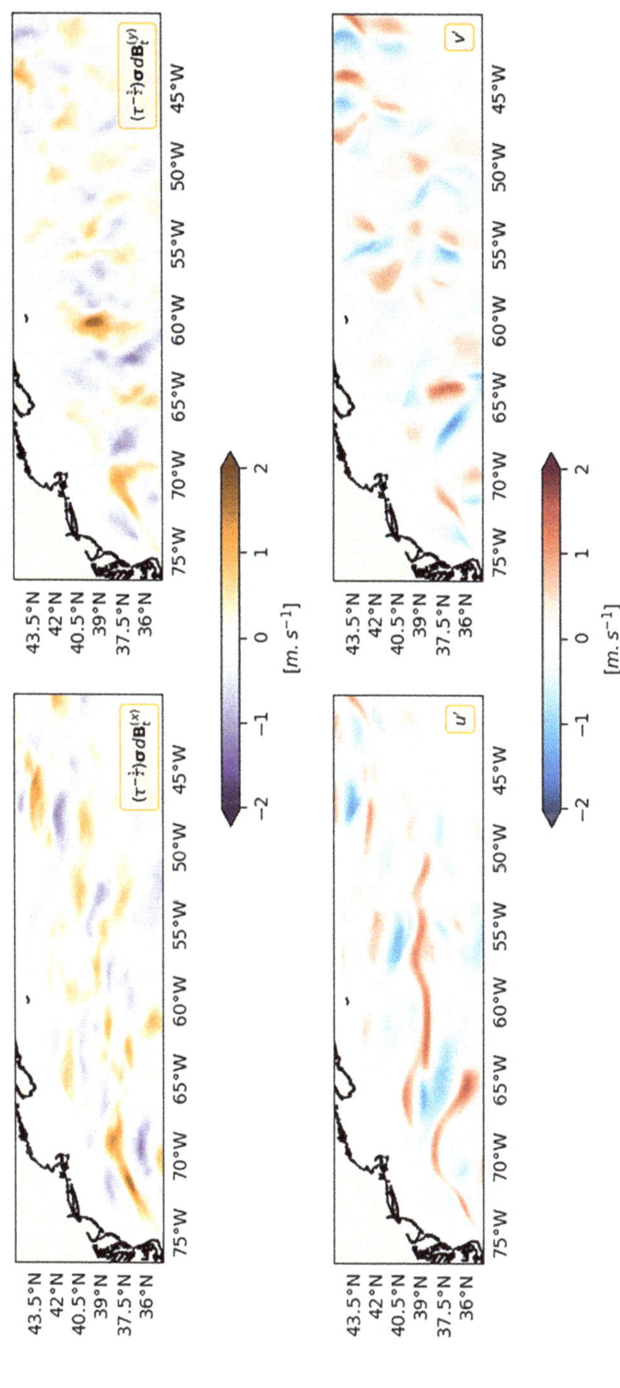

Fig. 8.1 Example of a realization of the x (left) and y (right) components of the noise normalized by $\sqrt{\tau}$ (top panel), and of the velocity fluctuations u' and v' (bottom panel). All maps are shown at $z = 94$ m on January 1, 2010

The components of the variance tensor can be directly inferred from the spectral decomposition on the m modes via

$$a^{xx}(x) = \tau\lambda_k\phi_k^x(x)\phi_k^x(x), \quad a^{xy}(x) = \tau\lambda_k\phi_k^x(x)\phi_k^y(x), \quad a^{yy}(x) = \tau\lambda_k\phi_k^y(x)\phi_k^y(x),$$

(8.22)

where $a^{xy} = a^{yx}$ by symmetry.

An example of the variance tensor components is shown in Fig. 8.2. Although a^{xy} has a smaller amplitude than a^{xx} and a^{yy}, it is clearly non-negligible, indicating a correlation between the two fluctuation components. Both the noise field and its variance are strongly anisotropic and spatially inhomogeneous. Within the Gulf Stream path, the variance tensor reaches values on the order of $\mathcal{O}(100 \text{ m}^2\text{s}^{-1})$, roughly one order of magnitude larger than the explicit horizontal viscosity coefficient $A_h = 20 \text{ m}^2\text{s}^{-1}$ used in our $\frac{1}{12}^\circ$ configuration. Nevertheless, these values remain smaller than typical subgrid-scale viscosities in non-eddy-resolving models (i.e., with horizontal resolution $\sim \mathcal{O}(1°)$), where coefficients often reach $\mathcal{O}(1000 \text{ m}^2\text{s}^{-1})$. The fact that the dissipation induced by our stochastic representation falls between these two regimes supports the interpretation that the noise effectively models the unresolved eddy feedback on the large-scale mean flow.

8.2.4.3 Statistical Diagnosis

Thanks to the ease of generating samples of the stochastic fluctuations, empirical estimates of the order-p statistical moment of the DADV term can be computed as

$$\mathbb{E}[(DADV - \overline{DADV})^p,] \approx \frac{1}{n-1}\sum_{i=1}^{n}\left[\left(\rho_0\overline{u_h} \cdot ((\sigma dB_t^{(i)} \cdot \nabla)(\sigma dB_t^{(i)} \cdot \nabla)\overline{u}h) - \overline{DADV}\right)^p, \right],$$

(8.23)

with the mean term defined as

$$\overline{DADV} = \frac{1}{n}\sum_{i=1}^{n}\left[\rho_0\overline{u_h} \cdot \left((\sigma dB_t^{(i)} \cdot \nabla)(\sigma dB_t^{(i)} \cdot \nabla)\overline{u}_h\right)\right],$$

(8.24)

where n is the number of realizations. A sufficiently large number of samples ensures statistical convergence. In the next section, we compare these diagnostic quantities to the empirical ensemble-mean estimate of the MEC term.

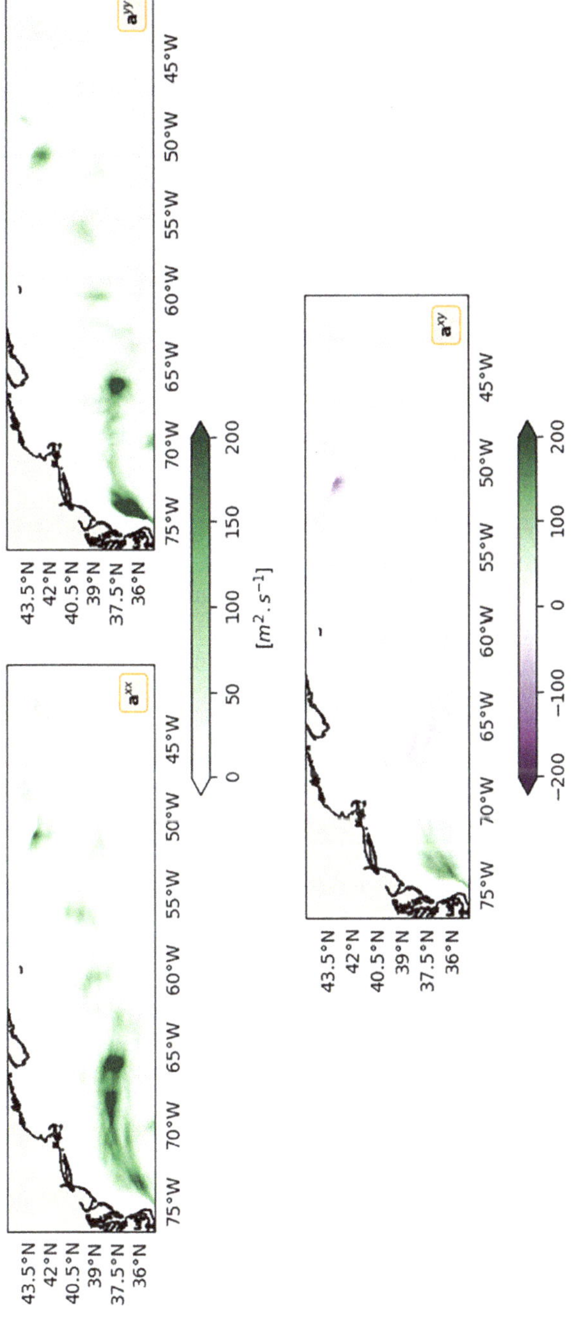

Fig. 8.2 Variance tensor components a^{xx} and a^{yy} (top panel), and a^{xy} (bottom panel), at $z = 94$ m on January 1, 2010

8.3 Numerical Results

8.3.1 Ensemble Simulations

We apply the statistical diagnostics introduced above to an ensemble of simulations generated using the MITgcm ocean model (Marshall et al. 1997), coupled to the idealized atmospheric boundary layer model CheapAML (Deremble et al. 2013). The ensemble consists of 48 eddy-resolving (1/12°) simulations of the North Atlantic subtropical gyre, covering the region from 20°S to 55°N. Each simulation spans the period 1963–2012 and is forced by realistic atmospheric and open boundary conditions. Open boundaries at 20°S, 55°N and the Strait of Gibraltar are derived from the ORCA12.L46-MJM88 global ocean configuration based on NEMO (Molines et al. 2014; Sérazin et al. 2015), and linearly interpolated onto the model grid.

At the surface, the ocean is forced by 6-hourly atmospheric winds, precipitation, and radiative fluxes from the Drakkar forcing set (Brodeau et al. 2010; Dussin et al. 2016). Turbulent heat and freshwater exchanges are computed using the COARE3 bulk algorithm (Fairall et al. 2003). CheapAML simulates the horizontal dynamics of air temperature and humidity via a simplified advection–diffusion equation. This setup avoids excessive damping of upper-ocean variability, as compared to more traditional bulk forcing approaches (Jamet et al. 2019). Further documentation and references are available at: https://github.com/quentinjamet/chaocean. The 48-member ensemble is built by integrating the model from slightly perturbed initial conditions, following the *micro*-initial condition perturbation framework (Stainforth et al. 2007). This approach captures a range of plausible ocean states under a common external forcing. Unless otherwise noted, diagnostics are computed at $z = -94$ m, within the winter mixed layer (Uchida et al. 2022), and based on snapshot outputs from January 1, 2010. Some diagnostics have also been extended to other times of the year to assess seasonal variability.

In this numerical study, we specifically focus on the deterministic eddy–mean kinetic energy transfer term MEC and its stochastic counterparts: the dissipation term $DADV$ and the Itô–Stokes drift. Figure 8.3 compares the spatial structures of these contributions. The Itô–Stokes drift, driven by the divergence of the noise covariance tensor ($\nabla \cdot \mathbf{a}$), appears negligible in comparison with $DADV$, despite evident spatial heterogeneity in the noise structure (Fig. 8.2). Consequently, this term is not considered further in this study. In contrast, $DADV$ exhibits similar magnitudes and spatial patterns to MEC, supporting its interpretation as a meaningful stochastic proxy for eddy–mean energy exchanges. An illustrative snapshot of the deterministic MEC field on January 1, 2010 is shown in the top panel of Fig. 8.3. It is important to note, however, that unlike $DADV$, the MEC term exhibits poor convergence due to the limited ensemble size, as will be demonstrated. Although MEC is not converged, a qualitative comparison between the two expressions can still be performed.

The structure of the MEC and $DADV$ terms reflects several characteristic patterns of energy transfer. Notably, the Gulf Stream separation acts as a persistent sink of energy for the mean flow, with slowly varying oscillations in the meridional (y)

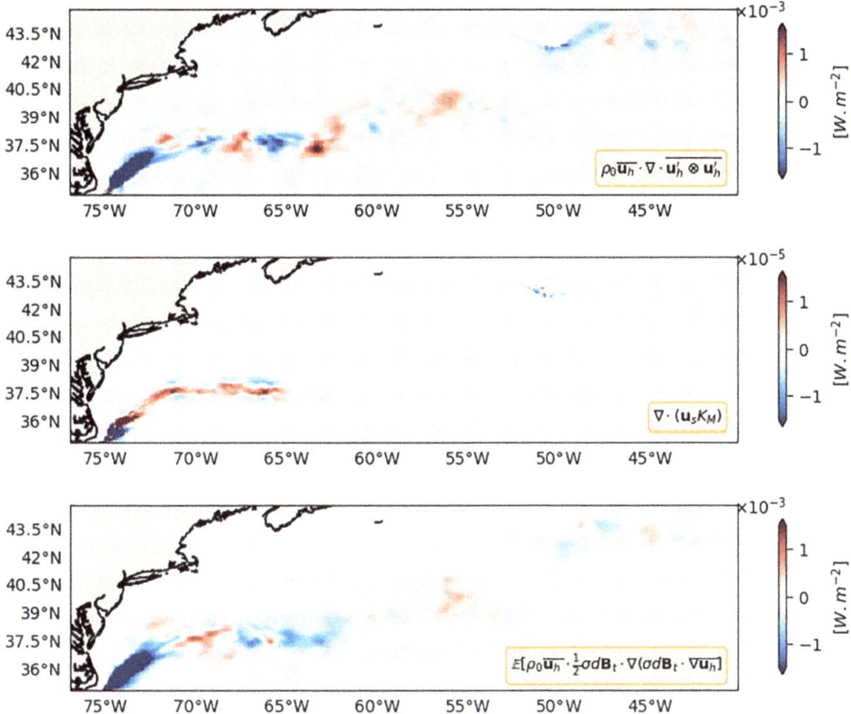

Fig. 8.3 Illustration of the mean–eddy kinetic energy transfer term MEC (upper panel), along with its stochastic equivalent associated with Itô–Stokes drift (center panel) and dissipation ($DADV$, bottom panel). Horizontal maps are taken at $z = 94$ m on January 1, 2010. Note the different scale for the Itô–Stokes drift component

direction. Along the current path, alternating bands of negative and positive energy transfer also appear, indicating a complex spatial organization of eddy–mean interactions. As indicated previously , to ensure consistency between the deterministic and stochastic representations of the mean–eddy kinetic energy transfer term, the decorrelation timescale τ—which governs the stochastic noise and its associated covariance tensor (see Eqs.(8.21) and (8.22))—has been locally adjusted via a procedure described in Appendix 8.4. The value of τ reflects the desired degree of smoothness of the stochastic representation and is here estimated over neighborhoods of 20×20 grid points. Figure 8.4 presents the resulting map of τ.

8.3.2 Convergence of the Statistical Moments

We begin by assessing the convergence of ensemble statistics for the mean kinetic energy (K_M) and eddy kinetic energy (K_E), as illustrated in Fig. 8.5. These

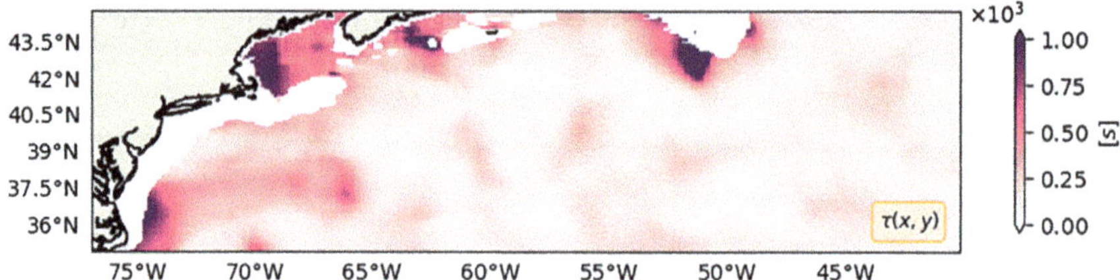

Fig. 8.4 Time scale map τ. This time scale is computed to align the amplitude of $DADV$ and MEC (8.19) (Appendix 8.4). The time scale is piecewise constant in space and has been computed over local neighborhoods of 20 by 20 grid points

diagnostics show that ensemble-averaged values of K_M and K_E stabilize after approximately 20 and 30 members, respectively. The analysis is repeated for four different dates to account for potential seasonal variability. While some minor fluctuations are observed, all curves exhibit convergence, confirming that the 48-member ensemble is sufficient for reliable estimation of K_M and K_E. However, achieving convergence for more complex diagnostics, such as the eddy–mean flow energy transfer terms, is substantially more difficult. As we will show, adopting a stochastic framework significantly improves this convergence behavior, in comparison to direct empirical statistics performed with a small ensemble.

In nonlinear systems governed by the primitive equations, the underlying attractor is typically high-dimensional and complex. Consequently, the associated statistical distributions often depart from Gaussianity. Capturing such non-Gaussian features— including skewness (asymmetry) and kurtosis (heavy tails—requires the analysis of higher-order statistical moments. Robust estimation of these moments is thus essential for accurately characterizing variability and uncertainty.

Figure 8.6 presents a comparison of the convergence behavior of the first four statistical moments—mean, variance, skewness, and kurtosis—for both the deterministic MEC term and its stochastic counterpart, $DADV$. For MEC, statistics are computed directly from progressively larger subsets of the ensemble of 48 members. For $DADV$, convergence is evaluated by increasing the number of stochastic samples drawn from the underlying random velocity field. This field is reconstructed using the full set of empirical orthogonal functions (EOFs) obtained from the ensemble, thereby preserving all the original statistical information.

To test the robustness of the stochastic approach, four different sizes of EOF bases are considered ($N_m = 12, 24, 36$, and 48). In all cases, the statistics of $DADV$ exhibit reliable convergence, confirming the method still remains consistent for a restricted number of modes.

The results show that, even with 48 members, the MEC moments remain noisy, with convergence errors on the order of 5%. In contrast, the stochastic method achieves full convergence for the first two moments, and provides more accurate estimates for skewness and kurtosis than the deterministic approach. Beyond the mean error level, the slope of convergence—as measured by the RMSE versus the

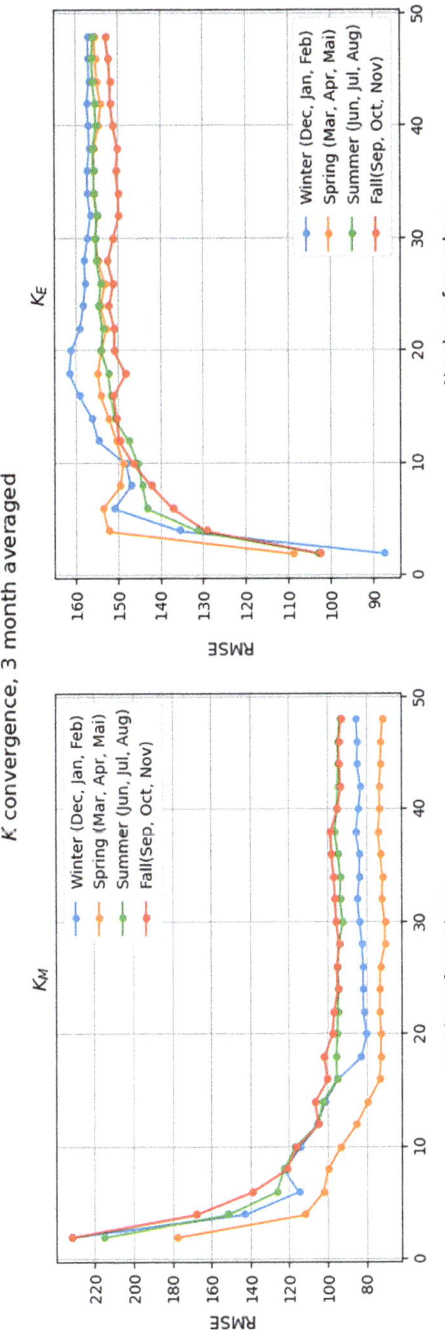

Fig. 8.5 Convergence of ensemble estimates of mean kinetic energy (K_M) and eddy kinetic energy (K_E). RMS values are computed for an increasing number of members, averaged horizontally over the domain (35°N–44°N; 75°W–40°W) at $z = 94$ m, for January 1, 2010

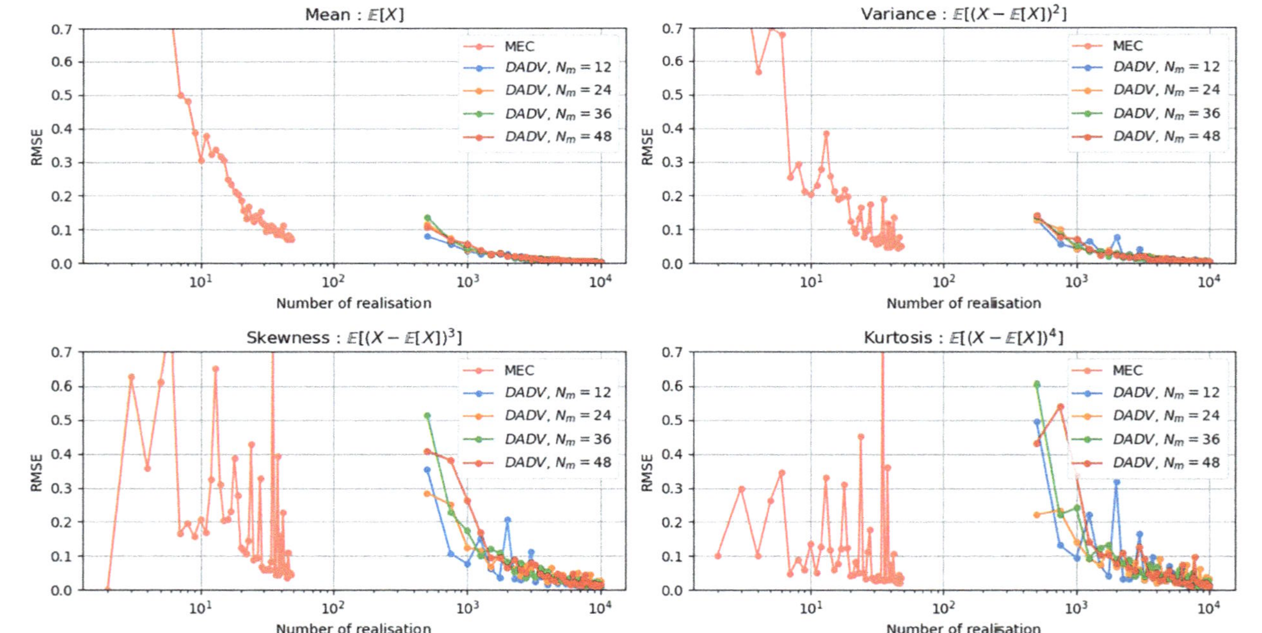

Fig. 8.6 Convergence of the first four statistical moments (mean, variance, skewness, and kurtosis) for the stochastic ($DADV$) and deterministic (MEC) terms. The vertical axis represents the absolute RMSE of each moment. For MEC, the number of ensemble members ranges from 2 to 48. For $DADV$, the number of stochastic realizations ranges from 250 to 10,000, and results are shown for four different EOF basis sizes: $N_m = 12$ 24, 36, and 48 (full set). The x-axis is in logarithmic scale. All diagnostics are computed over the domain (35°N–44°N, 75°W–40°W) at $z = 94$ m on January 1, 2010, with horizontal averaging

number of samples—indicates that the MEC statistical moments are still far from being fully converged.

From a computational standpoint, the stochastic method is highly efficient: random samples can be generated on demand, requiring storage only for the EOF basis. By comparison, improving the deterministic diagnostics would necessitate significantly larger ensembles, incurring substantial computational and storage costs.

Figure 8.7 displays the spatial structure of the first four moments of the $DADV$ term. These diagnostics highlight key regions of variability and intermittency, notably near the Gulf Stream detachment and around 65°W. In these areas, pronounced skewness and kurtosis reveal significant departures from Gaussianity, indicating the presence of extreme and intermittent energy transfer events.

8.3.3   Temporal Evolution of Energy Transfer Terms

Figure 8.8 presents the temporal evolution of the deterministic MEC and stochastic $DADV$ energy transfer terms, integrated over three spatially distinct regions along the Gulf Stream: the detachment zone (blue), a downstream segment (green), and the tail of the current (red). In the detachment region, both diagnostics exhibit a relatively steady and consistent energy loss from the mean flow, with $DADV$ and MEC closely overlapping. This agreement suggests that the deterministic MEC term is well converged in this specific region, and both frameworks yield similar interpretations. Downstream (green box), the behavior of the two terms begins to diverge. The $DADV$ diagnostic indicates a persistent energy gain, while the MEC term becomes unstable, oscillating between positive and negative values with no clear trend. In the far downstream region (red box), $DADV$ again points to a consistent energy loss, whereas MEC remains erratic and unstable.

These oscillations in the MEC diagnostic seem to indicate a lack of statistical convergence, particularly in the downstream regions, suggesting that they may stem from numerical noise or sampling artifacts rather than genuine physical variability. In contrast, the $DADV$ term, derived within the Location Uncertainty framework, exhibits a clearer spatial delineation between regions of energy gain and loss. Nevertheless, the time series of energy transfer across all three regions remains highly variable. Despite the improved statistical sampling enabled by the stochastic approach, the underlying dynamics of eddy–mean flow energy exchanges are still not fully understood. Further research is required to evaluate the robustness of LU-based diagnostics and to establish whether they can provide a reliable and interpretable representation of these complex interactions.

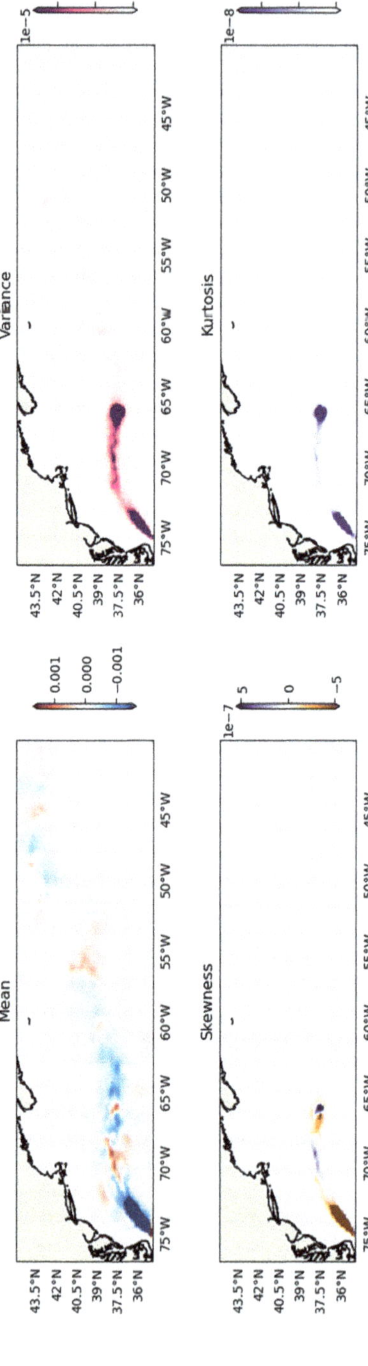

Fig. 8.7 Spatial distribution of the first four statistical moments (mean, variance, skewness, kurtosis) of the $DADV$ term on January 1, 2010

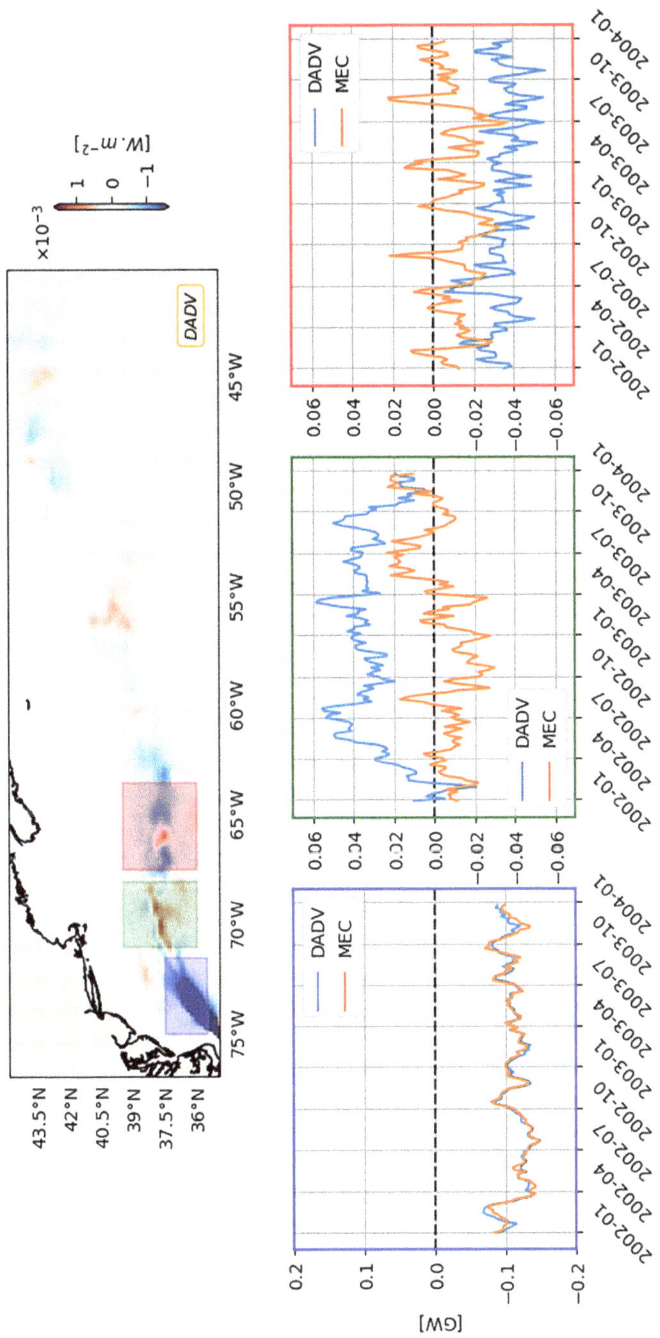

Fig. 8.8 Time series of the *DADV* (stochastic) and *MEC* (deterministic) terms, spatially integrated over three subregions: near the Gulf Stream detachment (blue), a downstream region (green), and the far tail of the current (red)

8.4 Conclusion

In this study, we investigated eddy–mean flow interactions in large-scale ocean circulation, focusing on the limitations of ensemble-based Reynolds averaging in General Circulation Models (GCMs). To address these challenges, we employed the Location Uncertainty (LU) framework, introducing a stochastic approach that enhances the statistical representation of kinetic energy exchanges between mean and eddy components.

Our results demonstrate that the stochastic formulation enables more robust and statistically convergent diagnostics of the Reynolds stress work. Specifically, the classical computation of the mean–eddy coupling term (MEC) using a 48-member ensemble suffers from significant convergence issues, limiting its reliability for energy transfer diagnostics. In contrast, the LU-based stochastic approach reconstructs eddy fluctuations through a reduced-order model—here based on Proper Orthogonal Decomposition (POD)—while preserving the full statistical information contained in the original ensemble. By incorporating stochastic diffusion and the Itô–Stokes drift, this method offers a physically consistent representation of eddy–mean flow interactions diagnostic.

The analysis of statistical moments further highlights the advantages of the LU approach, especially for higher-order metrics like skewness and kurtosis, which are crucial for capturing the non-Gaussian characteristics of energy transfer. A key strength of this framework is its ability to generate large ensembles at minimal additional computational cost, thereby improving statistical convergence and enhancing the accuracy of diagnostics. Spatial and temporal analyses reveal that energy exchanges are highly localized, exhibiting intense and coherent transfer patterns—for example, at the Gulf Stream detachment. The LU-based diagnostic, $DADV$, captures these localized phenomena more distinctly than its deterministic counterpart, despite persistent strong temporal variability. Moreover, these regions display heavy-tailed distributions, indicating the presence of rare and extreme events that deviate from classical Gaussian assumptions.

Looking ahead, future work will focus on refining the stochastic noise parameterization, investigating a wider range of decorrelation timescales, and extending the LU framework to additional energy transfer diagnostics. While this study concentrated on the mean-to-eddy energy transfer via Reynolds stress work (MEC), other critical processes—such as those involved in the eddy kinetic energy budget—warrant further investigation, including both local and non-local contributions (Jamet et al. 2022). A natural progression will be to develop a consistent stochastic definition of eddy kinetic energy, which will facilitate the characterization of further transfer mechanisms. By improving statistical sampling and capturing complex flow variability, the LU framework provides a promising avenue toward more accurate and physically consistent diagnostics of energy transfers in ocean circulation models.

Acknowledgements This research was supported by the ERC EU project 856408-STUOD and the French national LEFE program (Les Enveloppes Fluides et l'Environnement), under the SNOEMI project.

Appendix A: Proper Orthogonal Decomposition (POD)

Proper Orthogonal Decomposition (POD) is a widely used technique in fluid dynamics for extracting a set of orthogonal modes that optimally represent the dataset's variance (Holmes et al. 1996). As a dimensionality-reduction method, POD identifies patterns that capture the energy-dominant features of the data. It achieves this by decomposing the dataset into a set of orthogonal basis functions (modes) modulated by corresponding temporal coefficients. These modes are ranked by the amount of variance they explain in the data, enabling the construction of a reduced-order representation by retaining only the most significant modes. The decomposition works on a set of velocity fluctuation $\boldsymbol{u}' = (u', v')$ organized in a matrix $\boldsymbol{U} \in \mathbb{R}^{n \times m}$, where n is twice the number of spatial points (to include both u' and v') and m is the number of snapshots. Typically, these snapshots are part of a time series, but in this work, the decomposition is carried across the ensemble dimension. This means we work at a fixed time with $m = 48$, corresponding to the 48 ensemble members. Each column of $\boldsymbol{U}$ contains the velocity field $\boldsymbol{u}'_i \in \mathbb{R}^n$ for a specific ensemble member. Thus, the data matrix is formed as

$$\boldsymbol{U} = [\boldsymbol{u}'_1, \boldsymbol{u}'_2, \ldots, \boldsymbol{u}'_m] \in \mathbb{R}^{n \times m}. \tag{8.25}$$

The "direct" method involves then computing the spatial covariance matrix $\boldsymbol{C} \in \mathbb{R}^{n \times n}$ of the data and solving the associated eigenvalue problem:

$$\boldsymbol{C} = \frac{1}{m-1} \boldsymbol{U}^T \boldsymbol{U}, \quad \boldsymbol{C}\phi = \lambda\phi. \tag{8.26}$$

This method is efficient for velocity field of small dimension. Once the eigenvectors ϕ are computed, the original data can be projected onto the new basis to compute the POD coefficients:

$$A = \boldsymbol{U}\phi. \tag{8.27}$$

When the number of snapshots m (or the ensemble dimension here) is much smaller than the number of spatial points n, the snapshot method (Sirovich 1987; Wang et al. 2016) is preferred. It reduces computational intensity by constructing the covariance matrix in the snapshot (ensemble) dimension:

$$\boldsymbol{C}_s = \frac{1}{m-1} \boldsymbol{U}\boldsymbol{U}^T. \tag{8.28}$$

The eigenvalue problem and the spatial modes are derived as

$$\boldsymbol{C}_s \boldsymbol{A}_s = \lambda \boldsymbol{A}_s, \quad \phi_s = \boldsymbol{U}^T \boldsymbol{A}_s. \tag{8.29}$$

Although the snapshot method is computationally faster, it introduces a normalization factor between modes and coefficients. To ensure equivalence with the direct method, normalization is applied:

$$\phi = \frac{\phi_s}{||\phi_s||}, \quad A = ||\phi_s||A_s. \tag{8.30}$$

The system can be approximately reconstructed using a truncated sum of the most significant POD modes. The reconstructed velocity field $\tilde{u}$ is then given by

$$\tilde{u}(x, t) \approx \sum_{k=1}^{N_{modes}} A_k(t)\phi_k(x) = \sum_{k=1}^{N_{modes}} A_{s,k}(t)\phi_{s,k}(x). \tag{8.31}$$

Degree of approximation of the flow can be estimated by comparing the eddy kinetic energy over the domain as $EKE = \int u' \cdot u' dx$ and compare it to the sum of eigenvalues $\sum \lambda_k$, which represents the energy of the approximation.

Appendix B: Stochastic Evolution Equation for K_M

$$\rho_0 \overline{u_h} \cdot \overline{d_t u_h} = \rho_0 \overline{u_h} \cdot \overline{\left(-\nabla \cdot (u_h \otimes (u^* dt + \sigma d B_t)) + \frac{1}{2}\nabla \cdot (a\nabla u_h)dt - fk \times u_h dt - \frac{1}{\rho_0}\nabla d P\right)}$$

$$d_t K_M = -\rho_0 \overline{u_h} \cdot \left(\nabla \cdot \overline{(u_h \otimes (u^* dt + \sigma d B_t))}\right) + \rho_0 \overline{u_h} \cdot \frac{1}{2}\nabla \cdot (a\nabla \overline{u_h})dt$$
$$- \underbrace{\overline{\rho_0 u_h \cdot fk \times u_h}}_{=0} dt - \overline{u_h} \cdot \nabla d P$$

$$d_t K_M = -\rho_0 \overline{u_h} \cdot \left(\nabla \cdot \overline{(u_h \otimes u^*)}dt + \nabla \cdot \underbrace{\overline{(u_h \otimes \sigma d B_t)}}_{=0}\right) + \rho_0 \overline{u_h} \cdot \left(\frac{1}{2}\nabla \cdot (a\nabla \overline{u_h})dt\right)$$
$$- \nabla \overline{u}\,\overline{P}dt - \overline{w}\overline{b}dt \tag{8.32}$$

With all the terms in $d B_t$ cancelling in mean, we obtain the time derivative $\partial_t K_M$:

$$\partial_t K_M = -\rho_0 \overline{u_h} \cdot \left(\nabla \cdot (\overline{u_h} \otimes \overline{u})\right) + \rho_0 \overline{u_h} \cdot \left(\nabla \cdot (\overline{u_h} \otimes u_s)\right) + \rho_0 \overline{u_h} \cdot \left(\frac{1}{2}\nabla \cdot (a\nabla \overline{u_h})\right) - \nabla \overline{u}\,\overline{P} - \overline{w}\overline{b}, \tag{8.33}$$

which simplifies as

$$\partial_t K_M = -\nabla \cdot (\overline{u} K_M) + \nabla \cdot (u_s K_M) + \rho_0 \overline{u_h} \cdot \left(\frac{1}{2}\nabla \cdot (a\nabla \overline{u_h})\right) - \nabla \overline{u}\,\overline{P} - \overline{w}\overline{b}. \tag{8.34}$$

Appendix C: Milstein-Type Scheme for Transport Noise

In this section, we describe the time-stepping scheme in the stochastic case, focusing on the noise advection and diffusion terms in the stochastic transport equation for a scalar quantity q (i.e., the resolved advection velocity, $\boldsymbol{u}^*$, is here assumed to be null without loss of generalization). In compact form, the equation is expressed as follows:

$$d_t q = f(q, t)\, dt + \sum_{i=1}^{m} g^i(q, t)\, d\beta_t^i, \tag{8.35}$$

$$g^i(q, t) = -\boldsymbol{\varphi}_t^i \cdot \nabla q, \tag{8.36}$$

$$f(q, t) = \frac{1}{2}\nabla \cdot (\boldsymbol{a}_t \nabla q) = \frac{1}{2}\sum_{i=1}^{m} g^i \circ g^i(q, t), \tag{8.37}$$

where $(\boldsymbol{\varphi}_t^i)_{i=1,\dots,m}$ represents a set of non-divergent ($\nabla \cdot \boldsymbol{\varphi}_t^i = 0$) basis functions at time t, and $(\beta^i)_{i=1,\dots,m}$ is a sequence of independent standard Brownian motions. The last equality in (8.37) derives from the definition of the diffusion tensor, $\boldsymbol{a}_t = \sum_{i=1}^{m} \boldsymbol{\varphi}_t^i (\boldsymbol{\varphi}_t^i)^{\mathsf{T}}$, along with the divergence-free property of $\boldsymbol{\varphi}_t^i$. We now consider the semi-discrete form of this equation:

$$dq_h(t) = f_h\big(q_h(t), t\big)\, dt + \sum_{i=1}^{m} g_h^i\big(q_h(t), t\big)\, d\beta_t^i. \tag{8.38}$$

Here, the subscript h denotes quantities or operators that have been discretized in space. Applying the functional extension of the Itô formula (Cont and Fournie 2010) to both f_h and g_h^i, we obtain

$$q_h(t) = q_h(s) + \int_s^t f_h\big(q_h(r), r\big)\, dr + \int_s^t \sum_{i=1}^{m} g_h^i\big(q_h(r), r\big)\, d\beta_r^i$$

$$+ \int_s^t \sum_{j=1}^{m} \dot{g}_h^j\left(\int_s^r \sum_{i=1}^{m} g_h^i\big(q_h(p), p\big)\, d\beta_p^i, r \right) d\beta_r^j + R, \tag{8.39}$$

where R denotes higher-order terms in the Milstein scheme, and $\dot{g}_h^i$ denotes the derivative of g_h^i with respect to function q_h. From (8.36), we find that $\dot{g}_h^i = -\boldsymbol{\varphi}^i \cdot \nabla = g_h^i$. Consequently, we obtain the following first-order approximation of the above integral problem:

$$q_h^{n+1} = q_h^n + f_h(q_h^n, t_n)\,\Delta t + \sum_{i=1}^{m} g_h^i(q_h^n, t_n)\,\Delta\beta^i$$

$$+ \sum_{i,j=1}^{m} g_h^j \circ g_h^i(q_h^n, t_n) \underbrace{\int_{t_n}^{t_{n+1}} \int_{t_n}^{r} \mathrm{d}\beta_p^i \mathrm{d}\beta_r^j}_{:=I_{ij}}, \tag{8.40}$$

where q^n is the value of q at time t_n, $\Delta t = t_{n+1} - t_n$ is the time step, and $\Delta\beta \sim \sqrt{\Delta t}\,\mathcal{N}(0, 1)$ represents the Brownian motion increment. The double integral I_{ij} can be decomposed into a symmetric and an anti-symmetric part:

$$I_{ij} = \frac{1}{2}\underbrace{(I_{ij} + I_{ji})}_{:=S_{ij}} + \frac{1}{2}\underbrace{(I_{ij} - I_{ji})}_{:=A_{ij}}, \tag{8.41}$$

where the symmetric component expands as follows:

$$\begin{aligned}
S_{ij} &= \int_{t_n}^{t_{n+1}} \int_{t_n}^{s} \mathrm{d}\beta_r^i \mathrm{d}\beta_s^j + \int_{t_n}^{t_{n+1}} \int_{t_n}^{s} \mathrm{d}\beta_r^j \mathrm{d}\beta_s^i \\
&= \int_{t_n}^{t_{n+1}} \beta_s^i\,\mathrm{d}\beta_s^j + \int_{t_n}^{t_{n+1}} \beta_s^j\,\mathrm{d}\beta_s^i - \int_{t_n}^{t_{n+1}} \beta_{t_n}^i\,\mathrm{d}\beta_s^j - \int_{t_n}^{t_{n+1}} \beta_{t_n}^j\,\mathrm{d}\beta_s^i \\
&= \beta_{t_{n+1}}^i \beta_{t_{n+1}}^j - \beta_{t_n}^i \beta_{t_n}^j - \delta_{ij}\Delta t - \beta_{t_n}^i \beta_{t_{n+1}}^j + \beta_{t_n}^i \beta_{t_n}^j - \beta_{t_{n+1}}^i \beta_{t_n}^j + \beta_{t_n}^i \beta_{t_n}^j \\
&= \underbrace{(\beta_{t_{n+1}}^j - \beta_{t_n}^j)}_{\Delta\beta^j} \underbrace{(\beta_{t_{n+1}}^i - \beta_{t_n}^i)}_{\Delta\beta^i} - \delta_{ij}\Delta t
\end{aligned} \tag{8.42}$$

using the Itô's integration-by-part formula (Kunita 1990) for two independent Brownian motions. Therefore, I_{ij} can be rewritten as

$$I_{ij} = \frac{1}{2}\big(\Delta\beta^i \Delta\beta^j - \delta_{ij}\Delta t + A_{ij}\big). \tag{8.43}$$

Substituting this into (8.40) and recalling (8.37), we derive the Milstein scheme for the semi-discrete problem (8.38):

$$q_h^{n+1} = q_h^n + \sum_{i=1}^{m} g_h^i(q_h^n, t_n)\,\Delta\beta^i + \frac{1}{2}\sum_{i,j=1}^{m} g_h^j \circ g_h^i(q_h^n, t_n)\big(\Delta\beta^j \Delta\beta^i + A_{ij}\big). \tag{8.44}$$

Accordingly, the diffusion term in the continuous formulation is precisely balanced by the diagonal component of the higher-order double integral. The anti-symmetric component, A_{ij}, known as the Lévy area process, is notably challenging to simulate. However, as numerically supported in recent studies (Boulvard and Mémin 2023; Fiorini et al. 2023), for LU types dynamics, the Milstein scheme, without Lévy area, significantly reduces approximation error relative to the Euler scheme, though the strong convergence order remains the same. Consequently, we choose to disregard A_{ij} in practical applications. Additionally, as illustrated in Mémin et al. (2024), for linear waves, the proposed scheme enables to implement effectively the energy balance between the noise and the diffusion through an implicit discretization of the anisotropic diffusion operator involved (via the iterative advection term).

With the discrete noise term given by $\boldsymbol{\sigma}^n \Delta \boldsymbol{B} = \sum_{i=1}^{m} \boldsymbol{\varphi}_{t_n}^i \, \Delta \beta^i$, generated at each time step, the previous scheme can be expressed in a more compact form as follows:

$$q_h^{n+1} = q_h^n - (\boldsymbol{\sigma}^n \Delta \boldsymbol{B} \cdot \nabla q^n)_h + \frac{1}{2}\big(\boldsymbol{\sigma}^n \Delta \boldsymbol{B} \cdot \nabla (\boldsymbol{\sigma}^n \Delta \boldsymbol{B} \cdot \nabla q^n)\big)_h. \tag{8.45}$$

Incorporating transport noise into the Leapfrog scheme used in the current version of NEMO, we propose the following multi-scale approach:

$$\widetilde{q}_h = q_h^{n-1} + F_h(q_h^n, t_n)\Delta t$$

$$q_h^{n+1} = \widetilde{q}_h + F_h(q_h^n, t_n)\Delta t - (\boldsymbol{\sigma}^n \Delta \boldsymbol{B} \cdot \nabla q^n)_h + \frac{1}{2}(\boldsymbol{\sigma}^n \Delta \boldsymbol{B} \cdot \nabla (\boldsymbol{\sigma}^n \Delta \boldsymbol{B} \cdot \nabla q^n))_h. \tag{8.46}$$

In this scheme, F gathers the classical resolved advection together with the advection of q by the Itô-Stokes drift $\boldsymbol{v}_s = \frac{1}{2}\nabla \cdot \boldsymbol{a}$ in the stochastic formulation. This approach is also adaptable to other schemes, such as Adams–Bashforth or Runge–Kutta, widely used in OGCMs, where transport noise would be included in the final correction.

Appendix D: Decorrelation Timescale τ

A decorrelation timescale τ is required for an accurate representation of the energy transfer. To that end, the stochastic term should be representative of the MEC as written in (8.19):

$$-\rho_0 \overline{\boldsymbol{u}_h} \cdot \nabla \cdot (\overline{\boldsymbol{u}_h' \otimes \boldsymbol{u}_h'}) = \nabla \cdot (\boldsymbol{u}_s K_M) + \rho_0 \overline{\boldsymbol{u}_h} \cdot \mathbb{E}\big((\boldsymbol{\sigma} d\boldsymbol{B}_t \cdot \nabla)(\boldsymbol{\sigma} d\boldsymbol{B}_t \cdot \nabla)\overline{\boldsymbol{u}}_h\big),$$

which reads integrating in space:

$$\int_\Omega -\rho_0 \overline{\boldsymbol{u}_h} \cdot \nabla \cdot (\overline{\boldsymbol{u}_h' \otimes \boldsymbol{u}_h'})\, d\boldsymbol{x} = \int_\Omega \nabla \cdot ((\frac{1}{2}\nabla \cdot \boldsymbol{a})K_M)$$
$$+ \rho_0 \overline{\boldsymbol{u}_h} \cdot \mathbb{E}\big((\boldsymbol{\sigma} d\boldsymbol{B}_t \cdot \nabla)(\boldsymbol{\sigma} d\boldsymbol{B}_t \cdot \nabla)\overline{\boldsymbol{u}}_h\big)\, d\boldsymbol{x},$$

and expressed in spectral form, where the decorrelation time is explicitly involved:

$$\int_\Omega -\rho_0 \overline{u_h} \cdot \nabla \cdot (\overline{u_h' \otimes u_h'})\, dx = \int_\Omega \nabla \cdot ((\frac{1}{2}\nabla \cdot (\sum_k \tau \phi_k \phi_k^T))K_M)$$

$$+ \rho_0 \overline{u_h} \cdot \mathbb{E}((\sum_k \sqrt{\tau}\phi_k dB_k \cdot \nabla)(\sum_k \sqrt{\tau}\phi_k dB_k \cdot \nabla)\overline{u_h})\, dx. \tag{8.47}$$

To infer a local expression of the time decorrelation, we consider any neighborhood $\nu(x)$ centered at (x, y), within which τ is assumed constant. This allows us to compute a piecewise approximation of $\tau(x, y)$, yielding the following expression for the timescale:

$$\tau(x) = A^{-1} \int_{\nu(x)} -\rho_0 \overline{u_h} \cdot \nabla \cdot (\overline{u_h' \otimes u_h'})\, dx, \tag{8.48}$$

where the factor reads

$$A = \int_{\nu(x)} \nabla \cdot ((\frac{1}{2}\nabla \cdot (\sum_k \phi_k \phi_k^T))K_M) + \rho_0 \overline{u_h} \cdot \mathbb{E}((\sum_k \phi_k dB_k \cdot \nabla)(\sum_k \phi_k dB_k \cdot \nabla)\overline{u_h})\, dx. \tag{8.49}$$

References

Adcroft AJ (1995) Numerical algorithms for use in a dynamical model of the ocean. PhD thesis, University of London

Aluie H, Hecht M, Vallis GK (2018) Mapping the energy cascade in the North Atlantic Ocean: the coarse-graining approach. 48(2):225–244

Bachman SD (2019) The GM+ E closure: a framework for coupling backscatter with the Gent and McWilliams parameterization. Ocean Model 136:85–106

Bauer W, Chandramouli P, Li L, Mémin E (2020a) Deciphering the role of small-scale inhomogeneity on geophysical flow structuration: a stochastic approach. J Phys Oceanogr 983–1003

Bauer W, Chandramouli P, Li L, Mémin E (2020b) Stochastic representation of mesoscale eddy effects in coarse-resolution barotropic models. Ocean Model 151:101646

Boulvard P-M, Mémin E (2023) Diagnostic of the Lévy area for geophysical flow models in view of defining high order stochastic discrete-time schemes. Found Data Sci 6(1):1–21

Brecht R, Li L, Bauer W, Mémin E (2021) Rotating shallow water flow under location uncertainty with a structure-preserving discretization. J Adv Model Earth Syst 13:e2021MS002492

Brodeau L, Barnier B, Treguier A-M, Penduff T, Gulev S (2010) An ERA40-based atmospheric forcing for global ocean circulation models. 31(3–4):88–104

Chapron B, Dérian P, Mémin E, Resseguier V (2018) Large-scale flows under location uncertainty: a consistent stochastic framework. Quart J Royal Meteorol Soc 144(710):251–260

Cont R, Fournie D (2010) A functional extension of the ITO formula. CR Math 348(1):57–61

Deremble B, Wienders N, Dewar WK (2013) CheapAML: a simple, atmospheric boundary layer model for use in ocean-only model calculations. 141(2):809–821

Dussin R, Barnier B, Brodeau L, Molines J-M (2016) The making of the Drakkar Forcing Set DFS5. DRAKKAR/MyOcean Rep. 01–04:16

Eden C, Greatbatch RJ (2008) Towards a mesoscale eddy closure. 20(3):223–239

Fairall CW, Bradley EF, Hare JE, Grachev AA, Edson JB (2003) Bulk parameterization of air–sea fluxes: updates and verification for the COARE algorithm. 16(4):571–591

Fiorini C, Boulvard P-M, Li L, Mémin E (2023) A two-step numerical scheme in time for surface quasi geostrophic equations under location uncertainty. In: Stochastic transport in upper ocean dynamics. Springer International Publishing, pp 57–67

Gent PR, McWilliams JC (1990) Isopycnal mixing in ocean circulation models. 20(1):150–155

Grooms I, Loose N, Abernathey R, Steinberg JM, Bachman SD, Marques G, Guillaumin AP, Yankovsky E (2021) Diffusion-based smoothers for spatial filtering of gridded geophysical data. J Adv Model Earth Syst 13(9):e2021MS002552

Grooms I, Nadeau L-P, Shafer Smith K (2013) Mesoscale eddy energy locality in an idealized ocean model. 43(9):1911–1923

Holmes P, Lumley JL, Berkooz G (1996) Turbulence, coherence structures, dynamical systems and symetry. Cambridge University Press

Jamet Q, Dewar WK, Wienders N, Deremble B (2019) Fast warming of the surface ocean under a climatological scenario. 46(7):3871–3879

Jamet Q, Leroux S, Dewar WK, Penduff T, Le Sommer J, Molines J-M, Gula J (2022) Non-local Eddy-mean kinetic energy transfers in submesoscale-permitting ensemble simulations. J Adv Model Earth Syst 14(10):e2022MS003057

Jansen MF, Adcroft A, Khani S, Kong H (2019) Toward an energetically consistent, resolution aware parameterization of ocean mesoscale eddies. 11(8):2844–2860

Kang D, Curchitser EN (2015) Energetics of eddy–mean flow interactions in the Gulf Stream region. 45(4):1103–1120

Kunita H (1990) Stochastic flows and stochastic differential equations. Cambridge University Press

Leutbecher M, Palmer TN (2008) Ensemble forecasting. J Comput Phys 227(7):3515–3539

Li L, Deremble B, Lahaye N, Mémin E (2023) Stochastic data-driven parameterization of unresolved eddy effects in a baroclinic quasi-geostrophic model. J Adv Model Earth Syst 15(2)

Marshall DP, Maddison JR, Berloff PS (2012) A framework for parameterizing eddy potential vorticity fluxes. 42(4):539–557

Marshall J, Adcroft A, Hill C, Perelman L, Heisey C (1997) A finite-volume, incompressible Navier Stokes model for studies of the ocean on parallel computers. 102(C3):5753–5766

Matsuta T, Masumoto Y (2021) Modified view of energy budget diagram and its application to the Kuroshio extension region. J Phys Oceanogr 51(4):1163–1175

Mémin E, Li L, Lahaye N, Tissot G, Chapron B (2024) Linear wave solutions of a stochastic shallow water model. In: Stochastic transport in upper ocean dynamics II. Springer International Publishing, pp 223–245

Molines J-M, Barnier B, Penduff T, Treguier A-M, Le Sommer J (2014) ORCA12. L46 climatological and interannual simulations forced with DFS4. 4: GJM02 and MJM88. Drakkar Group Experiment Rep. Technical report, GDRI-DRAKKAR-2014-03-19, 50 pp. http://www.drakkar-ocean.eu/publications/reports/orca12_reference_experiments_2014

Mémin E (2014) Fluid flow dynamics under location uncertainty. Geophys Astrophys Fluid Dyn 108(2):119–146

Palmer TN (2012) Towards the probabilistic earth-system simulator: a vision for the future of climate and weather prediction. Quart J Royal Meteorol Soc 138(665):841–861

Redi MH (1982) Oceanic isopycnal mixing by coordinate rotation. 12(10):1154–1158

Resseguier V, Mémin E, Chapron B (2017a) Geophysical flows under location uncertainty, Part I Random transport and general models. Geophys Astrophys Fluid Dyn 111:149–176

Resseguier V, Mémin E, Chapron B (2017b) Geophysical flows under location uncertainty, Part II Quasi-geostrophy and efficient ensemble spreading. Geophys Astrophys Fluid Dyn 111:177–208

Resseguier V, Mémin E, Chapron B (2017c) Geophysical flows under location uncertainty, Part III SQG and frontal dynamics under strong turbulence conditions. Geophys Astrophys Fluid Dyn 111:209–227

Resseguier V, Mémin E, Heitz D, Chapron B (2017d) Stochastic modelling and diffusion modes for proper orthogonal decomposition models and small-scale flow analysis. J Fluid Mech 826:888–917

Sérazin G, Penduff T, Grégorio S, Barnier B, Molines J-M, Terray L (2015) Intrinsic variability of sea level from global ocean simulations: spatiotemporal scales. 28(10):4279–4292

Sirovich L (1987) Turbulence and the dynamics of coherent structures. Q Appl Math 45:561–590

Stainforth DA, Allen MR, Tredger ER, Smith LA (2007) Confidence, uncertainty and decision-support relevance in climate predictions. Philos Trans Royal Soc A Math Phys Eng Sci 365(1857):2145–2161

Tucciarone FL, Mémin E, Li L (2023) Primitive equations under location uncertainty: analytical description and model development. In: Chapron B, Crisan D, Holm D, Mémin E, Radomska A (eds) Stochastic transport in upper ocean dynamics. Mathematics of planet earth, vol 10. Springer International Publishing, Cham, pp 287–300

Uchida T, Deremble B, Penduff T (2021) The seasonal variability of the ocean energy cycle from a quasi-geostrophic double gyre ensemble. Fluids 6(6):206

Uchida T, Jamet Q, Dewar WK, Le Sommer J, Penduff T, Balwada D (2022) Diagnosing the thickness-weighted averaged eddy-mean flow interaction from an eddying North Atlantic ensemble: the Eliassen-Palm Flux. 14(5):e2021MS002866

Uchida T, Jamet Q, Poje A, Dewar WK (2022) An ensemble-based eddy and spectral analysis, with application to the Gulf stream. J Adv Model Earth Syst 14(4):e2021MS002692

Wang Z, McBee B, Iliescu T (2016) Approximate partitioned method of snapshots for POD. J Comput Appl Math 307:374–384

Chapter 9
Sequential Importance Resampling
of Particle-in-Cell for Swell Dynamics

Tom Protin, Valentin Resseguier, Momme Hell, Bertrand Chapron, and Ronan Fablet

Abstract A method is proposed to model the propagation and directional spread of ocean surface swell systems. It follows the development of the Particle-in-Cell for Efficient Swell (PiCLES) wave modeling framework (Hell et al. 2024). In particular, we study how this model spreads an initial energy spectrum in space and time. An ensemble approach is considered to sample the energy spectrum with individual particles, i.e., wave packets. Sequential importance resampling techniques are then considered to control how the discrete distribution, i.e., the particle ensemble, can match the true distribution after a time step without increasing the ensemble size. Tests are performed and compared to analytical solutions driven by the expected diffuse geometric optic behavior.

9.1 Introduction

Storms over the ocean produce long surface gravity waves that propagate as swell out of their generation area. Away from that core, nonlinear interactions become negligible, and very long-period swells can be identified and followed, propagating over very large distances, up to halfway around the globe (Munk et al. 1963; Collard et al. 2009). Associated with a continuous directional spreading, a swell field may eventually cover a full ocean basin and has a lifetime that can extend over a few

T. Protin (✉) · B. Chapron
IFREMER, Plouzané, France
e-mail: tom.protin@ifremer.fr

V. Resseguier
INRAE, Rennes, France

M. Hell
Woods Hole Oceanographic Institution, Woods Hole, MA, USA

R. Fablet
IMT Atlantique, Plouzané, France

B. Chapron et al. (eds.), *Stochastic Transport in Upper Ocean Dynamics IV*, Mathematics of Planet Earth 15, https://doi.org/10.1007/978-3-032-12749-5_9

weeks. These widely spreading and persistent swell fields are made possible by a limited energy loss.

State-of-the-art wave forecast models are today very efficient and formulated in terms of a wave energy balance discretized into a large number of spectral bands. This can provide an explicit representation and evolution of all frequencies and directions. But in realistic conditions, especially for the description of highly directional swell systems, the state vector of spectral models is much sparser and not necessarily well represented within the prescribed spectral grid. To represent swell systems, spectral wave models may then appear to have a high computational cost, discretizing the spectrum in many wavenumbers and directions to explicitly solve the interactions between them. As an alternative to spectral models, the Particle-in-Cell (PiC) method was recently proposed to model surface waves (Hell et al. 2024). For the more specific problem of solving the propagation of the swell in the sphere, a comparable method was first proposed by Lavrenov and Onvlee (1993).

PiC is an efficient method that combines the Eulerian and Lagrangian frameworks to solve fluid dynamics equations in which the fluids undergo large deformations (Harlow 1957, 1988). PiC's general idea is to associate a fixed spatial Eulerian grid that computes the field-dependent terms with Lagrangian particles at the grid nodes, which simulate the advection-like dynamics. The timestep for the field-dependent terms and Lagrangian (advection) step alternate such that after the Lagrangian time step, the information advected by the particles is remeshed to the grid nodes from which new particles are launched to repeat the time advancement. The wave model PiCLES (Hell et al. 2024), currently devoted to wind-sea dynamics (Kudryavtsev et al. 2021), is based on this model framework. Here, it is extended to specifically model swell propagation. Unlike wind–sea systems, swell systems are mostly linear, dominated by the dispersion. The propagation obeys the laws of optical geometry, with the energy density decreasing with the distance traveled but also spreading in directions orthogonal to the main propagation direction. PiCLES (Hell et al. 2024) uses a mixed Lagrangian-Eulerian approach, which means that the particle information is stored either in a Lagrangian representation or at the Eulerian grid point at different steps of the algorithm. In this work, compared to PiCLES, the wave system is represented by an ensemble of particles. In PiCLES, the information contained in the particles is aggregated at the end of each time step at the mesh points. Here, particles are manipulated, copied, or destroyed, but we never completely remove all of them. A main advantage is to keep track of a diversity of particle states when remeshing. It enables to represent wave spectral distribution and, eventually, the geometrical optics behavior of wave dynamics. Without this change, only the energy and peak wavevector at each given grid point is represented, to only provide the main direction and wavenumber, losing the evolution of the energy distribution in the direction orthogonal to the main propagation direction.

A swell system and associated narrow-banded 2D spectrum is represented as an ensemble of particles, e.g., wave packets, at each grid point. Each particle propagates given its corresponding wavevector, controlling the wave packet group velocity. When remeshing the particles after a time step, the initial spectrum is modified, requiring a step involving importance resampling. Importance sampling is a probabilistic method for computing statistics from a ensemble of samples (Kloek and van Dijk 1978; Liu et al. 2001; Robert et al. 1999). The distribution of interest is generally untractable. Therefore, the ensemble is sampled from an approximate distribution instead, referred to as proposal distribution. To obtain unbiased statistics each sample is weighted such that more weight is given to the samples close to the target distribution. There are several ways to compute this weight. In our case, they are computed according to the distance between the original particle and the newly created particle. For more details on this part, see Eq. 9.8 and Sect. 9.2.2.3. Repeating the process many times, some samples lose weight every time, and thus some weights may become negligible. This means that we keep track of samples that almost carry no weight, and have very little impact on the distribution. It is very inefficient since computation time is still allocated to them. This problem is called weight degeneracy, or particle degeneracy. See Li et al. (2014) for more details. In Sequential Importance Resampling (SIR), the ensemble is resampled regularly according to the sample weights. It alleviates the weight degeneracy. This method is widely used in particle filters (Doucet et al. 2001, 2009; Cotter et al. 2020). Here we do not perform any particle filtering or data assimilation (see Sect. 9.4 for a discussion about the data assimilation potential of the method). We solely rely on SIR for a better use of our finite-size ensemble. We are interested in wave statistics (e.g., surface wave height, mean group velocity, or mean angular spreading) and the samples live in the four-dimensional space of spatial positions and wavevectors. Samples are initially randomly drawn from a given spectrum and evolve deterministically. The PiC remeshing biases the samples distribution, but the PiC energy reweighing shall correct it. To prevent weight degeneracy, we propose resampling the ensemble after each remeshing.

We will first cover the overview of the method, where the equations of swell dynamics are recalled, the modifications from Hell et al. (2024) are described, and the algorithm is explained. Then, we show the results we obtained when implementing the proposed solutions into PiCLES. Finally, a conclusion and perspectives are proposed.

9.2 Method

To account for the desired swell spreading behavior, an ensemble representation is performed at every point using a method similar to a Monte Carlo Markov Chain (MCMC). The ensemble of particles will represent an underlying probability distribution associated with the space-time wave energy spectrum.

9.2.1 Discrete Approximation of a Continuous Probability Density: An Ensemble Approach

The local wave spectrum, $E(t, \boldsymbol{x}, \boldsymbol{k}) \geq 0$, is the energy by unit of surface and by unit of wavevector surface. $\boldsymbol{x} = (x, y) \in S \subset \mathbb{R}^2$ and $\boldsymbol{k} = (k, l) \in P \subset \mathbb{R}^2 \setminus \{0_{\mathbb{R}^2}\}$ are concatenated in the 4D vector $\boldsymbol{Z}(t) = (\boldsymbol{x}, \boldsymbol{k})$ that lies in the bounded domain $D = S \times P \subset \mathbb{R}^4$. The spectrum E can be understood—up to a normalization constant—as the probability density function, $p_{\boldsymbol{Z}(t)}$, of the presence of a random particle, $\boldsymbol{Z}(t)$, in that 4D space:

$$p_{\boldsymbol{Z}(t)}(\boldsymbol{x}, \boldsymbol{k}) := \frac{E(t, \boldsymbol{x}, \boldsymbol{k})}{\int_D E(t, \boldsymbol{x}', \boldsymbol{k}')d\boldsymbol{x}'d\boldsymbol{k}'}. \tag{9.1}$$

Here, we consider an ensemble of M particles $\overline{\boldsymbol{Z}}(t) = \{\boldsymbol{Z}_m(t)\}_{1 \leq m \leq M} = \{(\boldsymbol{x}_m(t), \boldsymbol{k}_m(t)))\}_{1 \leq m \leq M}$ in the 4D space. We define $E_m(t)$ as the amount of energy carried by each particle. The ensemble $\overline{\boldsymbol{Z}}(t)$ defines a discrete distribution $\hat{p}_{\overline{\boldsymbol{Z}}(t)}$, and the normalized energies, $\frac{E_m(t)}{E_{\text{tot}}(t)}$, are used as probabilities:

$$\hat{p}_{\overline{\boldsymbol{Z}}(t)}(\boldsymbol{Z}) := \sum_{m=1}^{M} \frac{E_m(t)}{E_{\text{tot}}(t)} \delta(\boldsymbol{Z} - \boldsymbol{Z}_m(t)) \qquad \forall \boldsymbol{Z} \in D, \tag{9.2}$$

with $E_{\text{tot}}(t) := \sum_{m=1}^{M} E_m(t)$ the total energy. For convenient choices of $(\boldsymbol{Z}_m(t), E_m(t))$ we expect the discrete distribution $\hat{p}_{\overline{\boldsymbol{Z}}(t)}(\boldsymbol{x}, \boldsymbol{k})$ to be close to the continuous distribution $p_{\boldsymbol{Z}(t)}$ in a sense defined below (see Eq. (9.4) with $h(x) := x$). In some situations, we can also assume that all the weights are the same: $E_m(t) = E_1(t), \forall m$. In this case, we have

$$\hat{p}_{\overline{\boldsymbol{Z}}(t)}(\boldsymbol{Z}) = \sum_{m=1}^{M} \frac{1}{M} \delta(\boldsymbol{Z} - \boldsymbol{Z}_m(t)). \tag{9.3}$$

With either of those discrete measures, we can estimate the expectation of any function h associated with the measure $p_{\boldsymbol{Z}(t)}$ over the domain D using the discrete distribution $\hat{p}_{\overline{\boldsymbol{X}}}$:

$$\mathbb{E}_{\boldsymbol{Z}(t)}h(\boldsymbol{Z}(t)) \approx \hat{\mathbb{E}}_{\boldsymbol{Z}(t)}h(\boldsymbol{Z}(t)) = \mathbb{E}_{\hat{p}_{\overline{\boldsymbol{X}}}(t)}h(\boldsymbol{Z}(t)) = \sum_{m=1}^{M} \frac{E_m(t)}{E_{\text{tot}}(t)} h(\boldsymbol{Z}_m(t)). \tag{9.4}$$

Equation (9.4) is an integral estimation in the sense of importance sampling. With this tool, we can estimate all the statistics of $\boldsymbol{Z}(t) = (\boldsymbol{x}(t), \boldsymbol{k}(t))$ for our ensemble $\overline{\boldsymbol{Z}}$.

(a) Particles represented after the advection. Arrows indicate where they will be remeshed

(b) The set of particles after remeshing.

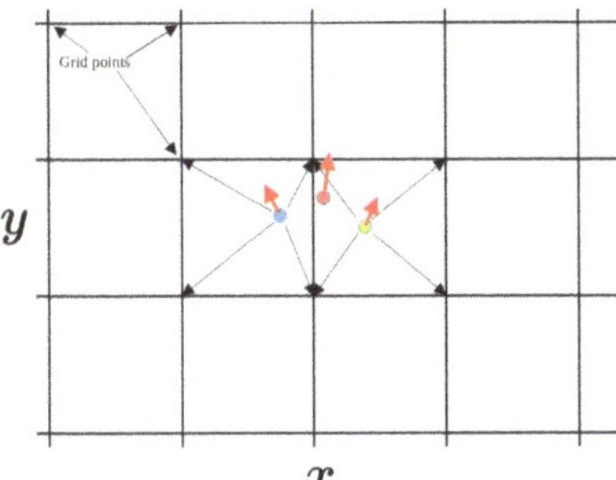

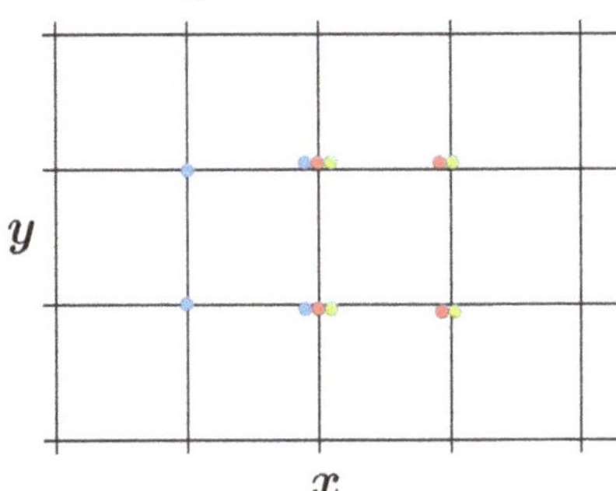

(c) The particles are numbered to allow for the importance resampling.

(d) Particles after resampling. The lines show their potential trajectories during the next advection.

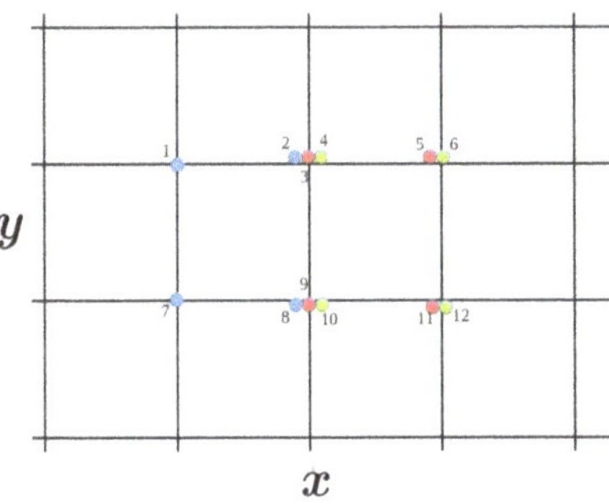

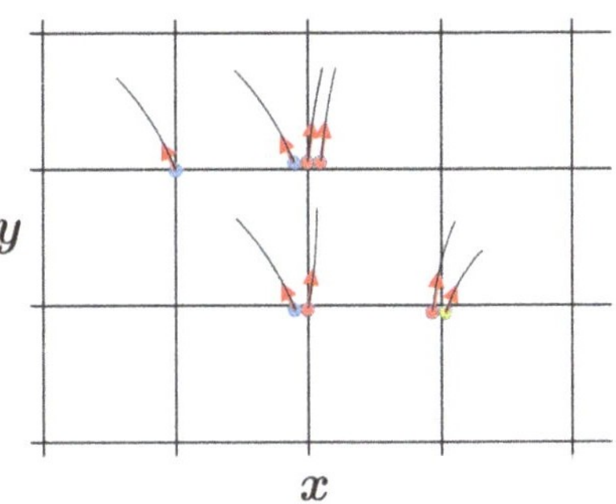

Fig. 9.1 Illustration of the steps of the remeshing and importance sampling

9.2.2 *Description of the Algorithm*

Figure 9.1 illustrates the main steps of the algorithm which are detailed in the following.

9.2.2.1 Physical Model: Time Stepping of the Particles

An ocean surface wave packet is dispersive. To first order over deep water and neglecting the currents, its frequency is

$$\omega^2 = gk \tag{9.5}$$

where g is the acceleration of gravity. The group velocity (the velocity at which the group energy propagates) and the evolution of the wavenumber are

$$\frac{d\boldsymbol{x}}{dt} = \frac{\partial \omega}{\partial \boldsymbol{k}} = \frac{1}{2}\sqrt{\frac{g}{k}}\boldsymbol{\kappa}, \tag{9.6}$$

$$\frac{d\boldsymbol{k}}{dt} = -\frac{\partial \omega}{\partial \boldsymbol{x}} = 0, \tag{9.7}$$

with the unit wavevector $\boldsymbol{\kappa} = \boldsymbol{k}/k$. Since surface currents are neglected here, the wave energy density, E, is conserved along the ray characteristics (Lavrenov 2013). System (9.6)–(9.7) is the kinematic equations describing the wave propagation. It takes the form of a transport equation and will be approximated by the advection of particles which carry the wavevector. Thus, the total derivative $\frac{d\boldsymbol{x}}{dt}$ is the evolution of the position of the propagated particle. This is the Lagrangian step of the algorithm and is the same in Hell et al. (2024). This Lagrangian representation is in four dimensions: two for the position in space and two for the position in the wavevector space. This step is shown in Fig. 9.1d. The transition from the Lagrangian to the Eulerian representation is done in the remeshing step, described in Sect. 9.2.2.2.

9.2.2.2   Lagrangian to Eulerian: Remeshing the Particles to the Grid

The remeshing step serves to transition from the Lagrangian to the Eulerian representation. While the Lagrangian step is used to propagate information, we can think of the Eulerian frame as a discretized sample of the 2D wavevector data, which we can use for resampling. The wavevector data at the grid points is therefore *indexed* by the grid point position.

At the end of the Lagrangian time step, the particles are distributed to the four neighboring cells, as shown in Fig. 9.1b. In total, the number of particles stored after remeshing is about four times the number of particles before. When remeshing one specific particle, the four new particles get a fraction of the original energy, so that the total energy remains conserved. The energy of each new particle is computed as a bilinear interpolation of the original particle. Suppose that the position of the particle after advection is $\boldsymbol{x}(t) = (x(t), y(t))$ and the positions of the four neighboring grid points are $\{\boldsymbol{x}_i\}_{i=1,2,3,4}$, with $\boldsymbol{x}_i = (x_i, y_i)$ $\forall k = 1, 2, 3, 4$. The distance between two grid points (the mesh size) is denoted Δx in the x-direction, and Δy in the y-direction. Assuming the energy of the original particle is $E_x(t)$, then the energy for each of the four remeshed particles is

$$E_i(t) := E_x(t)\underbrace{\left(1 - \frac{|x_i - x(t)|}{\Delta x}\right)}_{w_{i,x}(t)}\underbrace{\left(1 - \frac{|y_i - y(t)|}{\Delta y}\right)}_{w_{i,y}(t)}, \quad \forall i = 1, 2, 3, 4, \tag{9.8}$$

and $\sum_{i=1}^{4} E_i(t) = E_x(t)$.

The weights $w_{i,x}$ and $w_{i,y}$ are set such that energy is given to the neighbor cells using a bilinear interpolation on position. Doing this for every particle in the simulation, and storing all new particles in a unique set, we get the set R of particles after remeshing. Because each particle is divided into four fractions for each vertex, the size of R is approximately $4 \times M$.

The choice of bilinear interpolation is important to preserve the barycenter of the particles when remeshing. The particles' average position doesn't move when remeshing. In practice, this means that, on average, the particles travel at the desired speed, regardless of how they are moved around by the remeshing (see Appendix).

It should be mentioned that, after remeshing, all the particles present at a given grid point do not originate from the same pre-remeshing particle. This means that they may all have different wavevectors despite being at the same location. And thus, they can have different trajectories during the next time step.

The following section shows how this remeshing alters the resulting distribution and how to correct for this.

9.2.2.3 Eulerian to Lagrangian: Sequential Importance Resampling (SIR)

Importance sampling (Kloek and van Dijk 1978; Liu et al. 2001; Robert et al. 1999) is a technique for drawing samples from a hard-to-sample distribution called the **target probability distribution** $p(x)$. The idea is to draw N samples $\overline{Y} = \{y_i\}_{i \leq N}$ from an easier-to-sample distribution, called the **proposal distribution** $q(x)$. $q(x)$ has to cover the support of $p(x)$, $\text{supp}(p) \subset \text{supp}(q)$. The weights are computed such that they capture the "error" occurring from sampling q instead of p: $w_i = \frac{p(y_i)}{q(y_i)}$. Then these weights are used to draw M samples from $\overline{Y}$ with replacement to get $Y' = \{y_i'\}_{i \leq M}$. The $\overline{Y'}$ samples follow the target distribution.

Sequential importance resampling is built on importance sampling. Its goal is to correct the deformation of the spectrum induced by the remeshing step. It samples from a weighted ensemble to better represent the target probability distribution, and re-distributes the samples to the areas of higher probability while keeping the number of effective samples (or particles) constant. Doing it at each time step makes it a **sequential** importance resampling, preventing particle degeneracy. We can assimilate the underlying probability distribution before remeshing to the target distribution and the underlying probability distribution after remeshing to the proposal distribution. These two are different because the remeshing induces a modification of the particle positions, thereby modifying the distribution of energy and wave vector in space. The SIR allows for correction of this by resampling using the weights from the remeshing itself.

The set of particles R is used to draw and launch particles again from the grid points. We use their normalized remeshed energy $P = \{\frac{E_m}{\sum_{m'} E_{m'}}\}$ as a probability vector and do N successive draws with replacement, with N the total number of new particles to be launched in the whole simulation. In practice, $N = M$, and therefore the total number of particles remains constant. In Fig. 9.1c, the particles from the previous example are numbered from 1 to 12. A possible realization with $N = 8$ is shown in Fig. 9.1d, where the particles drawn are [1, 2, 3, 3, 8, 9, 11, 12]. We can see that some particles are lost (particles 4, 5, 6, 7, and 10), and one was selected twice (particle 3). The particles' trajectories during the next time step are then shown in black lines as an example in Fig. 9.1d.

SIR also keeps the number of particles constant in contrast to an ensemble particle approach that just re-meshes. Simple remeshing of particle ensembles would lead to a multiplication by 4 of the total number of particles each time step (ignoring boundaries), leading to a fast exponential growth of the number of particles (and thus memory space and compute time). By introducing importance resampling at each time step, we can control exactly the number of particles in the simulation. This is linked to particle degeneracy, see Sect. 9.1 for more details.

Applying these steps in order, particle time-stepping, remeshing, and resampling, is the core of the proposed algorithm. We expect the resampled particles to be assigned to the four-dimensional locations of most energy. As such, the four-dimensional distribution is expected to follow the laws of geometrical optics.

9.3 Results

This section shows some of the results we obtained when implementing the algorithm described in Sect. 9.2.2. The implementation was done in Julia 1.10.0. All the code can be found at https://doi.org/10.5281/zenodo.17037867.

9.3.1 Geometrical Optics Behavior

We set up a test case to verify if the propagating wave energy follows the expected laws of geometrical optics. The test domain is a rectangle 225 km $\times$ 112.5 km grid, without wind or currents, discretized in 151 $\times$ 46 points. This grids the domain into 1.5 km $\times$ 1.5 km cells. We initialize $1.5 \cdot 10^5$ particles close to the southern boundary at the point (112,500, 150,00). Although the wavenumber, and hence the particle velocity, is the same for all particles (22 m/s), the direction is distributed uniformly in angle between $\pm\frac{\pi}{2}$ from north, giving a total arc of π. The spatial distribution of the energy for three different moments in the simulation is shown in Fig. 9.2.

As expected, a uniform directional spreading appears. Since the total energy is conserved and the surface over which it is spread increases, the local energy in the northward direction will decay with the inverse of the distance traveled. This is shown in Fig. 9.3. The energy density seems to decay slightly faster than the expected rate $\frac{1}{r}$. This is likely the effect of the diffusion associated with the remeshing. Indeed,

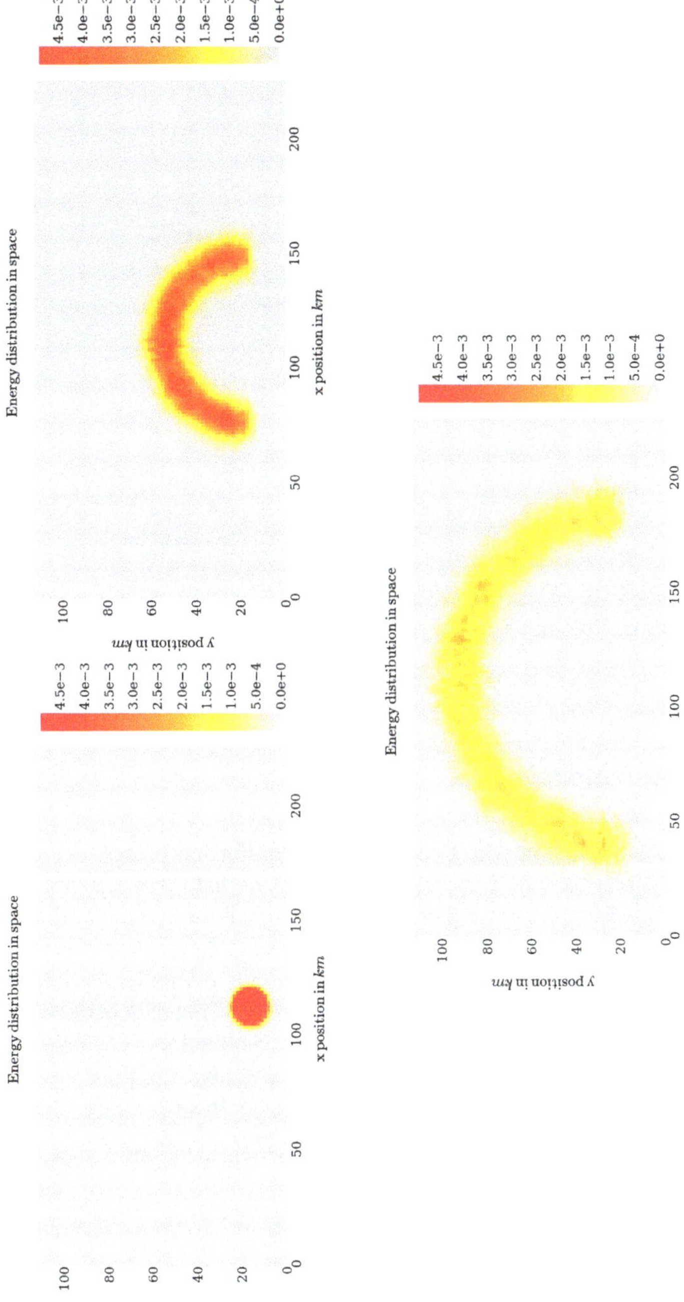

Fig. 9.2 From top left to bottom right, the spatial distribution of the energy at times $t = 0$ min, $t = 30$ and $t = 60$ min for the test case 1

Fig. 9.3 Energy density decaying with the distance traveled. A $\frac{1}{r}$ curve is superimposed for reference

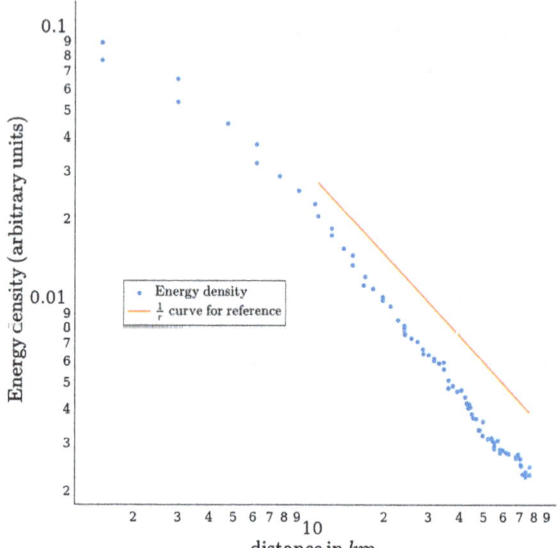

with diffusion, the thickness of the region where the energy is located also grows, decreasing the energy density compared to a non-diffusive situation.

9.3.2 Wave Dispersion

Next, the wave dispersion (Eq. (9.5)) is tested within the proposed propagation method. A test case is shown in Fig. 9.4. The test domain is a rectangle 168.750 km $\times$ 39.375 km discretized in 300×70 cells. This grids the domain roughly in 0.56 km $\times$ 0.56 km cells. 10^6 particles are initialized close to the western boundary at the point (4500, 18,750). The initial particle distribution is normally distributed in the 4D space $(X, C) \in \mathbb{R}^4$. Considering the group velocity C rather than the wavevector k, the distribution will remain normal throughout the simulation, as proved below. However, the mean (i.e., peak position and velocity) and covariance matrix (i.e., distribution spread in space and velocity) are changing. In such a case, an analytical solution can be derived and compared with the numerical solutions. Let $\boldsymbol{\mu}_0 = \begin{bmatrix} \boldsymbol{\mu}_{X_0} \\ \boldsymbol{\mu}_{C_0} \end{bmatrix}$ be the mean and $\boldsymbol{\Sigma}_0 = \begin{bmatrix} \boldsymbol{\Sigma}_{X_0} & \boldsymbol{0} \\ \boldsymbol{0} & \boldsymbol{\Sigma}_{C_0} \end{bmatrix}$ be the covariance matrix of the normal distribution at time $t = 0$ where we denote $\boldsymbol{0} = \begin{bmatrix} 0 & 0 \\ 0 & 0 \end{bmatrix}$. $\boldsymbol{\mu}_{X_0}$ describes the position of the wave packet at time $t = 0$ and $\boldsymbol{\mu}_{C_0}$ its peak velocity. $\boldsymbol{\Sigma}_{X_0}$ describes the spa-

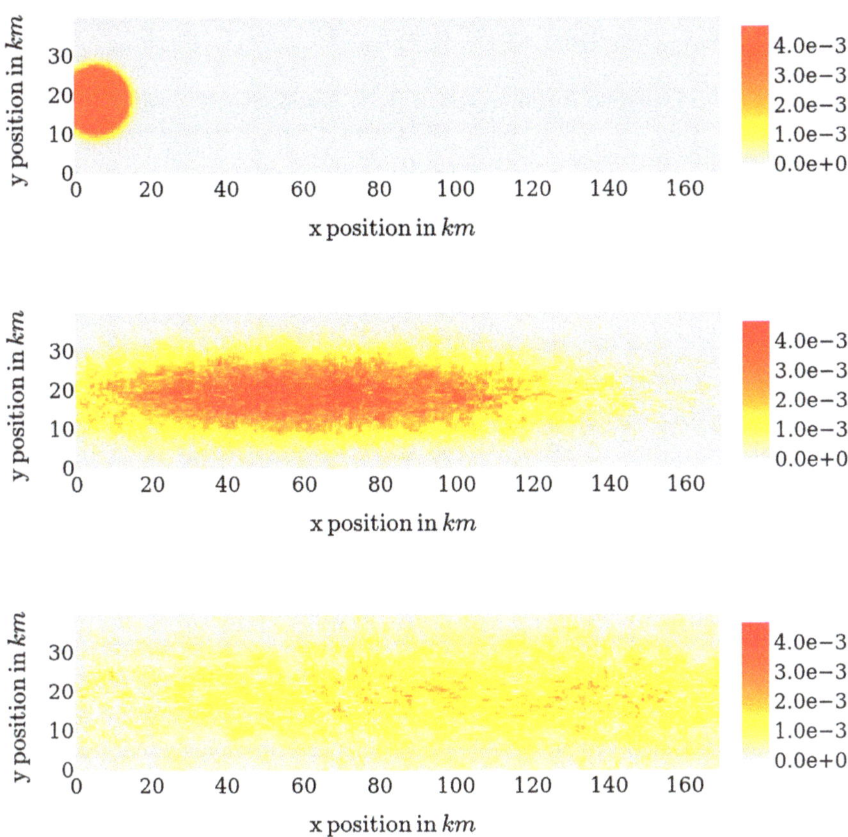

Fig. 9.4 From top to bottom, the spatial distribution of the energy at times $t = 0$ min, $t = 60$ and $t = 120$ min for the test case 2

tial extension of the wave packet and $\boldsymbol{\Sigma}_{C_0}$ the velocity distribution spread around the peak (related to the spectral width). Then, the normalized energy distribution is $\frac{1}{E_0} E(\boldsymbol{X}_0, \boldsymbol{C}_0, t = 0) = \mathcal{N}(\boldsymbol{X}_0, \boldsymbol{C}_0; \boldsymbol{\mu}_0, \boldsymbol{\Sigma}_0), \forall (\boldsymbol{X}_0, \boldsymbol{C}_0) \in \mathbb{R}^4$. Since the problem is entirely propagative and the energy along the ray is conserved, the method of the characteristics on $\{(9.6), (9.7)\}$ is used to obtain the solution at a later time t, with the energy $E(\boldsymbol{X}(t), \boldsymbol{C}(t), t) = E(\boldsymbol{X}_0, \boldsymbol{C}_0, t = 0)$.

$$\begin{bmatrix} \boldsymbol{X}(t) \\ \boldsymbol{C}(t) \end{bmatrix} = \begin{bmatrix} \boldsymbol{X}_0 + t\boldsymbol{C}_0 \\ \boldsymbol{C}_0 \end{bmatrix} \tag{9.9}$$

$$\begin{bmatrix} \boldsymbol{X}_0 \\ \boldsymbol{C}_0 \end{bmatrix} = \begin{bmatrix} \boldsymbol{X}(t) - t\boldsymbol{C}(t) \\ \boldsymbol{C}(t) \end{bmatrix}. \tag{9.10}$$

Therefore, for any $(\boldsymbol{X}, \boldsymbol{C}) \in \mathbb{R}^4$ we have

$$E(\boldsymbol{X}, \boldsymbol{C}, t) = E(\boldsymbol{X} - t\boldsymbol{C}, \boldsymbol{C}, 0) = E_0 \mathcal{N}(\boldsymbol{X} - t\boldsymbol{C}, \boldsymbol{C}; \boldsymbol{\mu}_0, \boldsymbol{\Sigma}_0), \qquad (9.11)$$

which is also a Gaussian function, i.e., $\boldsymbol{X}(t)$ is Gaussian. More explicitly, $\boldsymbol{X}_0 \sim \mathcal{N}(\boldsymbol{\mu}_0, \boldsymbol{\Sigma}_0)$ and from Eq. (9.9) $\boldsymbol{X}(t) = \underbrace{\begin{bmatrix} \boldsymbol{I}_2 & t\boldsymbol{I}_2 \\ 0 & \boldsymbol{I}_2 \end{bmatrix}}_{M(t)} \boldsymbol{X}_0$. It is a linear combination of multivariate normal random variables ($\boldsymbol{I}_2$ is the 2×2 identity matrix). Thus $\boldsymbol{X}(t) \sim \mathcal{N}(\boldsymbol{\mu}(t), \boldsymbol{\Sigma}(t))$ where

$$\boldsymbol{\mu}(t) := \begin{bmatrix} \boldsymbol{\mu}_X(t) \\ \boldsymbol{\mu}_C(t) \end{bmatrix} = M(t)\boldsymbol{\mu}_0 = \begin{bmatrix} t\boldsymbol{\mu}_{C_0} + \boldsymbol{\mu}_{X_0} \\ \boldsymbol{\mu}_{C_0} \end{bmatrix}, \qquad (9.12)$$

$$\boldsymbol{\Sigma}(t) := \begin{bmatrix} \boldsymbol{\Sigma}_X(t) & \boldsymbol{\Sigma}_{XC}(t) \\ \boldsymbol{\Sigma}_{XC}(t) & \boldsymbol{\Sigma}_C(t) \end{bmatrix} = M(t)\boldsymbol{\Sigma}_0^{\mathsf{T}} M^{\mathsf{T}}(t) = \begin{bmatrix} t^2 \boldsymbol{\Sigma}_{C_0} + \boldsymbol{\Sigma}_{X_0} & t\boldsymbol{\Sigma}_{C_0} \\ t\boldsymbol{\Sigma}_{C_0} & \boldsymbol{\Sigma}_{C_0} \end{bmatrix}. \qquad (9.13)$$

9.3.2.1 Effective Diffusion

In this section, we will explain how the remeshing induces a diffusion. This will also help us appropriately evaluate the model's performance. The remeshing step acts as a perturbation of the particle position.

Let $\Delta R = \begin{bmatrix} \Delta R_x \\ \Delta R_y \end{bmatrix}$ be the random perturbation in the particle position induced by the remeshing. Assuming that the particle falls between grid points positioned in x_i and x_{i+1}, Fig. 9.5 and Eq. 9.8 show that

$$\Delta R_x = \begin{cases} \Delta t(C_x \mod \frac{\Delta x}{\Delta t}) & \text{with probability } \mathbb{P}_1 = \frac{\Delta x - \Delta t(C_x \mod \frac{\Delta x}{\Delta t})}{\Delta x}; \\ \Delta x - \Delta t(C_x \mod \frac{\Delta x}{\Delta t}) & \text{with probability } \mathbb{P}_2 = 1 - \mathbb{P}_1 = \frac{\Delta t(C_x \mod \frac{\Delta x}{\Delta t})}{\Delta x} \end{cases}$$
$$(9.14)$$

and likewise for ΔR_y:

$$\Delta R_y = \begin{cases} \Delta t(C_y \mod \frac{\Delta y}{\Delta t}) & \text{with probability } \mathbb{P}_3 = \frac{\Delta y - \Delta t(C_y \mod \frac{\Delta y}{\Delta t})}{\Delta y}; \\ \Delta y - \Delta t(C_y \mod \frac{\Delta y}{\Delta t}) & \text{with probability } \mathbb{P}_4 = 1 - \mathbb{P}_3 = \frac{\Delta t(C_y \mod \frac{\Delta y}{\Delta t})}{\Delta y}. \end{cases}$$
$$(9.15)$$

We write $\tilde{C}_x = C_x \mod \frac{\Delta x}{\Delta t}$ and $\tilde{C}_y = C_y \mod \frac{\Delta y}{\Delta t}$.

This perturbation corresponds to the addition of a random walk process $\boldsymbol{R}$ on top of the 4D distribution evolution before remeshing. After a large number of steps, this converges to a centered normal distribution with covariance

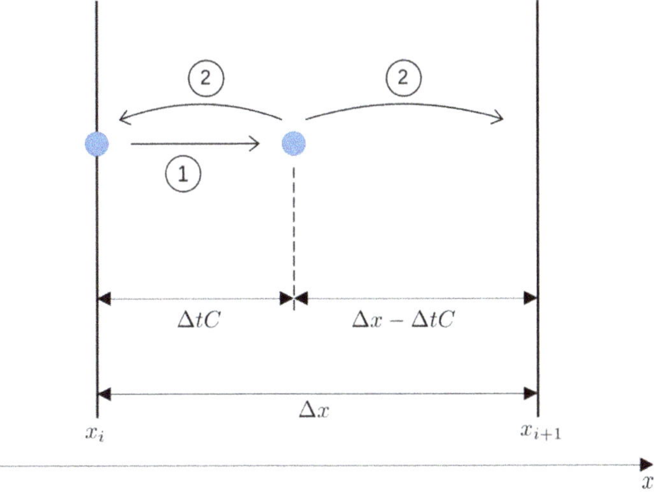

Fig. 9.5 Schematic of the remeshing step in x-direction. The circled 1 is the advection step, and the circled 2 is the remeshing step. The remeshing creates an unequal distribution of the charge (blue circle) to the grid point, impacting the effective diffusion of the scheme

$$COV = 2t\,\boldsymbol{D}(C) = t \begin{bmatrix} \frac{\mathrm{Var}(\Delta R_x)}{\Delta t} & \frac{\mathrm{Cov}(\Delta R_x, \Delta R_y)}{\Delta t} \\ \frac{\mathrm{Cov}(\Delta R_x, \Delta R_y)}{\Delta t} & \frac{\mathrm{Var}(\Delta R_y)}{\Delta t} \end{bmatrix} \tag{9.16}$$

$$= t \begin{bmatrix} \frac{\frac{\Delta x - \Delta t \tilde{C}_x}{\Delta x}(\Delta t \tilde{C}_x)^2 + \frac{\Delta t \tilde{C}_x}{\Delta x}(\Delta x - \Delta t \tilde{C}_x)^2}{\Delta t} & 0 \\ 0 & \frac{\frac{\Delta y - \Delta t \tilde{C}_y}{\Delta y}(\Delta t \tilde{C}_y)^2 + \frac{\Delta t \tilde{C}_y}{\Delta y}(\Delta y - \Delta t \tilde{C}_y)^2}{\Delta t} \end{bmatrix} \tag{9.17}$$

$$- t \begin{bmatrix} \Delta x \tilde{C}_x - \Delta t \tilde{C}_x^2 & 0 \\ 0 & \Delta y \tilde{C}_y - \Delta t \tilde{C}_y^2 \end{bmatrix}, \tag{9.18}$$

with D as a diffusivity term. In our case, we have $\Delta x = \Delta y$, and we know the distribution of C, so we can compute the overall value of $\boldsymbol{D}$ as follows:

$$\overline{\boldsymbol{D}} = \mathbb{E}_C[\boldsymbol{D}(C)]. \tag{9.19}$$

We estimate the value of $\overline{\boldsymbol{D}}$ using a trapezoidal integration scheme. Thus, the random walk can be modeled as

$$\boldsymbol{R}(t) \sim \mathcal{N}(0_{\mathbb{R}^2}, 2t\,\overline{\boldsymbol{D}}). \tag{9.20}$$

This yields a modified energy distribution with parameters:

$$\mu^{\mathrm{diff}}(t) = \begin{bmatrix} t\mu_{C_0} + \mu_{X_0} \\ \mu_{C_0} \end{bmatrix}, \tag{9.21}$$

$$\Sigma^{\mathrm{diff}}(t) = \begin{bmatrix} t^2\Sigma_{C_0} + 2t\overline{D} + \Sigma_{X_0} & t\Sigma_{C_0} \\ t\Sigma_{C_0} & \Sigma_{C_0} \end{bmatrix}. \tag{9.22}$$

9.3.2.2 Local Spectrum

From Eqs. (9.13) and (9.22), $X(t)$ and $C(t)$ are not independent. The correlation between $X(t)$ and $C(t)$ increases with time. This is expected since faster and slower waves should start detaching ahead and behind the peak energy packet, respectively. The spatial extent of the ensemble of wave packets also increases. This evolution of the correlation between position and group velocity is the statistical equivalent of the dispersion relation. $C(t)|\{X(t) = x\}$ can be computed analytically. It is also a normal random vector. Its distribution is a "slice" of the 4D distribution, interpreted as the local group velocity distribution observed at position x. Ignoring the diffusion, its mean and covariance are as follows (Eaton 2007):

$$\mu_{C|X=x}(t) := \mathbb{E}[C|X = x] \tag{9.23}$$
$$= \mu_C(t) + \Sigma_{CX}(t)(\Sigma_X(t))^{-1}(x - \mu_X(t)) \tag{9.24}$$
$$= \mu_{C_0} + t\Sigma_{C_0}(t^2\Sigma_{C_0} + \Sigma_{X_0})^{-1}(x - t\mu_{C_0} - \mu_{X_0}), \tag{9.25}$$
$$\Sigma_{C|X=x}(t) := Cov(C|X = x) \tag{9.26}$$
$$= \Sigma_C(t) - \Sigma_{CX}(t)(\Sigma_X(t))^{-1}\Sigma_{CX}^{\mathsf{T}}(t) \tag{9.27}$$
$$= \Sigma_C(t) - \Sigma_{XC}^{\mathsf{T}}(t)(\Sigma_X(t))^{-1}\Sigma_{XC}(t) \tag{9.28}$$
$$= \Sigma_{C_0} - t^2\Sigma_{C_0}(t^2\Sigma_{C_0} + \Sigma_{X_0})^{-1}\Sigma_{C_0} \tag{9.29}$$

using the Woodbury identity yields : $\tag{9.30}$

$$= \Sigma_{C_0} - \Sigma_{C_0}(\Sigma_{C_0}^{-1} - t^{-2}\Sigma_{C_0}^{-1}\Sigma_{X_0}(I_2 + t^{-2}\Sigma_{C_0}^{-1}\Sigma_{X_0})^{-1}\Sigma_{C_0}^{-1})\Sigma_{C_0} \tag{9.31}$$

$$= t^{-2}\Sigma_{X_0}(I_2 + t^{-2}\Sigma_{C_0}^{-1}\Sigma_{X_0})^{-1} \tag{9.32}$$
$$= (t^2\Sigma_{X_0}^{-1} + \Sigma_{C_0}^{-1})^{-1}. \tag{9.33}$$

Taking diffusion into account, a similar derivation leads to

$$\mu_{C|X=x}^{\mathrm{diff}}(t) = \mu_{C_0} + t\Sigma_{C_0}(t^2\Sigma_{C_0} + 2t\overline{D} + \Sigma_{X_0})^{-1}(x - t\mu_{C_0} - \mu_{X_0}), \tag{9.34}$$
$$\Sigma_{C|X=x}^{\mathrm{diff}}(t) = (\Sigma_{C_0}^{-1} + t^2(2t\overline{D} + \Sigma_{X_0})^{-1})^{-1}. \tag{9.35}$$

We can see that for large times, the diffused formula (9.35) behaves in $1/t$ while the non-diffused equation (9.26) predicts a $1/t^2$ behavior. In both cases, the local covariance matrix is narrowing around a new given peak velocity. The local spectrum—i.e., the distribution of $C(t)|\{X(t) = x\}$—also narrows and becomes more directional when the swell develops.

Using formulae (9.23) and (9.34), we can compare local values of peak velocities of our particles ensembles with the expected peak value. We can do the same with formulae (9.26) and (9.35) to compare local spectra of our ensemble with the diffused and non-diffused analytical expressions.

9.3.2.3 Local Peak Spectral Values

We first verify the local peak velocity (i.e., the most probable/energetic velocity for a given point in space) distribution in space for a given time using formulae (9.23) and (9.34). We thus look at the local peak velocity at different points of a line that runs through the distribution, along the direction of propagation (toward positive x, i.e., the east), at the final time, in Fig. 9.6.

To do so, we must estimate the peak velocity at a given position x in the domain using a cloud of particles with different speeds and positions. In general, an arbitrary $x \in \mathbb{R}^2$ does not correspond to a particle. Therefore, we cannot estimate moments from the natural importance sampling estimator (9.4) which can only give an estimation for the exact positions of the existing particles. A Gaussian kernel estimation method is used instead. The range of the kernel is set to the grid size Δx. This ensures that relevant particles lie within a radius coherent with the grid size. This kernel assigns to each particle $X_P(t)$ a weight $\lambda_P(t)$ based on its distance to x: $\lambda_P(t) = \mathcal{N}(X_P(t) - x; 0_{\mathbb{R}^2}, \Delta x^2 I_2)$. These weights are then normalized $\lambda'_P(t) = \lambda_P(t)/\sum_i \lambda_i(t)$. Finally, the estimated mean $\hat{\mu}_C$ is

$$\hat{\mu}_C(t) = \sum_i \lambda'_i(t) C_i(t). \tag{9.36}$$

Plotting this value for all the points on the line of interest gives the result shown in Fig. 9.6b, where we can see that the output of the model follows Eq. (9.34).

9.3.2.4 Local Spectrum Narrowing

Formulae (9.26) and (9.35) are now used to check the width of the local spectrum. We use the test case shown in Fig. 9.4 and tracked the time evolution of the local wave spectrum width resulting from the dispersion relation and importance resampling. Note that $\Sigma_{C|X=x}(t)$ and $\Sigma_{C|X=x}^{\text{diff}}(t)$ do not depend on the actual position x. In principle, we can choose any position x in the domain and compute the local spectrum. The peak speed would be different (as shown in Sect. 9.3.2.3), **but the covariance around that peak should be invariant**. To take advantage of this in the model experiment, we average the covariance estimates for several positions at each time step. The unbiased weighted covariance $\hat{\Sigma}_{C|X=x}(t)$ is then calculated at 15 points $X_i(t), \quad \forall i \in \{1, ..., 15\}$ using

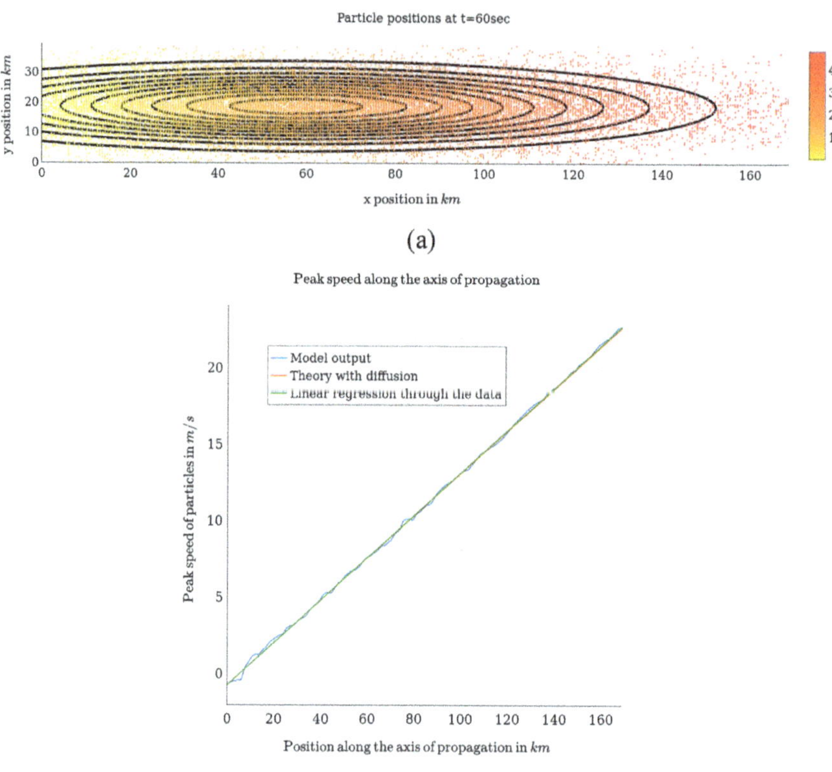

Fig. 9.6 Model output for the test case 2 at time $t = 60$ min (for Fig. 9.6a) and $t = 120$ min (for Fig. 9.6b). Panel **a** shows all the particles drawn as dots. Their color represents their speed, the colorbar on the right shows the correspondence between the speed and the color. The contours show the theoretical result given by Eqs. (9.12) and (9.13). Panel **b** shows the peak speed of the surrounding particles along a center line through the distribution ($x = y$ line) as function of distance from the origin

$$\hat{\boldsymbol{\Sigma}}_{C|X=x}(t) = \frac{1}{1 - \sum_i \lambda_i'^2(t)} \sum_i \lambda_i'(t)(\boldsymbol{C}_i(t) - \hat{\boldsymbol{\mu}}_C(t))(\boldsymbol{C}_i(t) - \hat{\boldsymbol{\mu}}_C(t))^\mathsf{T}, \quad (9.37)$$

and then averaged to get a value for a given time.

Plotting the local variance of the spectrum (i.e., the diagonal of $\hat{\boldsymbol{\Sigma}}_{C|X=x}(t)$) against time gives Fig. 9.7. The covariance closely follows the analytical solutions for the diffused dispersion, i.e., Eq. (9.35).

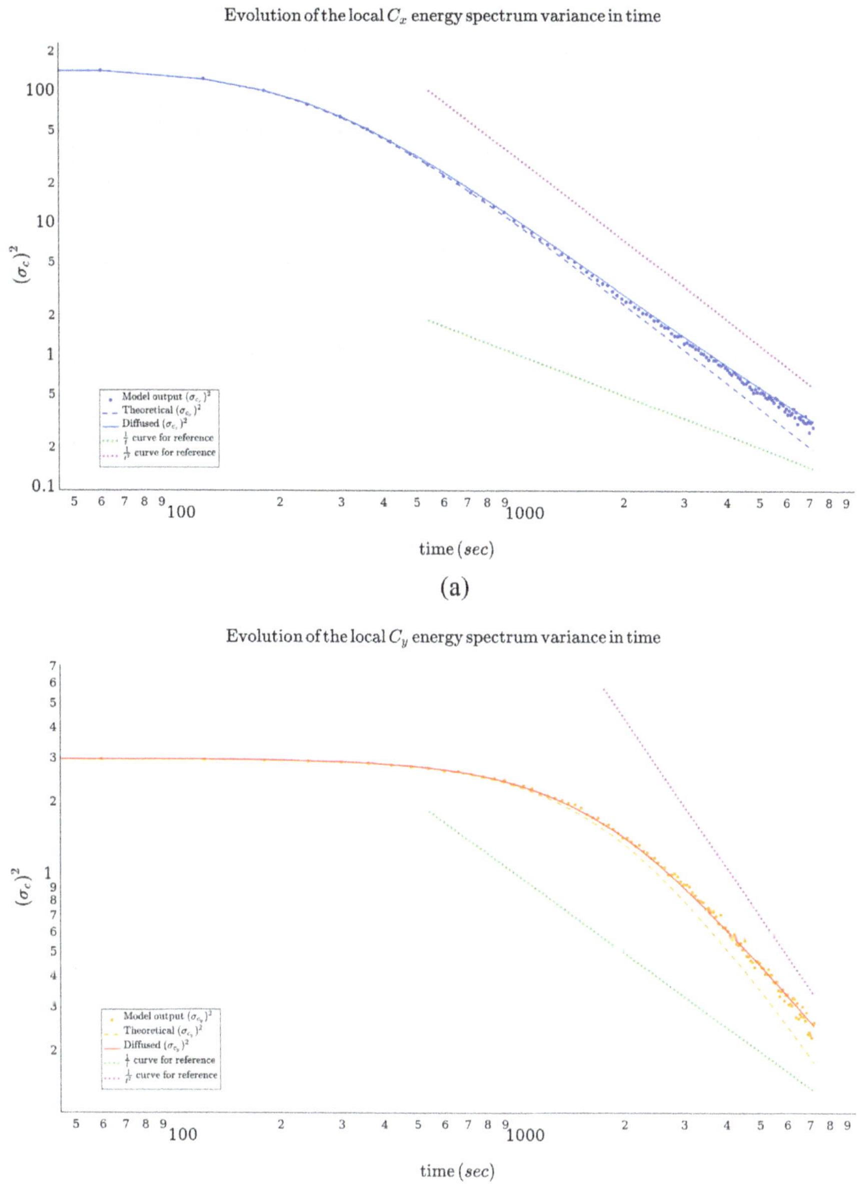

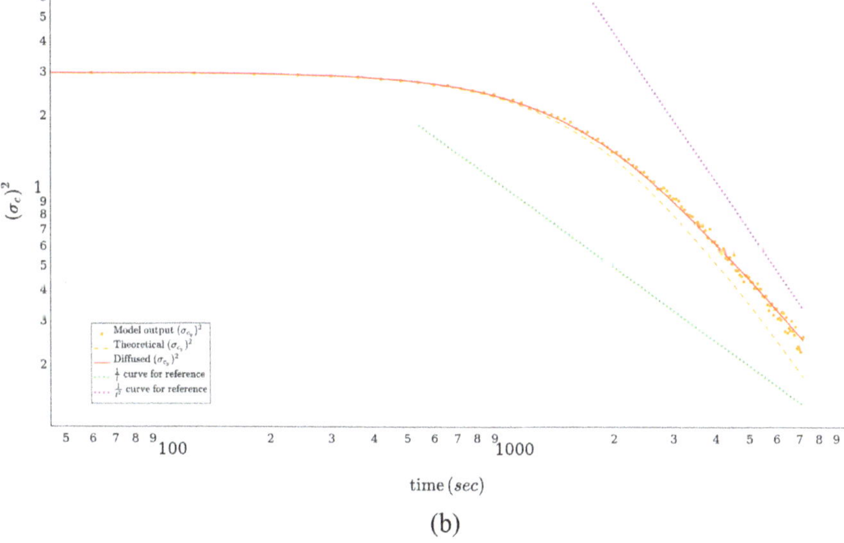

Fig. 9.7 Figures showing the evolution of the spectrum width, measured by $\hat{\sigma}_{c_x|X=x}(t)$ and $\hat{\sigma}_{c_y|Y=y}(t)$, with time. The markers represent the PiCLES output, the dashed line represents the expected behavior without diffusion (Eq. 9.26), and the solid line represents the expected behavior with diffusion (Eq. 9.35). $\frac{1}{t}$ and $\frac{1}{t^2}$ curves are represented in dotted lines for reference. The diffusivity term is computed using Eq. (9.19). In Fig. 9.7a, the diffusivity term is $\overline{D}_x = 442.187 \, \mathrm{m^2/s}$, and in Fig. 9.7b, it is $\overline{D}_y = 304.020 \, \mathrm{m^2/s}$

9.4 Conclusion

This paper describes an algorithm that expands the PiCLES model (Hell et al., 2024) to represent the swell dynamics. For that, we showed that the space-time evolution of an initial narrow-banded swell energy spectrum will evolve as a diffusion process, and tested this idea using an ensemble approach to sample the 4D (space-velocity) evolving distribution. We use a sequential importance resampling technique which is used to keep the discretized distribution (the particle ensemble) in line with the true distribution after the remeshing step of the PiC algorithm, without increasing the ensemble size. The geometrical optics behavior is well recovered (Figs. 9.2 and 9.3). The dispersion occurs at the expected rate (Fig. 9.6a, b), when diffusion is taken into account. For this analysis, we have compared to a Gaussian analytic case. Its local covariance evolution is derived in a predictable way (Fig. 9.7a, b).

This development enables future work on a parametric model to simplify the spectral distribution dynamics around the peak value, instead of relying on an ensemble assimilated to realizations of this distribution. Next, we will extend this framework to the dispersion of swell systems propagating through more realistic turbulent flows that characterize some highly dynamical ocean regions (Resseguier et al. 2024).

In future work, the order of accuracy of this implementation of the Particle-in-Cell scheme could also be studied both theoretically and experimentally. Here we have identified numerical diffusion as the main contributor to model error. Thus, we can expect the model error to evolve proportionally to the numerical diffusion, which is quantified in Eq. (9.18). This equation shows that the covariance of the random walk process associated to the position error of the particle evolves linearly with grid size (assuming that the time step is kept in line with the grid size $\Delta t \approx \alpha \Delta x$). This could lead to order-one error ($\sim \mathcal{O}(\Delta x^1)$), as is usual for Particle-in-Cell (Edwards and Bridson 2012; Jiang and Schroeder 2024). The order of accuracy could also be studied experimentally to confirm this hypothesis.

Finally, we remark that the Lagrangian step seems to be the most appropriate part for data assimilation, and maybe subject of future work. The observational data, from buoy, satellite, or both, provides different measurements for reconstructing of the 2D spectrum (frequency and direction) at a given location. This observed spectra can then inform the spectrum that our method implicitly assumes and tracks. However, during the Lagrangian step, only indirect information about the spectrum is available, in the form particles. They can be viewed as independent realizations of the spectrum interpreted as a probability density function. Here, data assimilation would update an ensemble of realizations using information on the true PDF. A possible way of comparison is to compute a likelihood for each particle (at least for the ones close to the observation) using the observed spectrum, and use this for the particle filter. In this way, we take full advantage of the ensemble nature of the Lagrangian step.

Appendix: Bilinear Interpolation Conserves Energy and the Energy-Weighted Barycenter of the Particles

In this appendix, using the notations in Fig. 9.8, we will show that

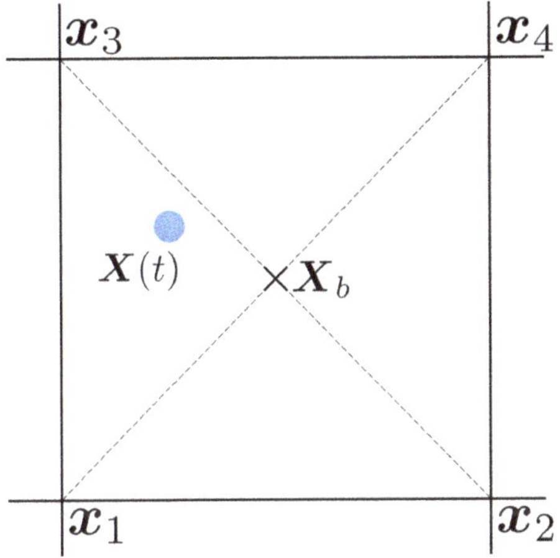

Fig. 9.8 Schematic of the remeshing step in two dimensions. $X(t) = [x(t), y(t)]^\mathsf{T}$ is the position of the particle before remeshing, $x_i = [x_i, y_i]^\mathsf{T}$ are the positions of the grid points surrounding the particle. A weight w_i is associated to each grid point i. $X_b = [X_b, Y_b]^\mathsf{T}$ is an example of the barycenter of the energy after remeshing if an equal weight is given to all grid points. In Appendix, we want to show that the bilinear interpolation shares the weights among the grid points such that $X_b = X(t)$, while keeping the energy constant

Proposition *Assume the interpolation scheme is a bilinear interpolation scheme*

$$w_i = \left(1 - \frac{|x_i - x(t)|}{\Delta x}\right)\left(1 - \frac{|y_i - y(t)|}{\Delta y}\right) \quad \forall i = 1, 2, 3, 4. \tag{9.38}$$

Then the interpolation scheme conserves total energy and the barycenter of the positions of the particles

$$\sum_{i=1}^{4} w_i = 1; \tag{9.39}$$

$$X_b := \sum_{i=1}^{4} w_i x_i = X(t). \tag{9.40}$$

Proof First, let's prove (9.39). Assume (9.38) is true. Then the sum of weights is

$$\sum_{i=1}^{4} w_i = \sum_{i=1}^{4} \left(1 - \frac{|x_i - x(t)|}{\Delta x}\right)\left(1 - \frac{|y_i - y(t)|}{\Delta y}\right) \tag{9.41}$$

remarking that $x_1 = x_3$; $x_2 = x_4$; $y_1 = y_2$ and $y_3 = y_4$,

$$= (1 - \frac{x(t) - x_1}{\Delta x} + 1 - \frac{x_2 - x(t)}{\Delta x})(1 - \frac{y(t) - y_1}{\Delta y} + 1 - \frac{y_3 - y(t)}{\Delta y}) \tag{9.42}$$

remarking that $x_2 = \Delta x + x_1$ and $y_3 = \Delta y + y_1$

$$= \left(1 - \frac{x(t) - x_1}{\Delta x} + 1 - \frac{\Delta x + x_1 - x(t)}{\Delta x}\right)\left(1 - \frac{y(t) - y_1}{\Delta y} + 1 - \frac{\Delta y + y_1 - y(t)}{\Delta y}\right) \tag{9.43}$$

$$= \left(1 - \frac{x(t) - \cancel{x_1}}{\cancel{\Delta x}} + \cancel{1} - \cancel{1} + \frac{x(t) - \cancel{x_1}}{\cancel{\Delta x}}\right)\left(1 - \frac{y(t) - \cancel{y_1}}{\cancel{\Delta y}} + \cancel{1} - \cancel{1} + \frac{y(t) - \cancel{y_1}}{\cancel{\Delta y}}\right) \tag{9.44}$$

$$\sum_{i=1}^{4} w_i = 1 \quad \Leftrightarrow \quad (9.39). \tag{9.45}$$

Thus, assuming (9.38), (9.39) **is true**.

Now, let's prove (9.40). Assume (9.38) is true. Then the barycenter of the grid points weighted by the weights (9.38) is

$$X_b = \sum_{i=1}^{4} w_i x_i = \sum_{i=1}^{4} \left(1 - \frac{|x_i - x(t)|}{\Delta x}\right)\left(1 - \frac{|y_i - y(t)|}{\Delta y}\right)\begin{bmatrix} x(t) \\ y(t) \end{bmatrix} \tag{9.46}$$

$$= \sum_{i=1}^{4} \begin{bmatrix} (x_i - \frac{x_i|x_i - x(t)|}{\Delta x})(1 - \frac{|y_i - y(t)|}{\Delta y}) \\ (1 - \frac{|x_i - x(t)|}{\Delta x})(y_i - \frac{y_i|y_i - y(t)|}{\Delta y}) \end{bmatrix} \tag{9.47}$$

$$= \begin{bmatrix} (x_1 - \frac{x_1(x(t) - x_1)}{\Delta x} + x_2 - \frac{x_2(x_2 - x(t))}{\Delta x})(1 - \frac{y(t) - y_1}{\Delta y} + 1 - \frac{y_3 - y(t)}{\Delta y}) \\ (1 - \frac{x(t) - x_1}{\Delta x} + 1 - \frac{x_2 - x(t)}{\Delta x})(y_1 - \frac{y_1(y(t) - y_1)}{\Delta y} + y_3 - \frac{y_3(y_3 - y(t))}{\Delta y}) \end{bmatrix} \tag{9.48}$$

using the same arguments as for the first part of the proof,

$$= \begin{bmatrix} (x_1 - \frac{x_1(x(t) - x_1)}{\Delta x} + x_2 - \frac{x_2(x_2 - x(t))}{\Delta x})(1) \\ (1)(y_1 - \frac{y_1(y(t) - y_1)}{\Delta y} + y_3 - \frac{y_3(y_3 - y(t))}{\Delta y}) \end{bmatrix} \tag{9.49}$$

$$= \begin{bmatrix} (x_1 - \frac{x_1(x(t) - x_1)}{\Delta x} + x_1 + \Delta x - \frac{(x_1 + \Delta x)(x_1 + \Delta x - x(t))}{\Delta x}) \\ (y_1 - \frac{y_1(y(t) - y_1)}{\Delta y} + y_1 + \Delta y - \frac{(y_1 + \Delta y)(y_1 + \Delta y - y(t))}{\Delta y}) \end{bmatrix} \tag{9.50}$$

$$= \begin{bmatrix} (2x_1 + \Delta x - \frac{x_1(x(t) - x_1)}{\Delta x} - \frac{(x_1^2 + 2x_1\Delta x - x_1 x(t) + \Delta x^2 - x(t)\Delta x)}{\Delta x}) \\ (2y_1 + \Delta y - \frac{y_1(y(t) - y_1)}{\Delta y} - \frac{(y_1^2 + 2y_1\Delta y - y_1 x(t) + \Delta y^2 - y(t)\Delta y)}{\Delta y}) \end{bmatrix} \tag{9.51}$$

$$= \begin{bmatrix} (2x_1 + \Delta x - \cancel{\frac{x_1(x(t) - x_1)}{\Delta x}} + \cancel{\frac{x_1(x(t) - x_1)}{\Delta x}} - \frac{(2x_1\Delta x + \Delta x^2 - x(t)\Delta x)}{\Delta x}) \\ (2y_1 + \Delta y - \cancel{\frac{y_1(y(t) - y_1)}{\Delta y}} + \cancel{\frac{y_1(y(t) - y_1)}{\Delta y}} - \frac{(2y_1\Delta y + \Delta y^2 - y(t)\Delta y)}{\Delta y}) \end{bmatrix} \tag{9.52}$$

$$= \begin{bmatrix} (\cancel{2x_1} + \cancel{\Delta x} - \cancel{2x_1} - \cancel{\Delta x} + x(t)) \\ (\cancel{2y_1} + \cancel{\Delta y} - \cancel{2y_1} - \cancel{\Delta y} + y(t)) \end{bmatrix} \tag{9.53}$$

$$X_b = \begin{bmatrix} x(t) \\ y(t) \end{bmatrix} \quad \Leftrightarrow \quad (9.40).$$

$$(9.54)$$

Thus, assuming (9.38), (9.40) **is true**.

References

Collard F, Ardhuin F, Chapron B (2009) Monitoring and analysis of ocean swell fields from space: new methods for routine observations. J Geophys Res Oceans 114(C7)

Cotter C, Crisan D, Holm DD, Pan W, Shevchenko I (2020) A particle filter for stochastic advection by lie transport: a case study for the damped and forced incompressible two-dimensional euler equation. SIAM/ASA J Uncertain Quantif 8(4):1446–1492

Doucet A, De Freitas N, Gordon NJ et al (2001) Sequential Monte Carlo methods in practice, vol 1. Springer

Doucet A, Johansen AM et al (2009) A tutorial on particle filtering and smoothing: fifteen years later. Handbook of nonlinear filtering 12(656–704):3

Eaton ML (2007) Multivariate statistics: a vector space approach. Inst Math Stat 53:i–512

Edwards E, Bridson R (2012) A high-order accurate particle-in-cell method. Int J Numer Methods Eng 90(9):1073–1088. https://onlinelibrary.wiley.com/doi/pdf/10.1002/nme.3356

Harlow FH (1957) Hydrodynamic problems involving large fluid distortions 4(2):137–142

Harlow FH (1988) PIC and its progeny 48(1):1–10

Hell M, Chapron B, Fox-Kemper B (2024) A particle-in-cell wave model for efficient sea-state and swell estimates in earth system models - PiCLES

Jiang H, Schroeder C (2024) Second order accurate particle-in-cell discretization of the Navier-Stokes equations 518:113302

Kloek T, van Dijk HK (1978) Bayesian estimates of equation system parameters: an application of integration by Monte Carlo. Econo Soc 46(1):1–19

Kudryavtsev V, Yurovskaya M, Chapron B (2021) Self-similarity of surface wave developments under tropical cyclones. J Geophys Res Oceans 126(4):e2020JC016916

Lavrenov I (2013) Wind-waves in oceans: dynamics and numerical simulations. Springer Science & Business Media

Lavrenov I, Onvlee J (1993) On the effects of limited spectral resolution in third-generation wave models. Koninklijk Nederlands Meteorologisch Instituut

Li T, Sun S, Sattar TP, Corchado JM (2014) Fight sample degeneracy and impoverishment in particle filters: a review of intelligent approaches. Expert Syst Appl 41(8):3944–3954

Liu JS, Chen R, Logvinenko T (2001) A theoretical framework for sequential importance sampling with resampling. In: Doucet A, Freitas N, Gordon N (eds) Sequential Monte Carlo Methods in Practice. Springer, New York, pp 225–246

Munk WH, Miller G, Snodgrass F, Barber NF (1963) Directional recording of swell from distant storms. Philos Trans Royal Soc Lond Ser A, Math Phys Sci 255(1062):505–584

Resseguier V, Hascoët E, Chapron B (2024) Wave propagation in random two-dimensional turbulence: a multiscale approach. J Fluid Mech 997:A51

Robert CP, Casella G, Casella G (1999) Monte Carlo statistical methods, vol 2. Springer

Author Index

© The Author(s) 2026
B. Chapron et al. (eds.), *Stochastic Transport in Upper Ocean Dynamics IV*, Mathematics
of Planet Earth 15, https://doi.org/10.1007/978-3 032-12749-5